BIRKHÄUSER

Advanced Courses in Mathematics
CRM Barcelona

Centre de Recerca Matemàtica

Managing Editor:
Manuel Castellet

Alfred Geroldinger
Imre Z. Ruzsa

Combinatorial Number Theory and Additive Group Theory

Birkhäuser
Basel · Boston · Berlin

Authors:

Alfred Geroldinger
Institute for Mathematics and
Scientific Computing
University of Graz
Heinrichstrasse 36
8010 Graz, Austria
e-mail: alfred.geroldinger@uni-graz.at

Imre Z. Ruzsa
Alfréd Rényi Institute of Mathematics
P.O. Box 127
1364 Budapest, Hungary
e-mail: ruzsa@renyi.hu

2000 Mathematical Subject Classification 11P70, 11B50, 11R27

Library of Congress Control Number: 2008941509

Bibliographic information published by Die Deutsche Bibliothek
Die Deutsche Bibliothek lists this publication in the Deutsche Nationalbibliografie; detailed bibliographic data is available in the Internet at <http://dnb.ddb.de>.

ISBN 978-3-7643-8961-1 Birkhäuser Verlag AG, Basel – Boston – Berlin

Printed on acid-free paper produced from chlorine-free pulp. TCF ∞

ISBN 978-3-7643-8961-1 ISBN 978-3-7643-8962-8 (eBook)

9 8 7 6 5 4 3 2 1 www.birkhauser.ch

Foreword

This book collects the material delivered in the 2008 edition of the *DocCourse in Combinatorics and Geometry* which was devoted to the topic of *Additive Combinatorics.*

The two first parts, which form the bulk of the volume, contain the two main advanced courses, *Additive Group Theory and Non-unique Factorizations*, by Alfred Geroldinger, and *Sumsets and Structure*, by Imre Z. Ruzsa.

The first part focusses on the interplay between zero-sum problems, arising from the Erdős–Ginzburg–Ziv theorem, and nonuniqueness of factorizations in monoids and integral domains.

The second part deals with structural set addition. It aims at describing the structure of sets in a commutative group from the knowledge of some properties of its sumset.

The third part of the volume collects some of the seminars which accompanied the main courses and covers several aspects of contemporary methods and problems in Additive Combinatorics: multiplicative properties of sumsets (Christian Elsholtz), a step further in the inverse $3k-4$-theorem (Gregory A. Freiman), the isoperimetric method (Yahya O. Hamidoune), new developments around Følner's theorem (Norbert Hegyvári), the polynomial method (Gyula Károlyi), a survey on open problems (Melvyn B. Nathanson), spectral techniques for the sum-product problem (Jozsef Solymosi), and multidimensional inverse problems (Yonutz V. Stanchescu).

We are grateful to Itziar Bardají and Lluís Vena for their careful proofreading of all its chapters.

This edition of the DocCourse has been supported by the Spanish project i-Math and by the Centre de Recerca Matemàtica, to which we express our gratitude. We particularly want to thank the director, Prof. Joaquim Bruna, and the staff of the Centre de Recerca Matemàtica for their excellent job in organizing this edition of the DocCourse.

Barcelona, July 2008
Javier Cilleruelo, Marc Noy and Oriol Serra
Coordinators of the DocCourse

Contents

Part I

Additive Group Theory and Non-unique Factorizations

Alfred Geroldinger

Introduction[1]

This is the extended and revised version of notes written for the Advanced Course in Combinatorics and Geometry: Additive Combinatorics. The course took place at the Centre de Recerca Matemàtica (CRM) at Barcelona in spring 2008. It gives a survey on the interaction between two, at first glance very disparate areas of mathematics: Non-Unique Factorization Theory (see [71, 70, 13, 87, 120]) and Additive Group Theory (see [103, 36, 104, 107, 23, 130, 52]). The main objective of factorization theory is a systematic treatment of phenomena related to the non-uniqueness of factorizations in monoids and integral domains. In the setting of Krull monoids (the main examples we have in mind are the multiplicative monoids of rings of integers of algebraic number fields) most problems can be translated into zero-sum problems over the class group. It will be a main aim of this course to highlight this relationship.

In Chapter 1 we introduce the basic concepts of factorization theory, point out that arithmetical questions in arbitrary Krull monoids can be translated into combinatorial questions on zero-sum sequences over the class group and formulate some main problems (Section 1.4). In Chapter 2 we study the Davenport constant and, using group algebras, we derive its precise value for p-groups (Theorem 2.2.6). In Chapter 3 we discuss the structure of sets of lengths (see Theorems 3.1.3, 3.2.4, 3.2.5 and 3.2.6). The characterization problem (Section 3.3) is a central topic. We give a proof in the case of cyclic groups and elementary 2-groups (Corollary 5.3.3), and this proof requires most of the results from additive group theory discussed in the previous parts. Chapter 4 starts with addition theorems, and then the Erdős–Ginzburg–Ziv constant and some of its variants are studied. We outline the power of the inductive method and determine the Davenport and the Erdős–Ginzburg–Ziv constant for groups of rank at most two (Theorem 4.2.10). Chapter 5 deals with inverse zero-sum problems. The focus is on cyclic groups and on groups of rank two.

[1]This work was supported by the Centre de Recerca Matemàtica (CRM) at Barcelona and by the Austrian Science Fund FWF, Project No. M1014-N13.

I would like to thank the co-ordinators of the DocCourse, Combinatorics and Geometry 2008, J. Cilleruelo, M. Noy and O. Serra, for their invitation and all their hospitality, and the CRM for providing an excellent working atmosphere.

A special thanks goes to W. Gao, D. Grynkiewicz, F. Halter-Koch, Y. ould Hamidoune and Wolfgang A. Schmid for all their suggestions and helpful comments on previous versions of this manuscript.

Notation

Our notation and terminology is consistent with [71]. We briefly gather some key notions. We denote by $\mathbb{N}$ the set of positive integers, and we put $\mathbb{N}_0 = \mathbb{N} \cup \{0\}$. For real numbers $a, b \in \mathbb{R}$ we set $[a, b] = \{x \in \mathbb{Z} \mid a \le x \le b\}$, and we define $\sup \emptyset = \max \emptyset = \min \emptyset = 0$.

Let $A, B \subset \mathbb{Z}$ be finite nonempty subsets. Then $A + B = \{a + b \mid a \in A, b \in B\}$ is their *sumset.* We denote by $\Delta(A)$ the *set of (successive) distances* of A, that is if $A = \{a_1, \ldots, a_t\}$ with $t \in \mathbb{N}$ and $a_1 < \cdots < a_t$, then $\Delta(A) = \{a_{\nu+1} - a_\nu \mid \nu \in [1, t-1]\}$. Moreover, we set $\Delta(\emptyset) = \emptyset$. A subset $P \subset \mathbb{Z}$ is called an *arithmetical progression* with *difference* $d \in \mathbb{N}$ if P is finite nonempty and $\Delta(P) \subset \{d\}$. If $A \subset \mathbb{N}$, we call

$$\rho(A) = \frac{\max A}{\min A} \in \mathbb{Q}_{\ge 1}$$

the *elasticity* of A, and we set $\rho(\{0\}) = 1$.

By a *monoid* we always mean a commutative semigroup with identity which satisfies the cancellation law (that is, if a, b, c are elements of the monoid with $ab = ac$, then $b = c$ follows). If R is an integral domain and $R^\bullet = R \setminus \{0\}$ its multiplicative semigroup of non-zero elements, then $R^\bullet$ is a monoid.

Throughout this paper, let H be a multiplicative monoid and G an additive finite abelian group.

Chapter 1

Basic concepts of non-unique factorizations

We denote by $H^\times$ the set of invertible elements of H, and we say that H is *reduced* if $H^\times = \{1\}$. Let $H_{\rm red} = H/H^\times = \{aH^\times \mid a \in H\}$ be the associated reduced monoid and $\mathsf q(H)$ a quotient group of H.

Let $a, b \in H$. We say that *a divides b* (and we write $a \,|\, b$) if there is an element $c \in H$ such that $b = ac$. We say that a and b are *associated* (and we write $a \simeq b$) if $a \,|\, b$ and $b \,|\, a$ (equivalently, $aH^\times = bH^\times$).

A monoid F is called *free* (*abelian, with basis* $P \subset F$) if every $a \in F$ has a unique representation in the form

$$a = \prod_{p \in P} p^{\mathsf v_p(a)} \quad \text{with} \quad \mathsf v_p(a) \in \mathbb N_0 \ \text{ and } \ \mathsf v_p(a) = 0 \ \text{ for almost all } \ p \in P\,.$$

In this case, F is (up to canonical isomorphism) uniquely determined by P, and conversely P is uniquely determined by F.

We set $F = \mathcal F(P)$ and call

$$|a|_F = |a| = \sum_{p \in P} \mathsf v_p(a) \quad \text{the } \textit{length} \text{ of } a\,.$$

An element $a \in H$ is called

- an *atom* (or an irreducible element) if $a \notin H^\times$ and, for all $b, c \in H$, $a = bc$ implies $b \in H^\times$ or $c \in H^\times$. We denote by $\mathcal A(H)$ the set of all atoms of H.
- a *prime* (or a prime element) if $a \notin H^\times$ and, for all $b, c \in H$, $a \,|\, bc$ implies $a \,|\, b$ or $a \,|\, c$.

The monoid H is called

- *atomic* if every $a \in H \setminus H^\times$ is a product of atoms.
- *factorial* if it satisfies one of the following equivalent conditions:
  (a) Every $a \in H \setminus H^\times$ is a product of primes.
  (b) H is atomic, and every atom is a prime.
  (c) Every $a \in H \setminus H^\times$ is a product of atoms, and this factorization is unique up to associates and the order of the factors.
  (d) $H_{\rm red}$ is free (in that case $H_{\rm red}$ is free with basis $\{pH^\times \mid p \in P\}$ where P denotes the set of primes of H).
  (e) $H = H^\times \times \mathcal{F}(P)$ for some subset $P \subset H$ (in that case P is a maximal set of pairwise non-associated primes of H).

Every prime is an atom, and every factorial monoid is atomic. An element $a \in H$ is an atom [a prime] of H if and only if $aH^\times$ is an atom [a prime] of $H_{\rm red}$. Thus $H_{\rm red}$ is atomic [factorial] if and only if H has this property.

By a *factorization* z of an element $a \in H$ we mean an equation of the form

$$z : a = u_1 \cdot \ldots \cdot u_l \text{ with } l \in \mathbb{N}_0 \text{ and } u_1, \ldots, u_l \text{ are atoms}.$$

The number of atoms l is called the length of the factorization, and two factorizations which differ only in the order of their factors and up to associates are considered as being equal. This concept can be formalized as follows.

The free monoid $\mathsf{Z}(H) = \mathcal{F}\big(\mathcal{A}(H_{\rm red})\big)$, whose basis is the set of atoms in $H_{\rm red}$, is called the *factorization monoid* of H. The homomorphism

$$\pi_H = \pi \colon \mathsf{Z}(H) \to H_{\rm red}, \quad \text{defined by} \quad \pi(z) = \prod_{u \in \mathcal{A}(H_{\rm red})} u^{\mathsf{v}_u(z)},$$

is called the *factorization homomorphism* of H. For $a \in H$, we set

$$\mathsf{Z}_H(a) = \mathsf{Z}(a) = \pi^{-1}(aH^\times) \subset \mathsf{Z}(H),$$

and we call the elements $z \in \mathsf{Z}(a)$ the *factorizations* of a. We say that a has *unique factorization* if $|\mathsf{Z}(a)| = 1$. For a factorization $z \in \mathsf{Z}(a)$, we call $|z|$ the *length* of z (clearly, this coincides with the above informal definition), and the set

$$\mathsf{L}_H(a) = \mathsf{L}(a) = \big\{|z| \bigm| z \in \mathsf{Z}(a)\big\} \subset \mathbb{N}_0$$

is called the *set of lengths* of a.

Note that $0 \in \mathsf{L}(a)$ if and only if $a \in H^\times$ and then $\mathsf{L}(a) = \{0\}$. We have $1 \in \mathsf{L}(a)$ if and only if a is an atom and then $\mathsf{L}(a) = \{1\}$. The monoid H is atomic if and only if $\mathsf{Z}(a) \neq \emptyset$ for all $a \in H$, and it is factorial if and only if $|\mathsf{Z}(a)| = 1$ for all $a \in H$. For every $b \in H$ we have

$$\mathsf{Z}(a)\mathsf{Z}(b) \subset \mathsf{Z}(ab) \quad \text{and} \quad \mathsf{L}(a) + \mathsf{L}(b) \subset \mathsf{L}(ab)\,.$$

Furthermore, the monoid H is called

- *half-factorial* if $|\mathsf{L}(a)| = 1$ for all $a \in H$.
- an FF-*monoid* (a finite factorization monoid) if $\mathsf{Z}(a)$ is finite and nonempty for all $a \in H$.
- a BF-*monoid* (a bounded factorization monoid) if $\mathsf{L}(a)$ is finite and nonempty for all $a \in H$.

Half-factorial monoids and domains have received a lot of attention in the literature (see [14, 20, 124] for recent surveys). Here is a first, very simple but important observation.

Lemma 1.0.1. *Let H be atomic but not half-factorial. Then for every $N \in \mathbb{N}$ there exists some $a \in H$ such that $|\mathsf{L}(a)| \geq N+1$.*

Proof. If $a = u_1 \cdot \ldots \cdot u_k = v_1 \cdot \ldots \cdot v_l$ with $k < l$ and $u_1, \ldots, u_k, v_1, \ldots, v_l \in \mathcal{A}(H)$, then

$$c = a^N = (u_1 \cdot \ldots \cdot u_k)^\nu (v_1 \cdot \ldots \cdot v_l)^{N-\nu} \quad \text{for all} \quad \nu \in [0, N]$$

whence $\{\nu k + l(N-\nu) \mid \nu \in [0,N]\} \subset \mathsf{L}(c)$. □

1.1 Arithmetical invariants

Most monoids studied so far in factorization theory are BF-monoids. In particular the multiplicative monoids of noetherian domains are BF-monoids (see [71, Theorem 2.2.9]). We call

$$\mathcal{L}(H) = \{\mathsf{L}(a) \mid a \in H\}$$

the *system of sets of lengths* of H. If H is a BF-monoid, then $\mathcal{L}(H)$ is a set of finite nonempty subsets of the non-negative integers, and apart from the trivial case of half-factoriality, for every $N \in \mathbb{N}$ there is an $L \in \mathcal{L}(H)$ such that $|L| > N$. In order to describe the structure of sets of lengths we introduce the following arithmetical invariants.

Definition 1.1.1. Let H be a BF-monoid.

1. For $a \in H$, we call $\rho(a) = \rho\big(\mathsf{L}(a)\big)$ the *elasticity* of a and

   $$\rho(H) = \sup\{\rho(a) \mid a \in H\} = \sup\{\rho(L) \mid L \in \mathcal{L}(H)\} \in \mathbb{R}_{\geq 1} \cup \{\infty\}$$

   the *elasticity* of H.

2. Let $k \in \mathbb{N}$. If $H = H^\times$, we set $\rho_k(H) = \lambda_k(H) = k$, and if $H \neq H^\times$, then we define

$$\rho_k(H) = \sup\{\max L \mid L \in \mathcal{L}(H),\ k \in L\} \in \mathbb{N} \cup \{\infty\} \quad \text{and}$$
$$\lambda_k(H) = \min\{\min L \mid L \in \mathcal{L}(H),\ k \in L\} \in [1, k]\,.$$

3. We call

$$\Delta(H) = \bigcup_{L \in \mathcal{L}(H)} \Delta(L) \ \subset \mathbb{N}$$

the *set of distances* of H.

Clearly, H is half-factorial if and only if $\Delta(H) = \emptyset$ if and only if $\rho_k(H) = k$ for all $k \in \mathbb{N}$. Furthermore, $|\Delta(H)| = 1$ if and only if all sets of lengths are arithmetical progressions with the same difference. Whereas the elasticity may be infinite in non-principal orders of algebraic number fields (see [71, Corollary 3.7.2]), we shall prove that it is finite in all Krull monoids with finite class group (thus, in particular, in all principal orders; see Theorems 1.3.5 and 2.3.1).

Lemma 1.1.2. *If* H *is a* BF*-monoid and* $\Delta(H)$ *is nonempty, then* $\min \Delta(H) = \gcd \Delta(H)$.

Proof. We set $d = \gcd \Delta(H)$. Clearly, it suffices to show that $d \in \Delta(H)$. There are $t \in \mathbb{N}, d_1, \ldots, d_t \in \Delta(H)$ and $m_1, \ldots, m_t \in \mathbb{Z} \setminus \{0\}$ such that $d = m_1 d_1 + \ldots + m_t d_t$. After renumbering, if necessary, there is some $s \in [1, t]$ such that $m_1, \ldots, m_s, -m_{s+1}, \ldots, -m_t$ are positive. For every $i \in [1, t]$, there are $x_i \in \mathbb{N}$ and $a_i \in H$ such that

$$\{x_i, x_i + d_i\} \subset \mathsf{L}(a_i) \quad \text{for every} \quad i \in [1, s]$$

and

$$\{x_i - d_i, x_i\} \subset \mathsf{L}(a_i) \quad \text{for every} \quad i \in [s+1, t]\,.$$

Then we get

$$\begin{aligned}
\Big\{k = \sum_{i=1}^{s} m_i x_i - \sum_{i=s+1}^{t} m_i x_i,\ l = \sum_{i=1}^{s} m_i (x_i + d_i) - \sum_{i=s+1}^{t} m_i (x_i - d_i)\Big\} \\
\subset \sum_{i=1}^{s} \{m_i x_i, m_i(x_i + d_i)\} + \sum_{i=s+1}^{t} \{(-m_i)(x_i - d_i), (-m_i) x_i\} \\
\subset \sum_{i=1}^{s} \mathsf{L}(a_i^{m_i}) + \sum_{i=s+1}^{t} \mathsf{L}(a_i^{-m_i}) \\
\subset \mathsf{L}\big(a_1^{|m_1|} \cdot \ldots \cdot a_t^{|m_t|}\big) = L\,.
\end{aligned}$$

Since $d \le \min \Delta(H)$, it follows that $d = l - k$ is a successive distance of L and hence $d \in \Delta(L) \subset \Delta(H)$. □

The structure of sets of lengths will be studied in detail in Chapter 3. We continue with concepts which consider factorizations in a more direct way and not only their lengths.

Definition 1.1.3. Let H be atomic and $z,\, z' \in \mathsf Z(H)$, say

$$z = u_1 \cdot \ldots \cdot u_l v_1 \cdot \ldots \cdot v_m \quad \text{and} \quad z' = u_1 \cdot \ldots \cdot u_l w_1 \cdot \ldots \cdot w_n\,,$$

where $l,\, m,\, n \in \mathbb N_0$, $u_1, \ldots, u_l,\, v_1, \ldots, v_m,\, w_1, \ldots, w_n \in \mathcal A(H_{\text{red}})$ and

$$\{v_1, \ldots, v_m\} \cap \{w_1, \ldots, w_n\} = \emptyset\,.$$

Then we call $\mathsf d(z, z') = \max\{m,\, n\} \in \mathbb N_0$ the *distance* between z and z'.

The distance function $\mathsf d \colon \mathsf Z(H) \times \mathsf Z(H) \to \mathbb N_0$ is a metric. The following observation is an analogue to Lemma 1.0.1.

Lemma 1.1.4. *Let H be atomic but not factorial. Then for every $N \in \mathbb N$ there exists some $a \in H$ such that $|\mathsf Z(a)| \ge N+1$, and there exist factorizations $z,\, z' \in \mathsf Z(a)$ such that $\mathsf d(z, z') \ge 2N$.*

This phenomenon motivates the following definition.

Definition 1.1.5. Let H be atomic.

1. We define the *catenary degree* $\mathsf c(a)$ for $a \in H$ to be the smallest $N \in \mathbb N_0 \cup \{\infty\}$ such that, for any two factorizations $z,\, z'$ of a, there exists a finite sequence $z = z_0\,, z_1\,, \ldots, z_k = z'$ of factorizations of a satisfying that $\mathsf d(z_{i-1}, z_i) \le N$ for all $i \in [1, k]$.

2. Globally, we define

$$\mathsf c(H) = \sup\{\mathsf c(a) \mid a \in H\} \in \mathbb N_0 \cup \{\infty\}\,,$$

   and we call $\mathsf c(H)$ the *catenary degree* of H.

The next lemma gathers some elementary properties. In particular, Lemma 1.1.6.1 shows that H is factorial if and only if the catenary degree $\mathsf c(H) = 0$.

Lemma 1.1.6. *Let H be atomic and $a \in H$.*

1. $\mathsf c(a) \le \sup \mathsf L(a)$, *and* $\mathsf c(a) = 0$ *if and only if* $|\mathsf Z(a)| = 1$.

2. *If $z,\, z' \in \mathsf Z(a)$ and $z \ne z'$, then $2 + \big||z| - |z'|\big| \le \mathsf d(z, z')$.*

3. *If $|\mathsf Z(a)| \ge 2$, then $2 + \sup \Delta\big(\mathsf L(a)\big) \le \mathsf c(a)$. In particular, $2 + \sup \Delta(H) \le \mathsf c(H)$.*

4. *If $\mathsf c(a) \le 2$, then $|\mathsf L(a)| = 1$, and if $\mathsf c(a) \le 3$, then $\mathsf L(a)$ is an arithmetical progression with difference* 1.

Proof. 1. If $z,\ z' \in \mathsf{Z}(a)$, then $\mathsf{d}(z,z') \le \max\{|z|,|z'|\} \le \sup \mathsf{L}(a)$. Hence $\mathsf{c}(a) \le \sup \mathsf{L}(a)$. The second assertion follows by the very definition of $\mathsf{c}(a)$.

2. Let $z,\ z' \in \mathsf{Z}(a)$ be distinct, $x = \gcd(z,z')$ and $z = xy,\ \ z' = xy'$, where $y,\ y' \in \mathsf{Z}(H)$. Then $|y| \ge 2,\ \ |y'| \ge 2$ and $\mathsf{d}(z,z') = \max\{|y|,|y'|\}$. Thus it follows that $2 + \big||z| - |z'|\big| = 2 + \big||y| - |y'|\big| \le \max\{|y|,|y'|\} = \mathsf{d}(z,z')$.

3. We may assume that $\Delta\big(\mathsf{L}(a)\big) \ne \emptyset$, and we must prove that $2+s \le \mathsf{c}(a)$ for every $s \in \Delta\big(\mathsf{L}(a)\big)$. If $s \in \Delta\big(\mathsf{L}(a)\big)$, then there exist factorizations $z,\ z' \in \mathsf{Z}(a)$ such that $|z'| = |z| + s$, and there is no factorization $z'' \in \mathsf{Z}(a)$ with $|z| < |z''| < |z'|$. By definition of $\mathsf{c}(a)$, there exist factorizations $z = z_0, z_1, \ldots, z_k = z' \in \mathsf{Z}(a)$ such that $\mathsf{d}(z_{i-1}, z_i) \le \mathsf{c}(a)$ for all $i \in [1,k]$. Thus there exists some $i \in [1,k]$ such that $|z_{i-1}| \le |z|$ and $|z_i| \ge |z'|$. Hence $2 + s \le 2 + |z_i| - |z_{i-1}| \le \mathsf{d}(z_{i-1}, z_i) \le \mathsf{c}(a)$.

4. This is obvious by 3. □

Next we consider local tameness. We start with the formal definition, and then we discuss the meaning of this concept in some detail.

Definition 1.1.7. Suppose that H is atomic.

1. For $a, b \in H$ let $\omega(a,b)$ denote the smallest $N \in \mathbb{N}_0 \cup \{\infty\}$ with the following property:

   For all $n \in \mathbb{N}$ and $a_1, \ldots, a_n \in H$, if $a = a_1 \cdot \ldots \cdot a_n$ and $b \mid a$, then there exists a subset $\Omega \subset [1,n]$ such that $|\Omega| \le N$ and

   $$b \,\Big|\, \prod_{\nu \in \Omega} a_\nu \,.$$

   In particular, if $b \nmid a$, then $\omega(a,b) = 0$. For $b \in H$ we define

   $$\omega(H,b) = \sup\big\{\omega(a,b) \,\big|\, a \in H\big\} \in \mathbb{N}_0 \cup \{\infty\}\,.$$

2. For $a \in H$ and $x \in \mathsf{Z}(H)$ let $\mathsf{t}(a,x) \in \mathbb{N}_0 \cup \{\infty\}$ denote the smallest $N \in \mathbb{N}_0 \cup \{\infty\}$ with the following property:

   If $\mathsf{Z}(a) \cap x\mathsf{Z}(H) \ne \emptyset$ and $z \in \mathsf{Z}(a)$, then there exists $z' \in \mathsf{Z}(a) \cap x\mathsf{Z}(H)$ such that $\mathsf{d}(z,z') \le N$.

   For subsets $H' \subset H$ and $X \subset \mathsf{Z}(H)$, we define

   $$\mathsf{t}(H',X) = \sup\big\{\mathsf{t}(a,x) \,\big|\, a \in H', x \in X\big\} \in \mathbb{N}_0 \cup \{\infty\}\,.$$

   H is called *locally tame* if $\mathsf{t}(H,u) < \infty$ for all $u \in \mathcal{A}(H_{\mathrm{red}})$.

Local tameness is a basic finiteness property in the theory of non-unique factorizations, in the sense that in many situations where the finiteness of an arithmetical invariant such as the catenary degree or the set of distances is studied, local tameness has to be proved first (see also the sketch of the proof of

Theorem 3.2.4). The closely related $\omega(H, \cdot)$-invariants, introduced in [68], are further well-established invariants in the theory of non-unique factorizations, which appear also in the context of direct-sum decompositions of modules (see [26, Remark 1.6]).

For simplicity of notation suppose that H is atomic and reduced, and let $u \in \mathcal{A}(H)$. Then u is a prime if and only if $\omega(H, u) = 1$. Thus $\omega(H, u)$ measures how far away is u from being a prime. Let $a \in H$. If $u \nmid a$, then $\mathsf{t}(a, u) = 0$ by definition. Suppose that $u \,|\, a$. Then $\mathsf{t}(a, u)$ is the smallest $N \in \mathbb{N}_0 \cup \{\infty\}$ with the following property: If $z = a_1 \cdot \ldots \cdot a_n$ is any factorization of a where $a_1, \ldots, a_n$ are atoms, then there exist a subset $\Omega \subset [1, n]$, say $\Omega = [1, k]$, and a factorization $z' = u u_2 \cdot \ldots \cdot u_l a_{k+1} \cdot \ldots \cdot a_n \in \mathsf{Z}(a)$, with atoms $u_2, \ldots, u_l$, such that $\max\{k, l\} \le N$. Thus $\mathsf{t}(a, u)$ measures how far away from any given factorization z of a there is a factorization z' of a which contains u, and if u is not a prime, then $\omega(H, u) \le \mathsf{t}(H, u)$. Suppose that u is a prime. Then every factorization of a contains u, we can choose $z' = z$ in the above definition, obtain that $\mathsf{d}(z, z') = \mathsf{d}(z, z) = 0$ and hence $\mathsf{t}(H, u) = 0$. Whereas in monoids, which satisfy the ascending chain condition for v-ideals, we have $\omega(H, u) < \infty$ for all atoms $u \in \mathcal{A}(H)$, this does not hold for the $\mathsf{t}(H, u)$ values (see [74, Theorems 3.6 and 4.4]).

1.2 Krull monoids

Krull monoids play a central role in factorization theory. We briefly summarize some of their main properties without giving any proofs. Then we discuss two main examples of Krull monoids: those stemming from domains and the monoid of zero-sum sequences over an abelian group. For more on the theory of Krull monoids we refer the reader to the monographs [88, 79, 71]. A detailed discussion of further examples may be found in [71, Examples 2.3.2 and 7.4.2].

Definition 1.2.1 (Krull monoids and class groups).

1. Let D be a monoid and $H \subset D$ a submonoid with $\mathsf{q}(H) \subset \mathsf{q}(D)$.
   (a) Then $H \subset D$ is called *saturated* if $\mathsf{q}(H) \cap D = H$ (that is, if $a, b \in H$ and a divides b in D, then a divides b in H).
   (b) For $a \in \mathsf{q}(D)$ we denote by $[a] = [a]_{D/H} = a\,\mathsf{q}(H) \in \mathsf{q}(D)/\mathsf{q}(H)$ the class containing a. We call $D/H = \{[a] \mid a \in D\} \subset \mathsf{q}(D)/\mathsf{q}(H)$ the *class group* of D modulo H.
2. H is called a *Krull monoid* if H_{red} is a saturated submonoid of a free monoid.
3. Let H be a Krull monoid and suppose that $H_{\mathrm{red}} \subset D = \mathcal{F}(P)$ is a saturated submonoid of a free monoid such that every $p \in P$ is the greatest common divisor of finitely many elements of H_{red}. Then we call D a monoid of *divisors* and P a set of *prime divisors* of H (for short, we refer to them as primes).

Let $H \subset D$ be as above. If $\mathsf{q}(D)/\mathsf{q}(H)$ is finite (this condition is fulfilled throughout the present article), then $D/H = \mathsf{q}(D)/\mathsf{q}(H)$. Class groups will be written additively whence $[1]$ is the zero element of D/H. Moreover, $H \subset D$ is saturated if and only if

$$H = \{a \in D \mid [a] = [1]\}.$$

Every Krull monoid possesses a monoid of divisors, and if D and D' are monoids of divisors of H, then there is a unique isomorphism $\Phi \colon D \to D'$ with $\Phi \mid H_{\text{red}} = \text{id}$. Hence the class group

$$\mathcal{C}(H) = D/H_{\text{red}} \quad \text{and the subset} \quad \{[p] \in \mathcal{C}(H) \mid p \in P\}$$

of all classes containing primes are uniquely determined by H (up to canonical isomorphism) and hence $\mathcal{C}(H)$ will be called the *class group* of the Krull monoid H.

Now we consider domains and outline when the multiplicative monoid of a domain is a Krull monoid (more details and proofs can be found in [71, Section 2.10]). Let R be a domain,

$$\mathcal{H}(R) = \{aR \mid a \in R^\bullet\}$$

the monoid of non-zero principal ideals and

$$\mathcal{I}^*(R) = \{I \lhd R \mid I \text{ is invertible}\}$$

the monoid of invertible ideals (recall that a non-zero ideal I of R is invertible if there is a non-zero ideal J of R such that their product IJ is a principal ideal). Then $(R^\bullet)_{\text{red}} \cong \mathcal{H}(R)$, the prime elements of the monoid $\mathcal{I}^*(R)$ are precisely the non-zero prime ideals of R, and $\mathcal{H}(R) \subset \mathcal{I}^*(R)$ is saturated.

Theorem 1.2.2. *Let R be an integral domain.*

1. *The following statements are equivalent:*
   (a) *$R^\bullet$ is a Krull monoid.*
   (b) *R is completely integrally closed and satisfies the ascending chain condition for v-ideals (also called divisorial ideals).*
   (c) *R satisfies the ascending chain condition for v-ideals and $R_{\mathfrak{m}}$ is a discrete valuation domain for all v-maximal v-ideals of R.*
2. *The following statements are equivalent:*
   (a) *R is integrally closed, noetherian and every non-zero prime ideal of R is maximal.*
   (b) *R is a one-dimensional Krull domain.*
   (c) *Every non-zero ideal is a product of prime ideals.*

A domain R is called a *Krull domain* if it satisfies the equivalent conditions of Theorem 1.2.2.1. In particular, every integrally closed noetherian domain is a Krull domain.

A domain R is called a *Dedekind domain* if it satisfies the equivalent conditions of Theorem 1.2.2.2. Suppose R is a Dedekind domain. Then $\mathcal{I}^*(R)$ is a monoid of divisors of $\mathcal{H}(R)$, the set of non-zero prime ideals is a set of prime divisors of $\mathcal{H}(R)$, and the class group of $\mathcal{H}(R) \subset \mathcal{I}^*(R)$ is the usual ideal class group of R. If K is an algebraic number field and $\mathfrak{o}_K$ is the ring of integers of K, then $\mathfrak{o}_K$ is a Dedekind domain with finite class group and every class contains infinitely many primes.

Next we discuss the monoid of zero-sum sequences over an abelian group, which will turn out to be a Krull monoid. It connects the theory of non-unique factorizations with additive group theory and combinatorial number theory.

Definition 1.2.3. Let $G_0 \subset G$ be a subset.

1. Let $\mathcal{F}(G_0)$ be the free (multiplicative) monoid with basis G_0. The elements of $\mathcal{F}(G_0)$ are called *sequences* over G_0. We write the sequences $S \in \mathcal{F}(G_0)$ in the form
$$S = \prod_{g \in G_0} g^{\mathsf{v}_g(S)} = g_1 \cdot \ldots \cdot g_l \in \mathcal{F}(G_0)\,,$$
where $\mathsf{v}_g(S) \in \mathbb{N}_0$.
2. If S is as above, then
$$\sigma(S) = \sum_{i=1}^{l} g_i = \sum_{g \in G} \mathsf{v}_g(S) g \in G \qquad \text{is called the } \textit{sum} \text{ of } S,$$
and we denote by $\mathcal{B}(G_0) = \{S \in \mathcal{F}(G_0) \mid \sigma(S) = 0\}$ the *monoid of zero-sum sequences* (*block monoid*) over G_0. The elements of $\mathcal{B}(G_0)$ are called *zero-sum sequences,* and the atoms of $\mathcal{B}(G_0)$ are called *minimal zero-sum sequences*.

For every arithmetical invariant $*(H)$ defined for the monoid H, we write $*(G_0)$ instead of $*\big(\mathcal{B}(G_0)\big)$ whenever the precise meaning is clear from the context. For example, we set $\mathcal{A}(G_0) = \mathcal{A}\big(\mathcal{B}(G_0)\big)$, $\mathcal{L}(G_0) = \mathcal{L}\big(\mathcal{B}(G_0)\big)$, $\Delta(G_0) = \Delta\big(\mathcal{B}(G_0)\big)$ and so on.

Proposition 1.2.4. *Let $G_0 \subset G$ be a nonempty subset.*

1. *$\mathcal{B}(G_0) \subset \mathcal{F}(G_0)$ is saturated and thus $\mathcal{B}(G_0)$ is a Krull monoid.*
2. *$\mathcal{A}(G_0)$ is finite and thus $\mathcal{B}(G_0)$ is finitely generated.*
3. *If $|G| \ne 2$, then $\mathcal{F}(G)$ is a monoid of divisors for $\mathcal{B}(G)$, $\mathcal{C}\big(\mathcal{B}(G)\big) \cong G$, and every class of $\mathcal{B}(G)$ contains exactly one prime.*

4. *The following statements are equivalent:*

   (a) $|G| \le 2$.

   (b) $\mathcal{B}(G)$ *is factorial.*

   (c) $\mathcal{B}(G)$ *is half-factorial.*

Proof. 1. This follows immediately from the definitions.

2. Every atom of $\mathcal{B}(G_0)$ divides the zero-sum sequence

$$B = \prod_{g \in G_0} g^{\mathrm{ord}(g)}$$

and hence there are only finitely many atoms.

3. Let $|G| \ne 2$. To verify that $\mathcal{F}(G)$ is a monoid of divisors for $\mathcal{B}(G)$, let $g \in G$ be given. If $\mathrm{ord}(g) = n \ge 3$, then $g = \gcd\bigl(g^n, g(-g)\bigr)$. If $\mathrm{ord}(g) = 2$, then there is an element $h \in G \setminus \{0, g\}$ and $g = \gcd\bigl(g^2, gh(g-h)\bigr)$.

The map $\sigma\colon \mathcal{F}(G) \to G$ is a monoid epimorphism. If $S, S' \in \mathcal{F}(G)$, then $\sigma(S) = \sigma(S')$ if and only if $S' \in [S] = S\mathsf{q}\bigl(\mathcal{B}(G)\bigr)$. Thus σ induces a group isomorphism $\Phi\colon \mathcal{F}(G)/\mathcal{B}(G) \to G$, defined by $\Phi([S]) = \sigma(S)$, and we have $[S] \cap G = \{g\}$. Thus the class $[S]$ contains exactly one prime.

4. (a) $\Rightarrow$ (b). If $G = \{0\}$, then $\mathcal{B}(G) = \mathcal{F}(G) \cong (\mathbb{N}_0, +)$ is factorial. Suppose that $G = \{0, e\}$. Then $\mathcal{A}(G) = \{0, e^2\}$, every atom is a prime and hence $\mathcal{B}(G)$ is factorial (indeed, $\mathcal{B}(G) \cong (\mathbb{N}_0^2, +)$).

(b) $\Rightarrow$ (c). This implication is obvious.

(c) $\Rightarrow$ (a). Suppose there is some $g \in G$ with $\mathrm{ord}(g) = n \ge 3$. Then $U = g^n$, $-U = (-g)^n$, $V = (-g)g$ are atoms of $\mathcal{B}(G)$ and $(-U)U = V^n$, a contradiction to half-factoriality. Thus $\mathrm{ord}(g) \le 2$ for all $g \in G$. Assume to the contrary that there are two distinct non-zero elements $e_1, e_2 \in G$ and set $e_0 = e_1 + e_2$. Then $U = e_0 e_1 e_2$ and $V_i = e_i^2$ are atoms of $\mathcal{B}(G)$ for $i \in [0, 2]$. But $U^2 = V_0 V_1 V_2$ is again a contradiction to half-factoriality. Thus G has no elements of order greater than or equal to 3, and at most one element of order 2 which implies $|G| \le 2$. □

1.3 Transfer principles

A central method in factorization theory is to study the arithmetic in auxiliary monoids and to shift the results to monoids and domains of arithmetical interest. We start with the crucial definition.

Definition 1.3.1. A monoid homomorphism $\theta\colon H \to B$ is called a *transfer homomorphism* if it has the following properties:

(T 1) $B = \theta(H)B^\times$ and $\theta^{-1}(B^\times) = H^\times$.

(T 2) If $u \in H$, $b, c \in B$ and $\theta(u) = bc$, then there exist $v, w \in H$ such that $u = vw$, $\theta(v) \simeq b$ and $\theta(w) \simeq c$.

Thus the strategy is to find, for a given monoid H, a simpler monoid B, to study the arithmetic in B, and then to shift the arithmetical results from B back to H. The next proposition shows that a shift back is possible.

Proposition 1.3.2. *Let $\theta\colon H \to B$ be a transfer homomorphism of atomic monoids and let $u \in H$.*

1. *If $n \in \mathbb{N}$, $b_1, \ldots, b_n \in B$ and $\theta(u) \simeq b_1 \cdot \ldots \cdot b_n$, then there exist $u_1, \ldots, u_n \in H$ such that $u \simeq u_1 \cdot \ldots \cdot u_n$ and $\theta(u_\nu) \simeq b_\nu$ for all $\nu \in [1, n]$.*
2. *u is an atom of H if and only if $\theta(u)$ is an atom of B.*
3. *$\mathsf{L}_H(u) = \mathsf{L}_B(\theta(u))$.*
4. *$\mathcal{L}(H) = \mathcal{L}(B)$. In particular, H is a* BF*-monoid if and only if B is a* BF*-monoid, and then we have $\rho(H) = \rho(B)$ and $\Delta(H) = \Delta(B)$.*

Proof. We suppose that H and B are reduced.

1. This follows by induction on n.

2. If $u \in \mathcal{A}(H)$ and $\theta(u) = bc$ for some $b, c \in B$, then there exist $v, w \in H$ such that $u = vw$, $\theta(v) = b$ and $\theta(w) = c$. Hence $v = 1$ or $w = 1$ and thus $b = 1$ or $c = 1$. If $\theta(u) \in \mathcal{A}(B)$ and $u = vw$ for some $v, w \in H$, then $\theta(u) = \theta(v)\theta(w)$ implies $\theta(v) = 1$ or $\theta(w) = 1$ and thus $v = 1$ or $w = 1$.

3. By **(T1)**, we have $u = 1$ if and only if $\theta(u) = 1$, and by definition we have $\mathsf{L}_H(1) = \{0\} = \mathsf{L}_B(\theta(u))$. Suppose that $u \ne 1$. If $k \in \mathsf{L}_H(u)$, then there are atoms $u_1, \ldots, u_k$ of H such that $u = u_1 \cdot \ldots \cdot u_k$. Then $\theta(u) = \theta(u_1) \cdot \ldots \cdot \theta(u_k)$. By 2., $\theta(u_1), \ldots, \theta(u_k)$ are atoms of B, and hence $k \in \mathsf{L}_B(\theta(u))$. Conversely, we pick $k \in \mathsf{L}_B(\theta(u))$. Then there are atoms $b_1, \ldots, b_k$ of B such that $\theta(u) = b_1 \cdot \ldots \cdot b_k$. By 1., there are $u_1, \ldots, u_k \in H$ such that $u = u_1 \cdot \ldots \cdot u_k$ and $\theta(u_\nu) = b_\nu$ for all $\nu \in [1, k]$. Thus by 2., $u_1, \ldots, u_k$ are atoms of H and hence $k \in \mathsf{L}_H(u)$.

4. This follows immediately from 3. □

We introduce the Davenport constant which will be investigated in detail in Chapter 2. Recall that $\mathcal{A}(G_0)$ is finite by Proposition 1.2.4.

Definition 1.3.3. Let $G_0 \subset G$ be a nonempty subset. Then

$$\mathsf{D}(G_0) = \max\{|U| \mid U \in \mathcal{A}(G_0)\} \in \mathbb{N}_0$$

is called the *Davenport constant* of G_0.

The next result gives the required link between factorization theory on the one side and additive group theory and combinatorial number theory on the other side.

Theorem 1.3.4. *Let H be a reduced Krull monoid with finite class group, $H \subset D = \mathcal{F}(P)$ a monoid of divisors and $G_0 = \{[p] \mid p \in P\} \subset G = D/H$ the set of classes containing primes. Let $\widetilde{\boldsymbol{\beta}}\colon D \to \mathcal{F}(G_0)$ be the unique homomorphism satisfying $\widetilde{\boldsymbol{\beta}}(p) = [p]$ for all $p \in P$.*

1. *For* $a \in D$ *we have* $\widetilde{\boldsymbol{\beta}}(a) \in \mathcal{B}(G_0)$ *if and only if* $a \in H$. *Thus* $\widetilde{\boldsymbol{\beta}}(H) = \mathcal{B}(G_0)$ *and* $\widetilde{\boldsymbol{\beta}}^{-1}(\mathcal{B}(G_0)) = H$.

2. *The restriction* $\boldsymbol{\beta} = \widetilde{\boldsymbol{\beta}}|H \colon H \to \mathcal{B}(G_0)$ *is a transfer homomorphism. In particular, we have* $\mathcal{L}(H) = \mathcal{L}(G_0)$.

3. $\mathsf{D}(G_0)$ *is the maximum of all* $l \in \mathbb{N}_0$ *with the following property: There exists an atom* $u \in H$ *such that* u *is the product of* l *primes of* D.

4. *We have* $\mathsf{c}(G_0) \le \mathsf{c}(H) \le \max\{\mathsf{c}(G_0), 2\}$.

Proof. 1. Let $a = p_1 \cdot \ldots \cdot p_l \in D$ where $l \in \mathbb{N}$ and $p_1, \ldots, p_l \in P$. Then

$$\widetilde{\boldsymbol{\beta}}(a) = [p_1] \cdot \ldots \cdot [p_l] \in \mathcal{F}(G_0) \quad \text{and} \quad \sigma\big([p_1] \cdot \ldots \cdot [p_l]\big) = [p_1] + \ldots + [p_l] = [a]\,.$$

Since $H \subset D$ is saturated, we have $[a] = 0 \in G$ if and only if $a \in H$, and thus all assertions follow.

2. By 1., $\boldsymbol{\beta} \colon H \to \mathcal{B}(G_0)$ is surjective and $\boldsymbol{\beta}^{-1}(1) = \{1\}$. Let $a = p_1 \cdot \ldots \cdot p_l \in H$, with $l \in \mathbb{N}$ and $p_1, \ldots, p_l \in P$, and suppose that $\boldsymbol{\beta}(a) = BC$, say $B = [p_1] \cdot \ldots \cdot [p_k]$ and $C = [p_{k+1}] \cdot \ldots \cdot [p_l]$. By 1., $b = p_1 \cdot \ldots \cdot p_k \in H$, $c = p_{k+1} \cdot \ldots \cdot p_l \in H$ and clearly we have $a = bc$. Therefore $\boldsymbol{\beta}$ is a transfer homomorphism, and thus Proposition 1.3.2 implies $\mathcal{L}(H) = \mathcal{L}(G_0)$.

3. Let $l \in \mathbb{N}_0$, $p_1, \ldots, p_l$ primes of D and $u = p_1 \cdot \ldots \cdot p_l$ an atom of H. Since $\boldsymbol{\beta}$ is a transfer homomorphism, $\boldsymbol{\beta}(u) = [p_1] \cdot \ldots \cdot [p_l] \in \mathcal{A}(G_0)$, and hence $\mathsf{D}(G_0) \ge |\boldsymbol{\beta}(u)| = l$. Conversely, let $U = g_1 \cdot \ldots \cdot g_l \in \mathcal{A}(G_0)$ with $\mathsf{D}(G_0) = |U| = l$. If $p_i \in G_0$ with $g_i = [p_i]$ for $i \in [1, l]$, then $u = p_1 \cdot \ldots \cdot p_l$ is an atom of H which is a product of l primes of D.

4. The proof is not difficult but requires concepts not introduced here (see [71, Theorem 3.4.10]). □

The homomorphism $\boldsymbol{\beta} \colon H \to \mathcal{B}(G_0)$ is called the *block homomorphism* of H. It transports arithmetical problems in H to zero-sum problems over G. In particular, if $a = p_1 \cdot \ldots \cdot p_l \in D$ is as above, then a is an atom of H if and only if $\boldsymbol{\beta}(a)$ is a minimal zero-sum sequence.

The next result states that, in a Krull monoid with finite class group all arithmetical invariants introduced so far are finite. The proof of finiteness is fairly simple. However, establishing the precise values of the invariants is a completely different task. It can be tackled with methods from additive group theory, and the remaining sections of this article will be devoted to that.

Theorem 1.3.5. *Let* H *be a Krull monoid with finite class group. Then* H *is a locally tame* FF*-monoid with finite catenary degree* $\mathsf{c}(H)$, *finite set of distances* $\Delta(H)$, *finite elasticity* $\rho(H)$ *and with* $\rho_k(H) < \infty$ *for all* $k \in \mathbb{N}$.

Proof. We may suppose that H is reduced. Let $D = \mathcal{F}(P)$ be a monoid of divisors of H, $G = D/H$ its class group and $G_0 \subset G$ the set of classes containing primes. We proceed in several steps.

H is an FF-*monoid.* If $a \in H$, then there are primes $p_1, \ldots, p_l \in P$ such that $a = p_1 \cdot \ldots \cdot p_l$, and this is the only factorization of a in D. Therefore every factorization $a = u_1 \cdot \ldots \cdot u_k$ of a into atoms of H corresponds uniquely to a partition

$$[1,l] = \bigcup_{\nu=1}^{k} I_\nu\,, \quad \text{where} \quad \sum_{j \in I_\nu} [p_j] = 0 \in G \quad \text{but} \quad \sum_{j \in I'_\nu} [p_j] \neq 0 \in G$$

for all nonempty proper subsets $I'_\nu \subset I_\nu$ and all $\nu \in [1,k]$. Thus a has only finitely many factorizations in H.

H is locally tame. Let $u = p_1 \cdot \ldots \cdot p_l \in \mathcal{A}(H)$ with $l \in \mathbb{N}$ and $p_1, \ldots, p_l \in P$. We assert that

$$\mathsf{t}(H,u) \le 1 + \frac{l(\mathsf{D}(G_0)-1)}{2} \le 1 + \frac{\mathsf{D}(G_0)(\mathsf{D}(G_0)-1)}{2}\,.$$

The second inequality follows from Theorem 1.3.4.3 and provides a global bound for all local tame degrees $\mathsf{t}(H,v)$ with $v \in \mathcal{A}(H)$. So we have to verify the first inequality. If u is a prime in H, then $\mathsf{t}(H,u) = 0$ by definition. Suppose that u is not a prime in H, that is, $u \notin P$. Then $\mathsf{D}(G_0) \ge l \ge 2$. We recall two notations. If $c \in D = \mathcal{F}(P)$, then $c = q_1 \cdot \ldots \cdot q_s$, where $s \in \mathbb{N}_0$ and $q_1, \ldots, q_s \in P$, and $|c|_D = s$. If $w \in \mathsf{Z}(H) = \mathcal{F}(\mathcal{A}(H))$, then $w = v_1 \cdot \ldots \cdot v_t$, where $t \in \mathbb{N}_0$ and $v_1, \ldots, v_t \in \mathcal{A}(H)$, and $|w| = t$.

Let $a \in H$, with $u \,|\, a$ and $z = u_1 \cdot \ldots \cdot u_r \in \mathsf{Z}(a)$, where $r \in \mathbb{N}$ and $u_1, \ldots, u_r \in \mathcal{A}(H)$. We must prove that there exists some $z' \in \mathsf{Z}(a) \cap u\mathsf{Z}(H)$ such that

$$\mathsf{d}(z,z') \le 1 + \frac{l(\mathsf{D}(G_0)-1)}{2}\,.$$

After renumbering, if necessary, we may assume that there exists some $k \in [1,r]$ such that $k \le l$, $u \,|\, u_1 \cdot \ldots \cdot u_k$, but u does not divide any proper subproduct of $u_1 \cdot \ldots \cdot u_k$ (in D and hence in H). Since $u \notin P$, it follows that $u_1, \ldots, u_k \notin P$, and thus $u_1 \cdot \ldots \cdot u_k$ is not divisible by any $p \in P \cap H$. If $u_1 \cdot \ldots \cdot u_k = uc$, where $c \in H$, and if $w \in \mathsf{Z}(c)$, then $|c|_D \ge 2|w|$ and

$$|w| \le \frac{|c|_D}{2} = \frac{|u_1|_D + \ldots + |u_k|_D - |u|_D}{2} \le \frac{k\mathsf{D}(G_0) - l}{2} \le \frac{l(\mathsf{D}(G_0)-1)}{2}\,.$$

Now it follows that $z' = uwu_{k+1} \cdot \ldots \cdot u_r \in \mathsf{Z}(a)$, and

$$\mathsf{d}(z,z') \le \max\{k, |w|+1\} \le \max\{l, |w|+1\} \le 1 + \frac{l\,(\mathsf{D}(G_0)-1)}{2}\,.$$

On the remaining invariants. By Theorem 1.3.4, we have $\Delta(H) = \Delta(G_0)$, $\mathsf{c}(H) \le \max\{\mathsf{c}(G_0), 2\}$, $\rho(H) = \rho(G_0)$ and $\rho_k(H) = \rho_k(G_0)$ for all $k \in \mathbb{N}$, and hence it suffices to consider $\mathcal{B}(G_0)$. Clearly, we have $\Delta(G_0) \subset \Delta(G)$, $\mathsf{c}(G_0) \le \mathsf{c}(G)$, $\rho(G_0) \le \rho(G)$ and $\rho_k(G_0) \le \rho_k(G)$ for all $k \in \mathbb{N}$. For these latter invariants we shall derive explicit upper bounds and in some cases even precise values in Section 2.3. □

1.4 Main problems in factorization theory

1. Which noetherian domains satisfy the main finiteness properties of factorization theory: local tameness, finiteness of the catenary degree and the Structure Theorem for Sets of Lengths? (See Definition 3.2.3 and the subsequent results.)

The goal is to derive explicit characterizations in terms of ring invariants. A prototype of such a result may be found in [96, Theorem 6.1]. It provides an explicit ring theoretical characterization of those finitely generated domains having finite elasticity.

2. If R is a ring of integers of an algebraic number field, then almost all elements of R have catenary degree at most 3 (see Corollary 3.2.8).

Which other domains have such a property? Of course, "almost all" has to be interpreted in a suitable way, i.e., for orders in global fields in the sense of Dirichlet density and for $\mathbb{Q}$-algebras in the sense of Zariski density.

3. Let H be a Krull monoid with finite class group G such that every class contains a prime (the multiplicative monoid of non-zero elements of a ring of integers of an algebraic number field is such a Krull monoid).

Find the precise values of arithmetical invariants of H (such as of $\mathsf{c}(H)$, $\Delta(H)$ and $\rho_k(H)$ for $k \in \mathbb{N}$) in terms of the group invariants of G (see [71, Chapter 6], and note that by the simple Theorem 1.3.5 all the invariants are finite). Results of this type are substantial for making progress on the Characterization Problem described in Section 3.3.

Let R be a noetherian domain. If R is integrally closed, then its multiplicative monoid $R^\bullet$ is a Krull monoid. Suppose that R is not integrally closed. If the integral closure $\overline{R}$ is a finitely generated R-module and some further natural finiteness conditions hold, then the arithmetic of R is studied via C-monoids and weakly C-monoids. These monoids play a similar role as the monoid of zero-sum sequences does for Krull monoids (see [71, Theorems 2.11.9 and 3.3.4] and [73]).

However, in the present article we restrict our attention to Krull monoids and focus on Problem **3** with respect to the invariants $\Delta(H)$, $\mathsf{c}(H)$ and $\rho_k(H)$ for all $k \in \mathbb{N}$.

Chapter 2

The Davenport constant and first precise arithmetical results

In this chapter we study the Davenport constant, a classical combinatorial invariant which has been investigated since the 1960s (see [115, 105, 31, 108, 103]). From the very beginning the investigation of this invariant was related also to arithmetical problems (it is reported in [108] that in 1966 H. Davenport asked for $\mathsf{D}(G)$, since it is the largest number of prime ideals occurring in the prime ideal decomposition of an irreducible integer in an algebraic number field with ideal class group G). However, it has turned out that this and related invariants occur in many branches of combinatorics, number theory and geometry (see [52] for a recent survey, and [99, 33] for the relationship with invariant theory).

We shall determine the precise value of the Davenport constant among others for p-groups and for groups of rank at most two (see Theorems 2.2.6 and 4.2.10 and Corollary 4.2.13). For general finite abelian groups the precise value is still unknown. After that we put the Davenport constant in connection with the arithmetical invariants introduced in Chapter 1.

2.1 The Davenport constant

We set $G^\bullet = G \setminus \{0\}$. Let $G_0 \subset G$ be a subset and let

$$S = \prod_{g \in G_0} g^{\mathsf{v}_g(S)} = g_1 \cdot \ldots \cdot g_l \in \mathcal{F}(G_0)$$

be a sequence over G_0. We call $\mathsf{v}_g(S)$ the *multiplicity* of g in S, and we say that S *contains* g if $\mathsf{v}_g(S) > 0$. S is called *squarefree* (in $\mathcal{F}(G)$) if $\mathsf{v}_g(S) \le 1$ for all $g \in G$. A sequence S_1 is called a *subsequence* of S if $S_1 \,|\, S$ in $\mathcal{F}(G)$ (equivalently, $\mathsf{v}_g(S_1) \le \mathsf{v}_g(S)$ for all $g \in G$), and it is called a *proper subsequence* of S if it is a subsequence with $1 \ne S_1 \ne S$. We call

$$|S| = l = \sum_{g \in G_0} \mathsf{v}_g(S) \in \mathbb{N}_0 \qquad \text{the } \textit{length} \text{ of } S,$$

$$\mathsf{h}(S) = \max\{\mathsf{v}_g(S) \mid g \in G\} \in [0, |S|]$$

the *maximum of the multiplicities* of S,

$$\operatorname{supp}(S) = \{g \in G \mid \mathsf{v}_g(S) > 0\} \subset G \qquad \text{the } \textit{support} \text{ of } S,$$

$$\Sigma_k(S) = \Big\{ \sum_{i \in I} g_i \;\Big|\; I \subset [1, l] \text{ with } |I| = k \Big\}$$

the *set of k-term subsums* of S, for all $k \in \mathbb{N}$,

$$\Sigma_{\le k}(S) = \bigcup_{j \in [1,k]} \Sigma_j(S), \qquad \Sigma_{\ge k}(S) = \bigcup_{j \ge k} \Sigma_j(S),$$

and

$$\Sigma(S) = \Sigma_{\ge 1}(S) \quad \text{the } \textit{set of (all) subsums} \text{ of } S.$$

We set $-S = (-g_1) \cdot \ldots \cdot (-g_l)$ and, for every $g \in G$, we set $g + S = (g + g_1) \cdot \ldots \cdot (g + g_l)$.

A sequence S is called *zero-sum free* if $0 \notin \Sigma(S)$, and we denote by $\mathcal{A}^*(G_0)$ the set of all zero-sum free sequences. Since every zero-sum free sequence S is a subsequence of

$$\prod_{g \in G_0} g^{\operatorname{ord}(g) - 1},$$

$\mathcal{A}^*(G_0)$ is finite. For convenience we introduce the following technical variant of the Davenport constant $\mathsf{D}(G_0)$ (introduced in Definition 1.3.3), and in Lemma 2.1.2.3 we give a straightforward characterization.

Definition 2.1.1. Let $G_0 \subset G$ be a nonempty subset. Then

$$\mathsf{d}(G_0) = \max\big\{ |S| \;\big|\; S \in \mathcal{F}(G_0) \text{ is zero-sum free} \big\} \in \mathbb{N}_0$$

is called the *little Davenport constant* of G_0.

Obviously, the map $\psi \colon \mathcal{A}^*(G_0) \to \mathcal{A}(G)$, defined by $S \mapsto (-\sigma(S))S$, is well defined, and $\mathsf{D}(G_0) \le 1 + \mathsf{d}(G_0)$. If $G_0 = G$, then ψ is surjective and $\mathsf{D}(G) = 1 + \mathsf{d}(G)$. The following two lemmas gather some elementary properties of the Davenport constant.

Lemma 2.1.2. 1. *If $S \in \mathcal{A}^*(G)$ has length $|S| = \mathsf{d}(G)$, then $\Sigma(S) = G^\bullet$ and $G = \langle \operatorname{supp}(S) \rangle$.*

2. $\mathsf{d}(G) = \max\big\{ |S| \;\big|\; S \in \mathcal{F}(G),\ \Sigma(S) = G^\bullet \big\}$.

3. *$\mathsf{D}(G)$ is the smallest integer $l \in \mathbb{N}$ such that every sequence $S \in \mathcal{F}(G)$ of length $|S| \ge l$ has a nontrivial zero-sum subsequence (that is, $S \notin \mathcal{A}^*(G)$).*

4. *If* $S \in \mathcal{A}^*(G)$, *then* $|S| \le |\Sigma(S)| \le |G| - 1$. *In particular,* $\mathsf{d}(G) \le |G| - 1$ *and* $\mathsf{D}(G) \le |G|$.

Proof. 1. Let $S \in \mathcal{A}^*(G)$ with $|S| = \mathsf{d}(G)$, and assume that there is some $h \in G^\bullet \setminus \Sigma(S)$. Then $T = (-h)S \in \mathcal{A}^*(G)$ and $|T| = 1 + |S|$, which contradicts the maximal choice of S. Clearly, $\Sigma(S) = G^\bullet$ implies $G = \langle \text{supp}(S) \rangle$.

2. If $S \in \mathcal{F}(G)$ and $\Sigma(S) = G^\bullet$, then $S \in \mathcal{A}^*(G)$, and thus $|S| \le \mathsf{d}(G)$. Conversely, if $S \in \mathcal{A}^*(G)$ and $|S| = \mathsf{d}(G)$, then $\Sigma(S) = G^\bullet$ by 1.

3. By definition we have $\mathsf{d}(G) = \max\{|S| \mid S \in \mathcal{A}^*(G)\}$. Hence $\mathsf{D}(G) = \mathsf{d}(G) + 1$ is the smallest integer $l \in \mathbb{N}$ such that every sequence $S \in \mathcal{F}(G)$ with $|S| \ge l$ does not lie in $\mathcal{A}^*(G)$.

4. If $S = g_1 \cdot \ldots \cdot g_l \in \mathcal{A}^*(G)$, then $C = \{g_1 + \ldots + g_k \mid k \in [1, l]\} \subset \Sigma(S) \subset G^\bullet$, and therefore $|S| = |C| \le |\Sigma(S)| \le |G| - 1$. Hence $\mathsf{d}(G) \le |G| - 1$, and thus $\mathsf{D}(G) \le |G|$. □

Let $r \in \mathbb{N}$. An r-tuple $(e_1, \ldots, e_r)$ of elements of $G^\bullet$ (resp., the elements $e_1, \ldots, e_r$) is said to be *independent* if for every $(m_i)_{i \in [1,r]} \in \mathbb{Z}^r$

$$\sum_{i=1}^{r} m_i e_i = 0 \quad \text{implies that} \quad m_1 e_1 = \cdots = m_r e_r = 0$$

(equivalently, $\langle e_1, \ldots, e_r \rangle = \langle e_1 \rangle \oplus \cdots \oplus \langle e_r \rangle$). Moreover, $(e_1, \ldots, e_r)$ is called a *basis* of G if $(e_1, \ldots, e_r)$ is independent and $\{e_1, \ldots, e_r\}$ generates G.

Suppose that $|G| > 1$. Then by the Structure Theorem of Finite Abelian Groups, we have

$$G \cong C_{n_1} \oplus \cdots \oplus C_{n_r},$$

where $1 < n_1 \mid \cdots \mid n_r$, $r = \mathsf{r}(G)$ is the *rank* of G and $n_r = \exp(G) = \text{lcm}\{\text{ord}(g) \mid g \in G\}$ is the *exponent* of G. We define

$$\mathsf{d}^*(G) = \sum_{i=1}^{r} (n_i - 1),$$

and we set $\mathsf{d}^*(\{0\}) = 0$. G is called an *(elementary) p-group* if $\exp(G)$ is a power of p (resp., $\exp(G) \mid p$).

Lemma 2.1.3. *Let* $\exp(G) = n \ge 2$.

1. *If* $e_1, \ldots, e_r \in G$ *are independent elements, then*

$$S = \prod_{i=1}^{r} e_i^{\text{ord}(e_i) - 1} \in \mathcal{A}^*(G).$$

2. *There exists a sequence* $S \in \mathcal{A}^*(G)$ *such that* $|S| = \mathsf{d}^*(G)$. *In particular,* $\mathsf{d}^*(G) \le \mathsf{d}(G)$.

Proof. 1. If $1 \neq T$ is a a subsequence of S, then $T = e_1^{k_1} \cdot \ldots \cdot e_r^{k_r}$ where $k_i \in [0, \mathrm{ord}(e_i) - 1]$ for all $i \in [1, r]$ and $k_i > 0$ for at least one $i \in [1, r]$. Hence $\sigma(T) = k_1 e_1 + \ldots + k_r e_r \neq 0$, and thus S is zero-sum free.

2. If $G \cong C_{n_1} \oplus \cdots \oplus C_{n_r}$ where $1 < n_1 \mid \cdots \mid n_r$ and $(e_1, \ldots, e_r)$ is a basis of G such that $\mathrm{ord}(e_i) = n_i$ for all $i \in [1, r]$, then $S = e_1^{n_1 - 1} \cdot \ldots \cdot e_r^{n_r - 1} \in \mathcal{A}^*(G)$ by 1., and hence $\mathsf{d}^*(G) = |S| \leq \mathsf{d}(G)$. □

Corollary 2.1.4.

1. *Let G be cyclic of order $n \geq 2$. A sequence $S \in \mathcal{F}(G)$ is zero-sum free of length $|S| = \mathsf{d}(G)$ if and only if $S = g^{n-1}$ for some $g \in G$ with $\mathrm{ord}(g) = n$. In particular, $\mathsf{d}(G) = \mathsf{d}^*(G) = n - 1$ and $\mathsf{D}(G) = n$.*

2. *Let G be an elementary 2-group. A sequence $S \in \mathcal{F}(G)$ is zero-sum free if and only if S is squarefree and $\mathrm{supp}(S)$ is an independent set. In particular, $\mathsf{d}(G) = \mathsf{d}^*(G) = \mathsf{r}(G)$.*

Proof. 1. By Lemmas 2.1.2 and 2.1.3, we have $n - 1 = \mathsf{d}^*(G) \leq \mathsf{d}(G) \leq |G| - 1 = n - 1$ and thus $\mathsf{d}(G) = n - 1$ and $\mathsf{D}(G) = n$. Obviously, if $g \in G$ with $\mathrm{ord}(g) = n$, then $S = g^{n-1} \in \mathcal{A}^*(G)$. Conversely, assume to the contrary that $S = g_1 \cdot \ldots \cdot g_{n-1} \in \mathcal{A}^*(G)$ and $g_1 \neq g_2$. If $\Sigma = \{g_1 + \ldots + g_k \mid k \in [1, n-1]\}$, then $|\Sigma| = n - 1$ and $g_2 \notin \Sigma$, a contradiction.

2. If S is squarefree and $\mathrm{supp}(S)$ is independent, then S is zero-sum free by Lemma 2.1.3.1. Conversely, if $S \in \mathcal{A}^*(G)$, then $\mathsf{v}_g(S) < \mathrm{ord}(g) \leq 2$ for all $g \in \mathrm{supp}(S)$. Hence S is squarefree, and $0 \notin \Sigma(S)$ implies that $\mathrm{supp}(S)$ is independent.

Thus we get $\mathsf{d}(G) = \mathsf{r}(G)$, and by the very definitions, it follows that $\mathsf{d}^*(G) = \mathsf{r}(G)$. □

There is a weighted version of the Davenport constant, called the cross number, which plays a crucial role in factorization theory (in particular, in the investigations of half-factorial and minimal non-half-factorial subsets; see [71, Chapter 5], [110, 111] and [77, 76] for recent progress).

2.2 Group algebras

Group algebras $R[G]$ – over suitable commutative rings R – have turned out to be powerful tools for a growing variety of questions from combinatorics and number theory. We discuss the classical application of group algebras to the investigation of zero-sum free sequences over p-groups, which is due to P. van Emde Boas, D. Kruyswijk and J.E. Olson (see [30], [31], [108]). Theorem 2.2.6 provides the classical result that, for a p-group G we have $\mathsf{d}(G) = \mathsf{d}^*(G)$.

Let R be a commutative ring (throughout, we assume that R has a unit element $1 \neq 0$). The *group algebra* $R[G]$ of G over R is a free R-module with basis $\{X^g \mid g \in G\}$, where multiplication is defined by

$$\Big(\sum_{g\in G} a_g X^g\Big)\Big(\sum_{g\in G} b_g X^g\Big) = \sum_{g\in G}\Big(\sum_{h\in G} a_h b_{g-h}\Big)X^g .$$

In particular, $X^g X^h = X^{g+h}$ for all $g, h \in G$. Thus we can think of this as a generalization of a polynomial ring, where the exponents come from the group G rather than from $\mathbb{N}_0$. Moreover, we view R as a subset of $R[G]$ by means of $a = aX^0$ for all $a \in R$. The *augmentation map*

$$\varepsilon\colon R[G] \to R, \quad \text{defined by} \quad \varepsilon\Big(\sum_{g\in G} a_g X^g\Big) = \sum_{g\in G} a_g,$$

is an epimorphism of R-algebras, and its kernel $\operatorname{Ker}(\varepsilon) = I_G$ is called the *augmentation ideal.*

Definition 2.2.1. For a commutative ring R, let $\mathsf{d}(G, R)$ denote the largest integer $l \in \mathbb{N}$ having the following property:

There is some sequence $S = g_1 \cdot \ldots \cdot g_l$ of length l over G such that

$$(a_1 - X^{g_1}) \cdot \ldots \cdot (a_l - X^{g_l}) \ne 0 \in R[G] \quad \text{for all} \quad a_1, \ldots, a_l \in R^\bullet .$$

Lemma 2.2.2. *Let R be an integral domain, $S = g_1 \cdot \ldots \cdot g_l \in \mathcal{F}(G)$ a zero-sum free sequence and $a_1, \ldots, a_l \in R^\bullet$. If*

$$f = \prod_{i=1}^{l} (a_i - X^{g_i}) = \sum_{g\in G} c_g X^g \in R[G] \quad \textit{with} \quad c_g \in R \ \textit{ for all } g \in G\,,$$

then $c_0 \ne 0$, and hence $f \ne 0$. In particular, we have $\mathsf{d}(G) \le \mathsf{d}(G, R)$.

Proof. Since R is an integral domain and $0 \notin \Sigma(S)$, it follows that $c_0 = a_1 \cdot \ldots \cdot a_l \ne 0$. □

Definition 2.2.3. Let $S = g_1 \cdot \ldots \cdot g_l \in \mathcal{F}(G)$ be a sequence of length $|S| = l \in \mathbb{N}_0$ and let $g \in G$.

1. For every $k \in \mathbb{N}_0$ let

$$\mathsf{N}_g^k(S) = \Big|\Big\{ I \subset [1, l] \ \Big| \ \sum_{i\in I} g_i = g \text{ and } |I| = k \Big\}\Big|$$

denote the number of subsequences T of S having sum $\sigma(T) = g$ and length $|T| = k$ (counted with the multiplicity of their appearance in S).

2. We define

$$\mathsf{N}_g(S) = \sum_{k \ge 0} \mathsf{N}_g^k(S), \quad \mathsf{N}_g^+(S) = \sum_{k \ge 0} \mathsf{N}_g^{2k}(S) \quad \text{and} \quad \mathsf{N}_g^-(S) = \sum_{k \ge 0} \mathsf{N}_g^{2k+1}(S).$$

Thus $\mathsf{N}_g(S)$ denotes the number of subsequences T of S having sum $\sigma(T) = g$, $\mathsf{N}_g^+(S)$ denotes the number of all such subsequences of even length, and $\mathsf{N}_g^-(S)$ denotes the number of all such subsequences of odd length (each counted with the multiplicity of its appearance in S).

Lemma 2.2.4. *Let p be a prime and G a p-group. Then the following identities hold in $\mathbb{F}_p[G]$.*

1. *If $g \in G$ and $\operatorname{ord}(g) = m \ge 2$, then*

$$(1 - X^g)^m = 0 \in \mathbb{F}_p[G], \quad (1 - X^g)^{m-1} = \sum_{j=0}^{m-1} X^{jg} \in \mathbb{F}_p[G]$$

and

$$(1 - X^g)^{m-2} = \sum_{j=0}^{m-1} (j+1) X^{jg} \in \mathbb{F}_p[G].$$

2. *Let $(e_1, \ldots, e_r)$ be a basis of G and $\operatorname{ord}(e_i) = n_i \ge 2$ for all $i \in [1, r]$. Then*

$$\prod_{i=1}^{r} (1 - X^{e_i})^{n_i - 1} = \sum_{g \in G} X^g \in \mathbb{F}_p[G],$$

and if $m \in \mathbb{N}$ and $g_1, \ldots, g_m \in G$, then

$$\prod_{\mu=1}^{m} (1 - X^{g_\mu}) = \sum_{j=1}^{t} c_j \prod_{i=1}^{r} (1 - X^{e_i})^{l_{j,i}} \in \mathbb{F}_p[G],$$

where $t \in \mathbb{N}_0$, $c_j \in \mathbb{F}_p$, $l_{j,1}, \ldots, l_{j,r} \in \mathbb{N}_0$ and $l_{j,1} + \ldots + l_{j,r} \ge m$ for all $j \in [1, t]$.

Proof. 1. Since m is a power of p, we obtain $(1 - X^g)^m = 1 - X^{mg} = 0 \in \mathbb{F}_p[G]$. For $k \in \{1, 2\}$, we have

$$(1 - X^g)^{m-k} = \sum_{j=0}^{m-k} \binom{m-k}{j} (-1)^j X^{jg}.$$

We assert that, for every $j \in [0, m-1]$,

$$\binom{m-1}{j} (-1)^j \equiv 1 \mod p \quad \text{and} \quad \binom{m-2}{j} (-1)^j \equiv (j+1) \mod p.$$

Indeed, in the polynomial ring $\mathbb{F}_p[T]$ we have

$$\sum_{j=0}^{m-1} \binom{m-1}{j} (-1)^j T^j = (1-T)^{m-1} = \frac{(1-T)^m}{1-T} = \frac{1-T^m}{1-T} = \sum_{j=0}^{m-1} T^j ,$$

whence the first assertion follows. Since

$$\binom{m-2}{j} = \binom{m-1}{j} \frac{m-j-1}{m-1} \equiv (j+1) \binom{m-1}{j} \mod p ,$$

the second assertion follows.

2. Every $g \in G$ has a unique representation of the form $g = \nu_1 e_1 + \ldots + \nu_r e_r$, where $\nu_i \in [0, n_i - 1]$ for all $i \in [1, r]$. Therefore 1. implies that

$$\prod_{i=1}^{r} (1 - X^{e_i})^{n_i - 1} = \prod_{i=1}^{r} \sum_{\nu_i = 0}^{n_i - 1} X^{\nu_i e_i} = \sum_{g \in G} X^g .$$

For the proof of the second identity we define, for every $\boldsymbol{l} = (l_1, \ldots, l_r) \in \mathbb{N}_0^r$,

$$g_{\boldsymbol{l}} = \prod_{i=1}^{r} (1 - X^{e_i})^{l_i} .$$

The augmentation ideal I_G is generated by $\{g_{\boldsymbol{l}} \mid \boldsymbol{0} \ne \boldsymbol{l} \in \mathbb{N}_0^r\}$. For $\mu \in [1, m]$ we have $1 - X^{g_\mu} \in I_G$ and therefore

$$1 - X^{g_\mu} = \sum_{\boldsymbol{0} \ne \boldsymbol{l} \in \mathbb{N}_0^r} c_{\mu, \boldsymbol{l}} g_{\boldsymbol{l}}$$

with coefficients $c_{\mu, \boldsymbol{l}} \in \mathbb{F}_p$. Hence

$$\prod_{\mu=1}^{m} (1 - X^{g_\mu}) = \sum_{\boldsymbol{0} \ne \boldsymbol{l}_1, \ldots, \boldsymbol{l}_m \in \mathbb{N}_0^r} c_{1, \boldsymbol{l}_1} \cdot \ldots \cdot c_{m, \boldsymbol{l}_m} g_{\boldsymbol{l}_1 + \ldots + \boldsymbol{l}_m} ,$$

and $|\boldsymbol{l}_1 + \ldots + \boldsymbol{l}_m| \ge m$ for all $\boldsymbol{l}_1, \ldots, \boldsymbol{l}_m \in \mathbb{N}_0^r \setminus \{\boldsymbol{0}\}$. □

Lemma 2.2.5. *Let p be a prime, G a p-group, $S = g_1 \cdot \ldots \cdot g_l \in \mathcal{F}(G)$, and*

$$f = \prod_{i=1}^{l} (1 - X^{g_i}) = \sum_{g \in G} c_g(S) X^g \in \mathbb{F}_p[G] .$$

1. *For every $g \in G$, we have $c_g(S) = \mathsf{N}_g^+(S) - \mathsf{N}_g^-(S) + p\mathbb{Z} \in \mathbb{F}_p$. In particular, if $c_0(S) = 0$, then $0 \in \Sigma(S)$, and if $g \in G^\bullet$ and $c_g(S) \ne 0$, then $g \in \Sigma(S)$.*

2. *For* $i \in [1, l]$, *let* $g_i = p^{m_i} g_i'$ *with* $g_i' \in G$ *and* $m_i \in \mathbb{N}_0$, *and define*

$$m = \sum_{i=1}^{l} p^{m_i} .$$

If $m > \mathsf{d}^*(G)$, *then* $c_g(S) = 0$ *for all* $g \in G$, $0 \in \Sigma(S)$, *and in particular* $\mathsf{N}_g^+(S) \equiv \mathsf{N}_g^-(S) \mod p$ *for all* $g \in G$.

Proof. 1. For $g \in G$, we set

$$\Omega_g = \Big\{ I \subset [1, l] \ \Big| \ \sum_{i \in I} g_i = g \Big\} .$$

Then $\emptyset \in \Omega_0$ and

$$c_g(S) = \sum_{J \in \Omega_g} (-1)^{|J|} + p\mathbb{Z} \ = \ \mathsf{N}_g^+(S) - \mathsf{N}_g^-(S) + p\mathbb{Z} \in \mathbb{F}_p .$$

Hence $c_0(S) = 0$ implies $0 \in \Sigma(S)$, and if $g \in G^\bullet$ is such that $c_g(S) \neq 0$, then $g \in \Sigma(S)$.

2. We shall repeatedly make use of Lemma 2.2.4. Let $(e_1, \ldots, e_r)$ be a basis of G, $\operatorname{ord}(e_i) = n_i$ for all $i \in [1, r]$, and $1 < n_1 \mid \cdots \mid n_r$. Then

$$\mathsf{d}^*(G) = (n_1 - 1) + \ldots + (n_r - 1) .$$

For $i \in [1, r]$ we have $(1 - X^{g_i'})^{p^{m_i}} = 1 - X^{p^{m_i} g_i'} = 1 - X^{g_i}$, and therefore

$$f \ = \ \prod_{i=1}^{l} (1 - X^{g_i'})^{p^{m_i}} = \sum_{j=1}^{t} c_j \prod_{i=1}^{r} (1 - X^{e_i})^{l_{j,i}}$$

for some $t \in \mathbb{N}_0$, $c_1, \ldots, c_t \in \mathbb{F}_p$, $l_{j,i} \in \mathbb{N}_0$ and $l_{j,1} + \ldots + l_{j,r} \geq m$ for all $j \in [1, t]$. If $j \in [1, t]$ and $l_{j,i} \geq n_i$ for some $i \in [1, r]$, then

$$\prod_{i=1}^{r} (1 - X^{e_i})^{l_{j,i}} = 0 \in \mathbb{F}_p[G] .$$

Hence we may assume that $l_{j,i} < n_i$ for all $i \in [1, r]$ and $j \in [1, t]$, and then either $t = 0$ or $m \leq l_{j,1} + \ldots + l_{j,r} \leq \mathsf{d}^*(G)$ for all $j \in [1, t]$.

If $m > \mathsf{d}^*(G)$, then $t = 0$, hence $f = 0$, and thus $c_g(S) = 0$ for all $g \in G$. The remaining assertions follow by 1. □

Theorem 2.2.6. *If* G *is a* p*-group, then* $\mathsf{d}^*(G) = \mathsf{d}(G) = \mathsf{d}(G, \mathbb{F}_p)$.

Proof. Suppose that G is a p-group. Lemmas 2.1.3.2 and 2.2.2 imply that $\mathsf{d}^*(G) \leq \mathsf{d}(G) \leq \mathsf{d}(G, \mathbb{F}_p)$. If $S = g_1 \cdot \ldots \cdot g_l \in \mathcal{F}(G)$ with $|S| = l > \mathsf{d}^*(G)$, then Lemma 2.2.5.2 (with $m_1 = \cdots = m_l = 0$) implies that

$$(1 - X^{g_1}) \cdot \ldots \cdot (1 - X^{g_l}) = 0 ,$$

and thus $\mathsf{d}(G, \mathbb{F}_p) \leq \mathsf{d}^*(G)$. □

An alternate proof of Theorem 2.2.6 was given by Zhi-Wei Sun who used covers of the integers (see [128, Corollary 2.1]). Here we briefly discuss some extensions of the classical approach via group algebras.

Let G' be a finite abelian group. Then, by G. Higman's theorem, $\mathbb{Z}[G] \cong \mathbb{Z}[G']$ implies that $G \cong G'$ (see [106, Corollary 3.5.6 and Theorem 9.1.4]). Therefore any combinatorial problem in G can, at least in principle, be tackled via the group algebra $\mathbb{Z}[G]$. Indeed, working over $\mathbb{Z}[G]$ allows to refine the congruences involving $\mathsf{N}_g^+(S)$ and $\mathsf{N}_g^-(S)$, as obtained in Lemma 2.2.5.2 (see [53]).

Let $\exp(G) = n$ and let K be a splitting field of G (that is, $|\{\zeta \in K \mid \zeta^n = 1\}| = n$). Following the ideas of P. van Emde Boas and using character theory, one obtains that

$$\mathsf{d}(G, K) \le (n-1) + n \log \frac{|G|}{n}$$

(see [71, Theorem 5.5.5]). In particular, for cyclic group this implies that $\mathsf{d}(G) = \mathsf{d}(G, K) = n - 1$. W. Gao conjectures that for every G there is a splitting field F such that $\mathsf{d}(G) = \mathsf{d}(G, F)$, and in [59] W. Gao and Y. Li showed that, for every splitting field K of $G = C_2 \oplus C_{2n}$ we have $\mathsf{d}(C_2 \oplus C_{2n}) = \mathsf{d}(C_2 \oplus C_{2n}, K)$ (see also [55]).

2.3 Arithmetical invariants again

Theorem 2.3.1. *Let H be a Krull monoid with class group G such that every class contains a prime and suppose that $|G| > 1$. Let $k \in \mathbb{N}$.*

1. *If $A = 0^m B \in \mathcal{B}(G)$, with $m \in \mathbb{N}_0$ and $B \in \mathcal{B}(G^\bullet)$, then*

$$2 \max \mathsf{L}(A) - m \le |A| \le \mathsf{D}(G) \min \mathsf{L}(A) - m(\mathsf{D}(G) - 1) \quad \text{and} \quad \rho(A) \le \frac{\mathsf{D}(G)}{2}.$$

2. *We have $k \le \rho_k(H) \le k\frac{\mathsf{D}(G)}{2}$ and $\rho(H)^{-1}k \le \lambda_k(H) \le k$.*
3. *$\rho_{2k}(H) = k\mathsf{D}(G)$ and $\rho(H) = \frac{\mathsf{D}(G)}{2}$.*
4. *If $j, l \in \mathbb{N}_0$ such that $l\mathsf{D}(G) + j \ge 1$, then*

$$2l + \frac{2j}{\mathsf{D}(G)} \le \lambda_{l\mathsf{D}(G)+j}(G) \le 2l + j.$$

In particular, $\lambda_{l\mathsf{D}(G)}(G) = 2l$ for every $l \in \mathbb{N}$.

Proof. By Theorem 1.3.4 it suffices to consider the block monoid $\mathcal{B}(G)$.

1. Let $A = 0^m U_1 \cdot \ldots \cdot U_l$ where $l, m \in \mathbb{N}_0$ and $U_1, \ldots, U_l \in \mathcal{A}(G^\bullet)$. Then $2 \le |U_\nu| \le \mathsf{D}(G)$ for all $\nu \in [1, l]$ and hence

$$m + 2l \le |A| \le m + l\mathsf{D}(G).$$

Choosing $l = \min \mathsf{L}(B)$ and $l = \max \mathsf{L}(B)$ we obtain the first inequalities, and then we get

$$\rho(A) = \frac{\max \mathsf{L}(A)}{\min \mathsf{L}(A)} = \frac{m + \max \mathsf{L}(B)}{m + \min \mathsf{L}(B)} \le \frac{\max \mathsf{L}(B)}{\min \mathsf{L}(B)} \le \frac{\mathsf{D}(G)}{2} .$$

2. By definition, we have $k \le \rho_k(G)$. If $A \in \mathcal{B}(G)$ with $k \in \mathsf{L}(A)$ and $\max \mathsf{L}(A) = \rho_k(G)$, then 1. implies that

$$\frac{\rho_k(G)}{k} \le \frac{\max \mathsf{L}(A)}{\min \mathsf{L}(A)} = \rho(A) \le \frac{\mathsf{D}(G)}{2} .$$

There is some $L \in \mathcal{L}(G)$ with $k, \lambda_k(G) \in L$, and hence it follows that

$$k \le \max L \le \rho(G) \min L = \rho(G)\lambda_k(G) .$$

3. By 1. and 2. it follows that $\rho_{2k}(G) \le k\mathsf{D}(G)$ and $\rho(G) \le \frac{\mathsf{D}(G)}{2}$. If $U = g_1 \cdot \ldots \cdot g_l \in \mathcal{A}(G)$ with $|U| = l = \mathsf{D}(G)$, then

$$(-U)^k U^k = \prod_{\nu=1}^{l} \big((-g_\nu)g_\nu\big)^k$$

shows that in both inequalities we actually have equality.

4. Let $j, l \in \mathbb{N}_0$ such that $l\mathsf{D}(G) + j \ge 1$. Then 2. and 3. imply that

$$2l + \frac{2j}{\mathsf{D}(G)} = \rho(G)^{-1}\big(l\mathsf{D}(G) + j\big) \le \lambda_{l\mathsf{D}(G)+j}(G) \le l\mathsf{D}(G) + j .$$

If $j = 0$, it follows that $\lambda_{l\mathsf{D}(G)}(G) = 2l$. □

Lemma 2.3.2.

1. *For $j \in \mathbb{N}_{\ge 2}$, the following statements are equivalent:*
   (a) *There exists some $L \in \mathcal{L}(G)$ with $\{2, j\} \subset L$.*
   (b) $j \le \mathsf{D}(G)$.
2. *Let $|G| \ge 3$ and $A \in \mathcal{B}(G)$. Then $\{2, \mathsf{D}(G)\} \subset \mathsf{L}(A)$ if and only if $A = U(-U)$ for some $U \in \mathcal{A}(G)$ with $|U| = \mathsf{D}(G)$.*

Proof. 1. (a) ⇒ (b). If $L \in \mathcal{L}(G)$ and $\{2, j\} \subset L$, then Theorem 2.3.1.3 implies that $j \le \sup L \le \rho_2(G) = \mathsf{D}(G)$.

(b) ⇒ (a). If $j \le \mathsf{D}(G)$, then there exists some $U \in \mathcal{A}(G)$ with $|U| = l \ge j$, say $U = g_1 \cdot \ldots \cdot g_l$. Then $V = g_1 \cdot \ldots \cdot g_{j-1}(g_j + \ldots + g_l) \in \mathcal{A}(G)$, and $\{2, j\} \subset \mathsf{L}\big(V(-V)\big)$.

2. If $\{2, \mathsf{D}(G)\} \subset \mathsf{L}(A)$, then there exist $U_1, U_2, V_1, \ldots, V_{\mathsf{D}(G)} \in \mathcal{A}(G)$ such that $A = U_1U_2 = V_1 \cdot \ldots \cdot V_{\mathsf{D}(G)}$, and clearly $0 \nmid A$, since otherwise $U_1 = 0$ or $U_2 = 0$ and $\mathsf{D}(G) = 2$. Theorem 2.3.1.1 implies that $\max \mathsf{L}(A) = \mathsf{D}(G)$ and $|A| = 2\mathsf{D}(G)$. Hence $|V_i| = 2$ for all $i \in [1, \mathsf{D}(G)]$, and $|U_1| = |U_2| = \mathsf{D}(G)$, which implies $U_2 = -U_1$. The converse is obvious. □

Lemma 2.3.3. *Suppose that $d \in \mathbb{N}$ has the following property:*

For all $U, V \in \mathcal{A}(G)$ with $\min\{|U|, |V|\} > d$ there exists a factorization $UV = W_1 \cdot \ldots \cdot W_k$ with $k \in [2,d]$ and $|W_1| \le d$.

Then $\mathsf{c}(G) \le d$.

Proof. We must prove that $\mathsf{c}(A) \le d$ for all $A \in \mathcal{B}(G)$. We proceed by induction on $|A|$, and we must prove that any two factorizations of A can be concatenated by a d-chain. Let z, z' be two factorizations of A, say

$$z = U_1 \cdot \ldots \cdot U_r \quad \text{and} \quad z' = V_1 \cdot \ldots \cdot V_s\,, \quad \text{where} \quad U_1, \ldots, U_r, V_1, \ldots, V_s \in \mathcal{A}(G)\,.$$

If $\max\{r, s\} \le d$, then $\mathsf{d}(z, z') \le d$ and we are done. Assume that $r > d$.

Case 1: $|V_i| \le d$ for some $i \in [1, s]$, say $|V_1| \le d$.

We may assume that $V_1 \,|\, U_1 \cdot \ldots \cdot U_{r-1}$, say $U_1 \cdot \ldots \cdot U_{r-1} = V_1 W_1 \cdot \ldots \cdot W_t$ with $t \in \mathbb{N}$ and $W_1, \ldots, W_t \in \mathcal{A}(G)$. By the induction hypothesis there is a d-chain of factorizations $y_0, \ldots, y_k$ concatenating $U_1 \cdot \ldots \cdot U_{r-1}$ and $V_1 W_1 \cdot \ldots \cdot W_t$, and there is a d-chain of factorizations $z_0, \ldots, z_l$ concatenating $W_1 \cdot \ldots \cdot W_t U_r$ and $V_2 \cdot \ldots \cdot V_s$. Then $z = y_0 U_r, \ldots, y_k U_r = z_0 V_1, \ldots, z_l V_1 = z'$ is a d-chain concatenating z and z'.

Case 2: $|V_i| > d$ for all $i \in [1, s]$.

By assumption there is a factorization $V_1 V_2 = W_1 \cdot \ldots \cdot W_k$, where $k \in [2, d]$ and $|W_1| \le d$. Then the factorization $z'' = W_1 \cdot \ldots \cdot W_k V_3 \cdot \ldots \cdot V_s$ of A satisfies $\mathsf{d}(z', z'') = \max\{2, k\} \le d$, and by *Case* 1 there is a d-chain of factorizations concatenating z and z''. □

Theorem 2.3.4. *Let H be a Krull monoid with class group G.*

1. $\mathsf{c}(H) \le \mathsf{D}(G)$.
2. *Suppose that $|G| \ge 3$. Then $\mathsf{c}(G) = \mathsf{D}(G)$ if and only if G is either cyclic or an elementary 2-group.*

Proof. 1. By Theorem 1.3.4 it suffices to show that $\mathsf{c}(G) \le \mathsf{D}(G)$. This follows immediately from Lemma 2.3.3 with $d = \mathsf{D}(G)$.

2. If G is cyclic, $g \in G$ with $\operatorname{ord}(g) = n = |G|$ and $U = g^n$, then $\mathsf{c}\big((-U)U\big) = n$ and hence $\mathsf{c}(G) = \mathsf{D}(G)$. If G is an elementary 2-group with basis $(e_1, \ldots, e_r)$, $e_0 = e_1 + \ldots + e_r$ and $U = e_0 \cdot \ldots \cdot e_r$, then $\mathsf{c}\big(U^2\big) = r+1 = \mathsf{D}(G)$ and hence $\mathsf{c}(G) = \mathsf{D}(G)$.

Assume now that G is neither cyclic nor an elementary 2-group. We shall prove that for all $U, V \in \mathcal{A}(G)$ with $|U| = |V| = \mathsf{D}(G)$ there exists some factorization $UV = W_1 \cdot \ldots \cdot W_k$ with $k \in [2, \mathsf{d}(G)]$ and $|W_1| \le \mathsf{d}(G)$. Then $\mathsf{c}(G) \le \mathsf{d}(G)$ by Lemma 2.3.3.

Let $U, V \in \mathcal{A}(G)$ with $|U| = |V| = \mathsf{D}(G)$. Then $\max \mathsf{L}(UV) \le \mathsf{D}(G)$, and equality holds if and only if $V = -U$ (cf. Lemma 2.3.2). Now we distinguish two cases.

Case 1: $V \neq -U$.

It is sufficient to prove that there exists some $W \in \mathcal{A}(G)$ such that $W \mid UV$ and $|W| < \mathsf{D}(G)$. Assume the contrary. Let $g \in \operatorname{supp}(U)$ and $V = h_1 \cdot \ldots \cdot h_l$ with $l = \mathsf{D}(G)$. For every $i \in [1, l]$, we consider the sequence $S_i = g h_i^{-1} V \in \mathcal{F}(G)$. Since $|S_i| = \mathsf{D}(G)$, there exists some $S_i' \in \mathcal{A}(G)$ such that $S_i' \mid S_i \mid UV$. By assumption, this implies $|S_i'| = \mathsf{D}(G)$, hence $S_i' = S_i$ and therefore $0 = \sigma(S_i) = g - h_i$. Thus $V = g^l$, and Lemma 2.1.2.1 implies $G = \langle \operatorname{supp}(V) \rangle = \langle g \rangle$, a contradiction.

Case 2: $V = -U$.

It is sufficient to prove that there exists some $W \in \mathcal{A}(G)$ such that $W \mid U(-U)$ and $2 < |W| < \mathsf{D}(G)$. Then we consider any factorization $U(-U) = W W_2 \cdot \ldots \cdot W_k$ with $W_2, \ldots, W_k \in \mathcal{A}(G)$, and we obtain that $k < \mathsf{D}(G)$.

By Lemma 2.1.2.1 we have $\langle \operatorname{supp}(U) \rangle = G$, and since G is not an elementary 2-group, there exists some $g_0 \in \operatorname{supp}(U)$ with $\operatorname{ord}(g_0) > 2$. We set $U = g_0^m g_1 \cdot \ldots \cdot g_l$ with $g_0 \notin \{g_1, \ldots, g_l\}$. Since $G = \langle \operatorname{supp}(U) \rangle$ is not cyclic, it follows that $l \geq 2$. If $W' = (-g_0)^m g_1 \cdot \ldots \cdot g_l$, then $W' \mid U(-U)$ and $|W'| = \mathsf{D}(G)$. Hence there exists some $W \in \mathcal{A}(G)$ with $W \mid W'$, and we shall prove that $2 < |W| < \mathsf{D}(G)$. Since $U \in \mathcal{A}(G)$, we have $W \nmid g_1 \cdot \ldots \cdot g_l$ and thus $-g_0 \mid W$. Since $g_0 \notin \{g_1, \ldots, g_l\}$ and $g_0 \neq -g_0$, it follows that $W \neq g_0(-g_0)$ and thus $|W| > 2$.

Assume to the contrary that $|W| = \mathsf{D}(G)$. Then $W = W'$, and $\sigma(U) = \sigma(W) = 0$ implies $2mg_0 = 0$ and thus $m > 1$. We consider the sequence $S = g_0^m g_1 \cdot \ldots \cdot g_{l-1}$. Since $S \in \mathcal{A}^*(G)$ and $|S| = \mathsf{d}(G)$, Lemma 2.1.2.1 implies $\Sigma(S) = G^\bullet$ and thus $(m+1)g_0 \in \Sigma(S)$, say

$$(m+1)g_0 = sg_0 + \sum_{i \in I} g_i \quad \text{with} \quad s \in [0, m] \quad \text{and} \quad I \subset [1, l-1] \,.$$

If $s = 0$, then

$$0 = 2mg_0 = (m-1)g_0 + \sum_{i \in I} g_i \ \in \Sigma(S),$$

a contradiction. If $s \geq 1$, then it follows that

$$T = (-g_0)^{m+1-s} \prod_{i \in I} g_i$$

is a proper zero-sum subsequence of W, a contradiction to $W \in \mathcal{A}(G)$. □

Corollary 2.3.5. *Let H be a Krull monoid with class group G such that every class contains a prime. Suppose that $|G| \geq 3$ and that $\exp(G) = n \geq 2$. Then*

$$[1, n-2] \subset \Delta(H) \subset [1, \mathsf{c}(G) - 2] \subset [1, \mathsf{D}(G) - 2] \,.$$

In particular, if G is cyclic, then $\Delta(H) = [1, n-2]$.

Proof. By Theorem 1.3.4, it suffices to consider the block monoid $\mathcal{B}(G)$. Lemma 1.1.6.3 implies that $\Delta(G) \subset [1, \mathsf{c}(G) - 2]$ and Theorem 2.3.4 implies that $\mathsf{c}(G) \leq \mathsf{D}(G)$.

Suppose that $n \geq 3$, pick $i \in [3, n]$ and $g \in G$ with $\operatorname{ord}(g) = n$. Then $T = g^n, U = (-g)^{i-1}\big((i-1)g\big), V = (-g)g$ and $W = g^{n-i+1}\big((i-1)g\big)$ are minimal zero-sum sequences. Then

$$TU = V^{i-1}W$$

shows that $\mathsf{L}(TU) = \{2, i\}$ whence $i - 2 \in \Delta\big(\mathsf{L}(TU)\big) \subset \Delta(G)$.

If G is cyclic, then Corollary 2.1.4 implies that $\mathsf{D}(G) = n$ and thus $\Delta(G) = [1, n-2]$. □

For all groups known so far it always holds that $\Delta(G) = [1, \mathsf{c}(G) - 2]$.

Corollary 2.3.6. *The following statements are equivalent:*

(a) *Every* $L \in \mathcal{L}(G)$ *with* $\{2, \mathsf{D}(G)\} \subset L$ *satisfies* $L = \{2, \mathsf{D}(G)\}$.

(b) $\{2, \mathsf{D}(G)\} \in \mathcal{L}(G)$.

(c) *G is either cyclic or an elementary 2-group.*

Proof. (a) ⇒ (b). By Lemma 2.3.2.1 there exists some $L \in \mathcal{L}(G)$ with $\{2, \mathsf{D}(G)\} \subset L$.

(b) ⇒ (c). If $L = \{2, \mathsf{D}(G)\} \in \mathcal{L}(G)$, then, by Lemma 1.1.6.3 and Theorem 2.3.4.1 we have $\mathsf{D}(G) \leq 2 + \sup \Delta(G) \leq \mathsf{c}(G) \leq \mathsf{D}(G)$, hence $\mathsf{c}(G) = \mathsf{D}(G)$, and the assertion follows by Theorem 2.3.4.2.

(c) ⇒ (a). Let $L \in \mathcal{L}(G)$ with $\{2, \mathsf{D}(G)\} \subset L$. By Lemma 2.3.2.2 we have $L = \mathsf{L}\big(U(-U)\big)$ for some $U \in \mathcal{A}(G)$ with $|U| = \mathsf{D}(G)$.

If G is cyclic of order $n \geq 3$, then Corollary 2.1.4 implies that $U = g^n$ for some $g \in G$ with $\operatorname{ord}(g) = n$. Since $\mathcal{A}(\{-g, g\}) = \{(-g)^n, g^n, g(-g)\}$, it follows that $\mathsf{L}\big(U(-U)\big) = \{2, \mathsf{D}(G)\}$.

If G is an elementary 2-group of rank $r \geq 2$ and $(e_1, \ldots, e_r)$ is a basis of G, then $U = e_1 \cdot \ldots \cdot e_r(e_1 + \ldots + e_r)$ by Corollary 2.1.4 and $\mathsf{L}\big(U(-U)\big) = \{2, r+1\} = \{2, \mathsf{D}(G)\}$. □

Chapter 3

The structure of sets of lengths

Sets of lengths in Krull monoids and in noetherian domains are finite and nonempty. Furthermore, either all sets of lengths are singletons or sets of lengths may become arbitrarily large (see Lemma 1.0.1).

3.1 Unions of sets of lengths

Definition 3.1.1. Let H be a BF-monoid and $k \in \mathbb{N}$. Let $\mathcal{V}_k(H)$ denote the set of all $m \in \mathbb{N}$ for which there exist $u_1, \ldots, u_k, v_1, \ldots, v_m \in \mathcal{A}(H)$ with $u_1 \cdot \ldots \cdot u_k = v_1 \cdot \ldots \cdot v_m$.

Lemma 3.1.2. *Let H be a BF-monoid with $H \neq H^\times$ and $k, l \in \mathbb{N}$.*

1. *$\mathcal{V}_1(H) = \{1\}$, $k \in \mathcal{V}_k(H)$ and*

$$\mathcal{V}_k(H) = \bigcup_{k \in L, L \in \mathcal{L}(H)} L\,.$$

   In particular, $\rho_k(H) = \sup \mathcal{V}_k(H)$ and $\lambda_k(H) = \min \mathcal{V}_k(H)$.

2. *$\mathcal{V}_k(H) + \mathcal{V}_l(H) \subset \mathcal{V}_{k+l}(H)$ and*

$$\lambda_{k+l}(H) \le \lambda_k(H) + \lambda_l(H) \le k + l \le \rho_k(H) + \rho_l(H) \le \rho_{k+l}(H)\,.$$

3. *We have $l \in \mathcal{V}_k(H)$ if and only if $k \in \mathcal{V}_l(H)$.*

Proof. This follows immediately from the definitions. □

Thus the sets $\mathcal{V}_k(H)$ are unions of sets of lengths. They were introduced by S.T. Chapman and W.W. Smith in 1990 (see [17]). The following result reveals that, in Krull monoids where every class contains a prime, the $\mathcal{V}_k(H)$ sets are intervals. This was first shown in [35]. The following simple proof is due to F. Halter-Koch.

Theorem 3.1.3. *Let H be a Krull monoid with finite class group G such that every class contains a prime. Then for every $k \in \mathbb{N}$ the set $\mathcal{V}_k(H)$ is an arithmetical progression with difference 1, and hence $\mathcal{V}_k(H) = [\lambda_k(H), \rho_k(H)]$.*

Proof. By Theorem 1.3.4, it suffices to consider the block monoid $\mathcal{B}(G)$.

If $|G| \le 2$, then $\mathcal{B}(G)$ is half-factorial by Proposition 1.2.4 whence the sets $\mathcal{V}_k(G)$ are singletons for all $k \in \mathbb{N}$. Let $|G| \ge 3$ and $k \in \mathbb{N}$. First, we point out that it suffices to prove that $[k, \rho_k(G)] \subset \mathcal{V}_k(G)$. Indeed, suppose that this is done, and let $l \in [\lambda_k(G), k]$. Then $l \le k \le \rho_l(G)$, hence $k \in \mathcal{V}_l(G)$ and consequently $l \in \mathcal{V}_k(G)$.

Thus let $l \in [k, \rho_k(G)]$ be minimal such that $[l, \rho_k(G)] \subset \mathcal{V}_k(G)$ and assume to the contrary that $l > k$. Let Ω be the set of all $A \in \mathcal{B}(G)$ such that $\{k, j\} \subset \mathsf{L}(A)$ for some $j \ge l$, and let $B \in \Omega$ be such that $|B|$ is minimal. Then $B = U_1 \cdot \ldots \cdot U_k = V_1 \cdot \ldots \cdot V_j$, where $j \ge l$ and $U_1, \ldots, U_k, V_1, \ldots, V_j \in \mathcal{A}(G)$. Since $j > k$, we have $B \ne 0^{|B|}$, and (after renumbering if necessary) we may assume that $U_k = g_1 g_2 U'$ and $V_{j-1} V_j = g_1 g_2 V'$, where $g_1, g_2 \in G$ and $U', V' \in \mathcal{F}(G)$. Then $U_k' = (g_1 + g_2) U' \in \mathcal{A}(G)$, and we suppose that $V_{j-1}' = (g_1 + g_2) V' = W_1 \cdot \ldots \cdot W_t$, where $t \in \mathbb{N}$ and $W_1, \ldots, W_t \in \mathcal{A}(G)$. If $B' = U_1 \cdot \ldots \cdot U_{k-1} U_k'$, then $|B'| < |B|$ and $B' = V_1 \cdot \ldots \cdot V_{j-2} W_1 \cdot \ldots \cdot W_t$. By the minimal choice of $|B|$, it follows that $j - 2 + t < l$, hence $t = 1$, $j = l$ and $l - 1 \in \mathcal{V}_k(G)$, a contradiction. □

The structure of unions of sets of lengths in much more general settings is studied in [48]. Here we stick to Krull monoids and determine the $\lambda_k(H)$-invariants with respect to the $\rho_k(H)$-invariants.

Corollary 3.1.4. *Let H be a Krull monoid with class group G such that every class contains a prime, and suppose that $|G| > 1$. Then for every $l \in \mathbb{N}_0$ we have*

$$\lambda_{l\mathsf{D}(G)+j}(H) = \begin{cases} 2l & \text{for } j = 0, \\ 2l+1 & \text{for } j \in [1, \rho_{2l+1}(G) - l\mathsf{D}(G)], \\ 2l+2 & \text{for } j \in [\rho_{2l+1}(G) - l\mathsf{D}(G) + 1, \mathsf{D}(G) - 1], \end{cases}$$

provided that $l\mathsf{D}(G) + j \ge 1$.

Proof. By Theorem 1.3.4, it suffices to consider the block monoid $\mathcal{B}(G)$. If $|G| = 2$, then $\mathcal{B}(G)$ is half-factorial, $\mathsf{D}(G) = 2$ and hence the assertion follows. Suppose that $|G| \ge 3$, and thus we get $\mathsf{D}(G) \ge 3$.

Let $l \in \mathbb{N}_0$ and $j \in [0, \mathsf{D}(G) - 1]$ such that $l\mathsf{D}(G) + j \ge 1$. For $j = 0$ the assertion follows from Theorem 2.3.1.4. Let $j \in [1, \mathsf{D}(G) - 1]$. By Theorem 2.3.1.4 we obtain that

$$2l + \frac{2j}{\mathsf{D}(G)} = \frac{l\mathsf{D}(G) + j}{\rho(G)} \le \lambda_{l\mathsf{D}(G)+j}(G) \le 2l + j\,.$$

Thus the assertion follows for $j = 1$, and hence from now on we may suppose that $j \ge 2$. Lemma 2.3.2.1 implies that $\{2, j\} \subset \mathsf{L}(B)$ for some $B \in \mathcal{B}(G)$ and thus

$\lambda_j(G) = 2$. Hence we obtain

$$\lambda_{l\mathsf{D}(G)+j}(G) \le \lambda_{l\mathsf{D}(G)}(G) + \lambda_j(G) = 2l + 2\,,$$

and therefore $\lambda_{l\mathsf{D}(G)+j}(G) \in \{2l+1, 2l+2\}$.

If $j \in [2, \rho_{2l+1}(G) - l\mathsf{D}(G)]$, then $l \ge 1$ and $l\mathsf{D}(G) + j \in \mathcal{V}_{2l+1}(G)$ by Theorem 3.1.3. Thus $\lambda_{l\mathsf{D}(G)+j}(G) \le 2l+1$ and hence $\lambda_{l\mathsf{D}(G)+j}(G) = 2l+1$.

If $j > \rho_{2l+1}(G) - l\mathsf{D}(G)$, then $l\mathsf{D}(G)+j > \rho_{2l+1}(G)$ and $l\mathsf{D}(G)+j \notin \mathcal{V}_{2l+1}(G)$. Thus $\lambda_{l\mathsf{D}(G)+j}(G) > 2l+1$ and hence $\lambda_{l\mathsf{D}(G)+j}(G) = 2l+2$. □

Corollary 3.1.5. *Let H be a Krull monoid whose class group G is an elementary 2-group and suppose that every class contains a prime. Then for every $k \in \mathbb{N}_{\ge 2}$ and every $l \in \mathbb{N}_0$ we have $\mathcal{V}_k(H) = [\lambda_k(H), \rho_k(H)]$,*

$$\rho_k(H) = \lfloor \frac{k\mathsf{D}(G)}{2} \rfloor \quad \text{and} \quad \lambda_{l\mathsf{D}(G)+j}(H) = \begin{cases} 2l+j & \text{for } j \in [0,1], \\ 2l+1 & \text{for } j \in [2, \mathsf{D}(G)/2] \text{ and } l \ge 1, \\ 2l+2 & \text{for } j \in [2, \mathsf{D}(G)-1] \text{ and} \\ & (\text{either } j > \mathsf{D}(G)/2 \text{ or } l = 0)\,, \end{cases}$$

provided that $l\mathsf{D}(G) + j \ge 1$.

Proof. By Theorem 1.3.4, it suffices to consider the block monoid $\mathcal{B}(G)$. By Theorem 3.1.3 we obtain that $\mathcal{V}_k(H) = [\lambda_k(G), \rho_k(G)]$. If $|G| = 2$, then $\mathcal{B}(G)$ is half-factorial by Proposition 1.2.4 whence for all $k \in \mathbb{N}$ we have $\lambda_k(G) = k = \rho_k(G)$.

Now suppose that $|G| \ge 4$ and hence $\mathsf{D}(G) > 2$. We first prove the assertion on $\rho_k(G)$ and then the assertion on $\lambda_{l\mathsf{D}(G)+j}(G)$.

1. Let $k \in \mathbb{N}$. If k is even, then the assertion follows from Theorem 2.3.1.3. Suppose we know that

$$\rho_3(G) \ge \lfloor \frac{3\mathsf{D}(G)}{2} \rfloor\,. \tag{$*$}$$

Then Theorem 2.3.1 and Lemma 3.1.2 imply that

$$\begin{aligned} \lfloor \frac{3\mathsf{D}(G)}{2} \rfloor + k\mathsf{D}(G) &\le \rho_3(G) + \rho_{2k}(G) \le \rho_{2k+3}(G) \\ &\le \lfloor \frac{(2k+3)\mathsf{D}(G)}{2} \rfloor = \lfloor \frac{3\mathsf{D}(G)}{2} \rfloor + k\mathsf{D}(G)\,, \end{aligned}$$

and hence the assertion follows. Thus it remains to prove $(*)$. We pick a basis $(e_1, \ldots, e_{\mathsf{r}(G)})$ of G and set $e_0 = e_1 + \ldots + e_{\mathsf{r}(G)}$.

First suppose that $\mathsf{r}(G) = 2s+1$ with $s \in \mathbb{N}$. Then

$$\begin{aligned} U &= e_1 \cdot \ldots \cdot e_{s+1} e_{s+2} \cdot \ldots \cdot e_{2s+1} e_0\,, \\ V &= e_1 \cdot \ldots \cdot e_{s+1} (e_1 + e_{s+2}) \cdot \ldots \cdot (e_s + e_{2s+1})(e_{s+1} + \ldots + e_{2s+1}) \quad \text{and} \\ W &= e_{s+2} \cdot \ldots \cdot e_{2s+1} e_0 (e_1 + e_{s+2}) \cdot \ldots \cdot (e_s + e_{2s+1})(e_{s+1} + \ldots + e_{2s+1}) \end{aligned}$$

are minimal zero-sum sequences of length $\mathsf{D}(G) = 2s + 2$. By construction, UVW may be written as a product of $3\mathsf{D}(G)/2$ minimal zero-sum sequences, and hence $(*)$ follows.

Second suppose that $\mathsf{r}(G) = 2s$ with $s \in \mathbb{N}$. Then

$$\begin{aligned} U &= e_1 \cdot \ldots \cdot e_s e_{s+1} \cdot \ldots \cdot e_{2s} e_0 \,, \\ V &= e_1 \cdot \ldots \cdot e_s (e_1 + e_{s+1}) \cdot \ldots \cdot (e_s + e_{2s})(e_{s+1} + \ldots + e_{2s}) \quad \text{and} \\ W &= e_{s+1} \cdot \ldots \cdot e_{2s} (e_1 + e_{s+1}) \cdot \ldots \cdot (e_s + e_{2s})(e_1 + \ldots + e_s) \end{aligned}$$

are minimal zero-sum sequences of length $\mathsf{D}(G) = 2s + 1$. By construction, UVW may be written as a product of $\lfloor 3\mathsf{D}(G)/2 \rfloor = 3s + 1$ minimal zero-sum sequences, and hence $(*)$ follows.

2. Since $\rho_k(G) = \lfloor \frac{k\mathsf{D}(G)}{2} \rfloor$, the assertion on $\lambda_{l\mathsf{D}(G)+j}(G)$ follows from Corollary 3.1.4. □

3.2 Almost arithmetical multiprogressions and the structure of sets of lengths

We start with four simple examples which show the variety of possible structures for sets of lengths.

Examples 3.2.1.

1. *Arithmetical progressions.* Let $d \in \mathbb{N}$. Let $g \in G$ with $\mathrm{ord}(g) = d + 2$, and set $B = (-g)^{d+2} g^{d+2}$. Then for every $l \in \mathbb{N}$ we obviously have

$$\mathsf{L}(B^l) = 2l + \{\nu d \mid \nu \in [0, l]\} \in \mathcal{L}(G) \,,$$

whence $\mathcal{L}(G)$ contains arithmetical progressions with difference d and any length l.

2. *Multidimensional arithmetical progressions.* Let $r \in \mathbb{N}$, $d_1, \ldots, d_r \in \mathbb{N}$ and $l_1, \ldots, l_r \in \mathbb{N}$. For every $i \in [1, r]$, let G_i be a finite abelian group and $B_i^{l_i} \in \mathcal{B}(G_i)$ as in 1., such that $\mathsf{L}(B_i^{l_i})$ is an arithmetical progression with difference d_i and length l_i. If $G_1 \oplus \cdots \oplus G_r \subset G$ and $B = B_1^{l_1} \cdot \ldots \cdot B_r^{l_r}$, then

$$\mathsf{L}(B) = \sum_{i=1}^{r} \mathsf{L}(B_i) \in \mathcal{L}(G)$$

is an r-dimensional arithmetical progression.

3. *Arithmetical progressions with gaps at their end.* Let $n \ge 2$, $e_1, e_2 \in G$ independent elements with $\mathrm{ord}(e_1) = 2$, $\mathrm{ord}(e_2) = 2n$ and set $e_0 = e_1 + ne_2$. If $G_0 = \{e_0, e_1, e_2, -e_2\}$ and $U = e_0 e_1 e_2^n$, then

$$\mathcal{A}(G_0) = \{g^{\mathrm{ord}(g)} \mid g \in G_0\} \cup \{((-e_2)e_2), U, -U\} \,.$$

For every $l \in \mathbb{N}$ we consider

$$B_l = e_0^2 e_1^2 \big((-e_2)e_2\big)^{n+2nl} .$$

Let $B_l = A_1 \cdot \ldots \cdot A_k$ with $A_1, \ldots, A_k \in \mathcal{A}(G_0)$, and let $I \subset [1,k]$ be the set of all $i \in [1,k]$ such that $\operatorname{supp}(A_i) \cap \{e_0, e_1\} \ne \emptyset$. Then $|I| = 2$, say $I = \{1,2\}$, and we set $C_l = \prod_{i=3}^k A_i$. There are the following four possibilities:

- $\{A_1, A_2\} = \{e_0^2, e_1^2\}$. Then $C_l = \big((-e_2)e_2\big)^{n+2nl}$ and
$$\mathsf{L}(C_l) = n + 2l + \{\nu(2n-2) \mid \nu \in [0,l]\} .$$
- $\{A_1, A_2\} = \{-U, U\}$. Then $C_l = \big((-e_2)e_2\big)^{2nl}$ and
$$\mathsf{L}(C_l) = 2l + \{\nu(2n-2) \mid \nu \in [0,l]\} .$$
- $\{A_1, A_2\} = \{U\}$. Then $C_l = (-e_2)^{2n}\big((-e_2)e_2\big)^{n+2n(l-1)}$ and
$$\mathsf{L}(C_l) = 1 + n + 2(l-1) + \{\nu(2n-2) \mid \nu \in [0,l-1]\} .$$
- $\{A_1, A_2\} = \{-U\}$. Then $C_l = e_2^{2n}\big((-e_2)e_2\big)^{n+2n(l-1)}$ and
$$\mathsf{L}(C_l) = 1 + n + 2(l-1) + \{\nu(2n-2) \mid \nu \in [0,l-1]\} .$$

Thus we obtain that

$$\mathsf{L}(B_l) = 2 + 2l + \Big(\big((\mathcal{D} + (2n-2)\mathbb{Z}) \cap [0,(2n-2)l]\big) \cup \{n + (2n-2)l\}\Big) ,$$

where $\mathcal{D} = \{0, n-1, n, 2n-2\}$. In terms of the following definition, $\mathsf{L}(B_l)$ is an almost arithmetical multiprogression with difference $2n-2$, period $\mathcal{D}$, length l and bound n. If in particular $n = 2$, then $\mathsf{L}(B_l)$ is an arithmetical progression with difference 2 and a gap at its end.

4. *Arithmetical multiprogressions.* It can be shown that for every finite subset $L \subset \mathbb{N}_{\ge 2}$ there is a finite abelian group G_1 such that $L \in \mathcal{L}(G_1)$ (see [71, Proposition 4.8.3]), say $L = \mathsf{L}(B_1) = x + \mathcal{D}$ where $x = \min L$, $\min \mathcal{D} = 0$, $\max \mathcal{D} = d$ and $B_1 \in \mathcal{B}(G_1)$. By 1., there is a group G_2 and a $B_2 \in \mathcal{B}(G_2)$ such that, for every $l \in \mathbb{N}$, $\mathsf{L}(B_2^l) = 2l + \{\nu d \mid \nu \in [0,l]\}$ is an arithmetical progression with difference d and length l. Thus for every $l \in \mathbb{N}$ we have

$$\begin{aligned}\mathsf{L}(B_1B_2^l) &= \mathsf{L}(B_1) + \mathsf{L}(B_2^l) \\ &= (x + 2l) + \mathcal{D} + \{\nu d \mid \nu \in [0,l]\} \\ &= \min \mathsf{L}(B_1B_2^l) + \Big(\mathcal{D} + d\mathbb{Z} \cap [0, \max \mathsf{L}(B_1B_2^l) - \min \mathsf{L}(B_1B_2^l)]\Big) .\end{aligned}$$

Definition 3.2.2. Let $d \in \mathbb{N}$, $l, M \in \mathbb{N}_0$ and $\{0, d\} \subset \mathcal{D} \subset [0, d]$. A subset $L \subset \mathbb{Z}$ is called an

- *arithmetical multiprogression* (AMP for short) with *difference* d, *period* $\mathcal{D}$ and *length* l, if L is an interval of $\min L + \mathcal{D} + d\mathbb{Z}$ (this means that L is finite nonempty and $L = (\min L + \mathcal{D} + d\mathbb{Z}) \cap [\min L, \max L])$), and l is maximal such that $\min L + ld \in L$.
- *almost arithmetical multiprogression* (AAMP for short) with *difference* d, *period* $\mathcal{D}$, *length* l and *bound* M, if

$$L = y + (L' \cup L^* \cup L'') \subset y + \mathcal{D} + d\mathbb{Z},$$

where L^* is an AMP with difference d (whence $L^* \neq \emptyset$), period $\mathcal{D}$ and length l such that $\min L^* = 0$, $L' \subset [-M, -1]$, $L'' \subset \max L^* + [1, M]$ and $y \in \mathbb{Z}$.

We call $y + L'$ the *initial part*, $y + L^*$ the *central part* and $y + L''$ the *end part* of L.

- *almost arithmetical progression* (AAP for short) with *difference* d, *bound* M and *length* l, if it is an AAMP with difference d, period $\{0, d\}$, bound M and length l.

Note that

- AMPs, AAMPs and AAPs are finite nonempty subsets of $\mathbb{Z}$.
- a set L is an AMP if and only if it is an AAMP with bound 0, and it is an arithmetical progression with difference d if and only if it is an AAP with difference d and bound 0.
- a set L is an AAMP if and only if the shifted set $y + L$ is an AAMP for any $y \in \mathbb{Z}$.
- $L^* = \big(\mathcal{D} + d\mathbb{Z}\big) \cap [0, \max L^*]$.

AAMPs, as defined above, were introduced in [38], and a slightly less restrictive notion was first defined in [67].

Definition 3.2.3. We say that *the Structure Theorem for Sets of Lengths holds for the monoid* H if H is atomic and there exist some $M^* \in \mathbb{N}_0$ and a finite nonempty set $\Delta^* \subset \mathbb{N}$ such that every $L \in \mathcal{L}(H)$ is an AAMP with some difference $d \in \Delta^*$ and bound M^*.

If the Structure Theorem for Sets of Lengths holds for the monoid H, then H is a BF-monoid with finite set of distances $\Delta(H)$ (note that the formulation in [71, Definition 4.7.1] is slightly different and erronous, since it was forgotten to require $\Delta^*(H)$ to be finite). We cite three key results on the structure of sets of lengths in Krull monoids (proofs can be found in [71, Section 4.7], [122] and [71, Theorem 7.6.9]).

Theorem 3.2.4. *Let H be a Krull monoid with finite class group G. Then the Structure Theorem for Sets of Lengths holds. Moreover, if $|G| \geq 3$ and every class contains a prime, then it holds with the set $\Delta^* = \{\min \Delta(G_0) \mid G_0 \subset G$ with $\Delta(G_0) \neq \emptyset\} \subset \Delta(G)$.*

Idea of the Proof. The proof splits into an abstract additive part and an ideal-theoretic part. Both steps are based on the concepts of *pattern ideals* and of *tamely generated ideals* which are defined as follows.

- For a finite nonempty set $A \subset \mathbb{Z}$ the pattern ideal $\Phi(A)$ is the set of all $a \in H$ for which there is some $y \in \mathbb{Z}$ such that $y + A \subset \mathsf{L}(a)$.
- A subset $\mathfrak{a} \subset H$ is called tamely generated if there exist a subset $E \subset \mathfrak{a}$ and a bound $N \in \mathbb{N}$ with the following property:

  For every $a \in \mathfrak{a}$ there exists some $e \in E$ such that $e \,|\, a$, $\sup \mathsf{L}(e) \leq N$ and $\mathsf{t}(a, \mathsf{Z}(e)) \leq N$.

In the additive part one proves that the Structure Theorem for Sets of Lengths holds for every BF-monoid H with finite set $\Delta(H)$ in which all pattern ideals are tamely generated. This is done in the spirit of additive number theory. To apply this additive result to a BF-monoid H, it must be proved that $\Delta(H)$ is finite and that all pattern ideals are tamely generated. This is fairly simple for finitely generated monoids (but far from being simple for C-monoids).

Let H be a Krull monoid as above and let $G_0 \subset G$ denote the set of classes containing primes. By Theorem 1.3.4, it suffices to prove the Structure Theorem for the monoid $\mathcal{B}(G_0)$ of zero-sum sequences over G_0. Since $\mathcal{B}(G_0)$ is finitely generated by Proposition 1.2.4, the assertion follows. □

Theorem 3.2.4 was recently generalized to Krull monoids with finite Davenport constant (see [69]). More classes of monoids and domains where the Structure Theorem for Sets of Lengths holds can be found in [71, Section 4.7]. The next theorem is a realization result stating that Theorem 3.2.4 is sharp.

Theorem 3.2.5. *Let $M \in \mathbb{N}_0$ and let $\Delta^* \subset \mathbb{N}$ be a finite nonempty set. Then there exists a Krull monoid H with finite class group such that the following holds: for every* AAMP *L with difference $d \in \Delta^*$ and bound M there is some $y_{H,L} \in \mathbb{N}$ such that*

$$y + L \in \mathcal{L}(H) \quad \textit{for all} \quad y \geq y_{H,L}.$$

Indeed, there exists an algebraic number field such that its ring of integers has this property.

The next result is in a certain contrast to the previous two. In terms of zero-sum sequences it states that whenever the underlying set $\text{supp}(S)$ of a given zero-sum sequence S is a group, then the factorizations of S are as nice as possible. Indeed, its catenary degree $\mathsf{c}(S)$ is bounded above by 3, and its set of lengths $\mathsf{L}(S)$ is an arithmetical progression with difference 1.

Theorem 3.2.6. *Let H be a Krull monoid with class group G and $a \in H$ such that*

$$\mathrm{supp}(\boldsymbol{\beta}(a)) \cup \{0\} \subset G \quad \text{is a subgroup}.$$

Then $\mathsf{c}(a) \le 3$ and hence the set of lengths $\mathsf{L}(a)$ is an arithmetical progression with difference 1.

The main tool in the proof of Theorem 3.2.6 is the following result on the structure of additively closed sequences (see [51] and [71, Theorem 7.5.2]).

Proposition 3.2.7. *Let S, B, $C \in \mathcal{F}(G^\bullet)$ be sequences such that $S = BC$, $|S| \ge 4$ and $|B| \ge |C|$. Suppose that, for all $g_1, g_2 \in G$,*

$$\text{if } g_1 g_2 \,|\, B \text{ or } g_1 g_2 \,|\, C, \text{ then } (g_1 + g_2) \,|\, S.$$

Then S has a proper zero-sum subsequence, apart from the following exceptions:

1. $|C| = 1$, *and we are in one of the following cases:*
   - $B = g^k$ *and* $C = 2g$ *for some* $k \ge 3$ *and* $g \in G$ *with* $\mathrm{ord}(g) \ge k+2$.
   - $B = g^k(2g)$ *and* $C = 3g$ *for some* $k \ge 2$ *and* $g \in G$ *with* $\mathrm{ord}(g) \ge k+5$.
   - $B = g_1 g_2 (g_1 + g_2)$ *and* $C = g_1 + 2g_2$ *for some* $g_1, g_2 \in G$ *with* $\mathrm{ord}(g_1) = 2$ *and* $\mathrm{ord}(g_2) \ge 5$.
2. $\{B, C\} = \{\, g(9g)(10g), (11g)(3g)(14g) \,\}$ *for some* $g \in G$ *with* $\mathrm{ord}(g) = 16$.

Theorem 3.2.6 and some analytic machinery are the main ingredients to obtain the following density result (see [71, Theorem 9.4.11]). It states that the factorizations and hence the sets of lengths of "almost all" elements in a ring of integers are "as nice as possible".

Corollary 3.2.8. *Let K be an algebraic number field with ring of integers R and ideal class group G, and let $H = \{aR \mid a \in R^\bullet\}$ denote the monoid of non-zero principal ideals. For a principal ideal $a \in H$ let $|a| = (R{:}a)$ denote its norm. Then for every $x \ge 2$ we have*

$$\frac{\big|\{a \in H \mid \mathsf{c}(a) \le 3\,, |a| \le x\}\big|}{\big|\{a \in H \mid |a| \le x\}\big|} = 1 + O\big((\log x)^{-1/|G|}\big)\,.$$

3.3 The characterization problem

Two reduced Krull monoids H and H' are isomorphic if and only if there is a group isomorphism $\Phi\colon \mathcal{C}(H) \to \mathcal{C}(H')$ such that for every class $g \in \mathcal{C}(H)$, the number of primes in g equals the number of primes in the class $\Phi(g) \in \mathcal{C}(H')$ (see [71, Theorem 2.5.4]). If H is the multiplicative monoid of the ring of integers of an algebraic number field, then the class group is finite and the set of primes in each class is denumerable. Thus the traditional idea in algebraic number theory, that the class group determines the arithmetic, is justified. Initiated by a problem of W. Narkiewicz in the 1970s (see Sections 7.1 and 7.2 in [71]), a huge variety of explicit results in this direction was derived.

If the class group of a Krull monoid H is finite and every class contains a prime, then, roughly speaking, the system of sets of factorizations $\mathcal{Z}(H) = \{\mathsf{Z}(a) \mid a \in H\}$ determines the class group (see [71, Sections 7.1 and 7.2]). The question arose, which is still wide open, whether the same is true for the system of sets of lengths. Clearly, if H and H' are reduced Krull monoids with isomorphic class groups G, G' and primes in all classes, then

$$\mathcal{L}(H) = \mathcal{L}(G) = \mathcal{L}(H'),$$

but H and H' need not be isomorphic. By Proposition 1.2.4.4 it follows that

$$\mathcal{L}(C_1) = \{\{k\} \mid k \in \mathbb{N}_0\} = \mathcal{L}(C_2),$$

and it is easy to check (details may be found in [71, Theorem 7.3.2]) that

$$\mathcal{L}(C_3) = \{y + 2k + [0, k] \mid y, k \in \mathbb{N}_0\} = \mathcal{L}(C_2 \oplus C_2).$$

Note that $\mathsf{D}(C_3) = \mathsf{D}(C_2 \oplus C_2) = 3$, and $C_1, C_2, C_2 \oplus C_2$ and C_3 are the only finite abelian groups G' with $\mathsf{D}(G') \le 3$. So the best we can hope for is a positive answer to the following question:

Given two finite abelian groups G and G' with $\mathsf{D}(G) \ge 4$ such that $\mathcal{L}(G) = \mathcal{L}(G')$, does it follow that $G \cong G'$?

Up to now there is known no pair of non-isomorphic groups (G, G') with $\mathsf{D}(G) \ge 4$ and $\mathcal{L}(G) = \mathcal{L}(G')$. We start with some simple observations and then we gather the results known so far.

Proposition 3.3.1.

1. $\mathcal{L}(G) = \{y + L \mid y \in \mathbb{N}_0,\ L \in \mathcal{L}(G^\bullet)\} \supset \{\{y\} \mid y \in \mathbb{N}_0\}$, *and equality holds if and only if* $|G| \le 2$.
2. *If* $G_0 \subset G$ *is a subset, then* $\mathcal{L}(G_0) \subset \mathcal{L}(G)$.
3. *Let* G' *be an abelian group with* $|G'| \ge 3$ *such that* $\mathcal{L}(G) = \mathcal{L}(G')$. *Then we have* $\rho_k(G) = \rho_k(G')$ *and* $\lambda_k(G) = \lambda_k(G')$ *for every* $k \in \mathbb{N}$, $\mathsf{D}(G) = \mathsf{D}(G')$ *and* $\Delta(G) = \Delta(G')$.
4. *There exist (up to isomorphisms) only finitely many finite abelian groups* G' *such that* $\mathcal{L}(G) = \mathcal{L}(G')$.

Proof. 1. Observe that $\mathcal{B}(G) = \{0^y B \mid B \in \mathcal{B}(G^\bullet),\ y \in \mathbb{N}_0\}$, and if $B \in \mathcal{B}(G^\bullet)$ and $y \in \mathbb{N}_0$, then $\mathsf{L}(0^y B) = y + \mathsf{L}(B)$. By definition, we have $|L| = 1$ for every $L \in \mathcal{L}(G)$ if and only if $\mathcal{B}(G)$ is half-factorial, and by Proposition 1.2.4 this is equivalent to $|G| \le 2$.

2. This is obvious.

3. By 1. we obtain $|G| \geq 3$. By the very definition we have $\Delta(G) = \Delta(G')$, $\lambda_k(G) = \lambda_k(G')$ and $\rho_k(G) = \rho_k(G')$ for every $k \in \mathbb{N}$, and hence $\mathsf{D}(G) = \rho_2(G) = \rho_2(G') = \mathsf{D}(G')$ by Theorem 2.3.1.3.

4. If G' is an abelian group with $\mathcal{L}(G) = \mathcal{L}(G')$ and $|G'| \geq 3$, then it follows that $\mathsf{D}(G) = \mathsf{D}(G') \geq 1 + \mathsf{d}^*(G')$ (see Lemma 2.1.3.2). By the very definition of $\mathsf{d}^*(\cdot)$, there are (up to isomorphisms) only finitely many finite abelian groups G' with $\mathsf{d}^*(G') < \mathsf{D}(G)$. □

Proposition 3.3.2. *Let G' be a finite abelian group with $\mathsf{D}(G') \in [4, 10]$. If $\mathcal{L}(G) = \mathcal{L}(G')$, then $G \cong G'$.*

Theorem 3.3.3. *Let G be a finite elementary p-group and let G' be a finite elementary q-group with $\mathsf{D}(G') \geq 4$ and with primes $p, q \in \mathbb{P}$. If $\mathcal{L}(G) = \mathcal{L}(G')$, then $G \cong G'$.*

Theorem 3.3.4. *Let G' be a finite abelian group with $\mathsf{D}(G') \geq 4$ and suppose that one of the following statements hold:*

1. *G is cyclic.*
2. *G is an elementary 2-group.*
3. *$G \cong C_2 \oplus C_{2n}$ with $n \geq 2$.*
4. *$G \cong C_n \oplus C_n$ with $n \geq 3$.*

If $\mathcal{L}(G) = \mathcal{L}(G')$, then $G \cong G'$.

The results given in Proposition 3.3.2 and Theorems 3.3.3 and 3.3.4 are mainly due to Wolfgang A. Schmid (see [123, 120, 119] and [71, Section 7.3]). They are based on solid investigations of the set $\Delta^*(G) = \{\min \Delta(G_0) \mid G_0 \subset G \text{ with } \Delta(G_0) \neq \emptyset\}$, which occurs in Theorem 3.2.4. We do not discuss these topics here, but using results on the $\rho_k(G)$-invariants we will be able to prove Theorem 3.3.4 for cyclic groups and for elementary 2-groups (see Corollary 5.3.3).

Chapter 4

Addition theorems and direct zero-sum problems

We start with the classical theorems of Kneser and Kemperman–Scherk which are fundamental in additive group theory. Having these results at our disposal we continue the investigation of the Davenport constant, of the Erdős–Ginzburg–Ziv constant and of related invariants in zero-sum theory.

4.1 The theorems of Kneser and of Kemperman-Scherk

Let $k \in \mathbb{N}$ and let $A, B, A_1, \ldots, A_k \subset G$ be nonempty subsets. Then

$$\mathrm{Stab}(A) = \{g \in G \mid g + A = A\}$$

denotes the *stabilizer* of A, which is a subgroup of A. For $g \in G$, let

$$\mathsf{r}_{A_1,\ldots,A_k}(g) = \Big| \Big\{(a_1, \ldots, a_k) \in A_1 \times \cdots \times A_k \mid g = a_1 + \ldots + a_k\Big\} \Big|$$

denote the number of representations of g as a sum of elements of $A_1, \ldots, A_k$. In particular, we have

$$r_{A,B}(g) = |\{(a, b) \in A \times B \mid g = a + b\}| = |A \cap (g - B)| \,.$$

In the 1950s M. Kneser proved the following addition theorem formulated in Theorem 4.1.1. Since a proof is given in the Part II "Sumsets and Structure" by Imre Z. Ruzsa, we do not give a proof here. Moreover, a variety of proofs and historical references may be found in each of the following monographs [104, Chapter 1], [107, Chapter 4], [71, Section 5.2], and [130, Theorem 5.5]. For some recent development around Kneser's theorem, and in particular on the isoperimetric approach, we refer the reader to [22, 83, 126, 97, 19, 81, 25, 24, 3, 89, 90, 93, 91, 80].

Theorem 4.1.2 was first proved by J.H.B. Kemperman (see [98]; the special case $\min\{ r_{A,B}(g) \mid g \in A+B\} = 1$ was settled before by P. Scherk answering a question of L. Moser [118]; a short proof of Scherk's theorem, which is not based on Kneser's theorem, may be found in [90]). Corollary 4.1.3 is crucial in many investigations on the structure of zero-sum free sequences (as, for example, in the proofs of Proposition 4.2.6 and Corollary 5.1.10).

Theorem 4.1.1 (Kneser). *Let* $K = \mathrm{Stab}(A+B)$ *be the stabilizer of* $A+B$.

1. *There exists a subgroup* $K' \subset K$ *such that* $|A+B| \ge |A| + |B| - |K'|$.
2. *There exists a subgroup* $K' \subset K$ *such that* $|A+B| \ge |A+K'|+|B+K'|-|K'|$.
3. $|A+B| \ge |A+K| + |B+K| - |K|$.
4. *Either* $|A+B| \ge |A| + |B|$ *or* $|A+B| = |A+K| + |B+K| - |K|$.

Theorem 4.1.2 (Kemperman–Scherk). *Let* $K = \mathrm{Stab}(A+B)$ *be the stabilizer of* $A+B$. *Then*

$$\begin{aligned}|A+B| &\ge |A| + |B| - \min\{ r_{(a+K)\cap A,(b+K)\cap B}(g) \mid a \in A,\ b \in B,\ g \in a+b+K\} \\ &\ge |A| + |B| - \min\{ r_{A,B}(g) \mid g \in A+B\}\,.\end{aligned}$$

Proof. If $a \in A$, $b \in B$ and $g \in a+b+K$, then $r_{(a+K)\cap A,(b+K)\cap B}(g) \le r_{A,B}(g)$, and therefore

$$\begin{aligned}&\min\{r_{(a+K)\cap A,(b+K)\cap B}(g) \mid a \in A,\ b \in B,\ g \in a+b+K\} \\ &\quad\le \min\{r_{A,B}(g) \mid a \in A,\, b \in B,\, g \in a+b+K\} \\ &\quad= \min\{r_{A,B}(g) \mid g \in A+B+K\} = \min\{r_{A,B}(g) \mid g \in A+B\}\,.\end{aligned}$$

Thus it suffices to prove the first inequality. We may assume that $|A+B| < |A|+|B|$, and then $|A+B| = |A+K| + |B+K| - |K|$ by Theorem 4.1.1.4.

Suppose that $a \in A$, $b \in B$ and $g \in a+b+K$. By definition, we have

$$r_{(a+K)\cap A,(b+K)\cap B}(g) = |C_1 \cap C_2|\,,$$

where $C_1 = (a+K)\cap A$ and $C_2 = g - [(b+K)\cap B]$, and thus we must prove that $|C_1 \cap C_2| \ge |A| + |B| - |A+B|$. Since $C_1 \cup C_2 \subset a+K$, we obtain

$$\begin{aligned}|C_1 \cap C_2| &= |C_1| + |C_2| - |C_1 \cup C_2| \ge |C_1| + |C_2| - |a+K| \\ &= |(a+K)\cap A| + |(b+K)\cap B| - |K| \\ &= |a+K| - |(a+K)\setminus A| + |b+K| - |(b+K)\setminus B| - |K| \\ &\ge |K| - |(A+K)\setminus A| - |(B+K)\setminus B| \\ &= |K| - |A+K| + |A| - |B+K| + |B| = |A| + |B| - |A+B|\,.\end{aligned}$$

□

Corollary 4.1.3. *Let* $k \in \mathbb{N}$.

1. *Let* $A_1, \ldots, A_k \subset G$ *be nonempty subsets. Then*

$$|A_1 + \ldots + A_k| \ge |A_1| + \ldots + |A_k| - (k-2) \\ - \min\{r_{A_1,\ldots,A_k}(g) \mid g \in A_1 + \ldots + A_k\}\,.$$

2. *If* $S = S_1 \cdot \ldots \cdot S_k \in \mathcal{A}^*(G)$, *then*

$$|\Sigma(S)| \ge |\Sigma(S_1)| + \ldots + |\Sigma(S_k)|\,.$$

Proof. 1. We proceed by induction on k. For $k = 1$ the assertion is clear. Suppose that $k \ge 2$. We start with the following assertion.

A. For every $g \in A_1 + \ldots + A_k$ we have

$$r_{A_1,\ldots,A_k}(g) \ge \min\{r_{A_1,\ldots,A_{k-1}}(a) \mid a \in A_1 + \ldots + A_{k-1}\} \\ + \min\{r_{A_1+\ldots+A_{k-1},A_k}(b) \mid b \in A_1 + \ldots + A_k\} - 1\,.$$

Proof of **A**. If $g \in A_1 + \ldots + A_k$, then we get

$$\begin{aligned} r_{A_1,\ldots,A_k}(g) &\ge \min\{r_{A_1,\ldots,A_{k-1}}(a) \mid a \in A_1 + \ldots + A_{k-1}\} \cdot r_{A_1+\ldots+A_{k-1},A_k}(g) \\ &\ge \min\{r_{A_1,\ldots,A_{k-1}}(a) \mid a \in A_1 + \ldots + A_{k-1}\} \\ &\quad \cdot \min\{r_{A_1+\ldots+A_{k-1},A_k}(b) \mid b \in A_1 + \ldots + A_k\} \\ &\ge \min\{r_{A_1,\ldots,A_{k-1}}(a) \mid a \in A_1 + \ldots + A_{k-1}\} \\ &\quad + \min\{r_{A_1+\ldots+A_{k-1},A_k}(b) \mid b \in A_1 + \ldots + A_k\} - 1\,. \end{aligned}$$

Now the induction hypothesis, Theorem 4.1.2 and **A** imply that

$$\begin{aligned} |A_1 + \ldots + A_k| &\ge |A_1 + \ldots + A_{k-1}| + |A_k| \\ &\qquad - \min\{r_{A_1+\ldots+A_{k-1},A_k}(b) \mid b \in A_1 + \ldots + A_k\} \\ &\ge |A_1| + \ldots + |A_{k-1}| - (k-3) \\ &\qquad - \min\{r_{A_1,\ldots,A_{k-1}}(a) \mid a \in A_1 + \ldots + A_{k-1}\} + |A_k| \\ &\qquad - \min\{r_{A_1+\ldots+A_{k-1},A_k}(b) \mid b \in A_1 + \ldots + A_k\} \\ &\ge |A_1| + \ldots + |A_k| - (k-2) \\ &\qquad - \min\{r_{A_1,\ldots,A_k}(g) \mid g \in A_1 + \ldots + A_k\}\,. \end{aligned}$$

2. For $i \in [1,k]$ we set $A_i = \Sigma(S_i) \cup \{0\}$. Then $r_{A_1,\ldots,A_k}(0) = 1$, and hence 1. implies that

$$|\Sigma(S)| \ge |A_1 + \ldots + A_k| - 1 \ge |A_1| + \ldots + |A_k| - k = |\Sigma(S_1)| + \ldots + |\Sigma(S_k)|\,. \quad \square$$

4.2 On the Erdős–Ginzburg–Ziv constant $\mathsf{s}(G)$ and on some of its variants

Definition 4.2.1. Let $\exp(G) = n$.

1. A sequence $S \in \mathcal{F}(G)$ is called *short* (in G) if $|S| \in [1, n]$.
2. We denote by $\eta(G)$ the smallest integer $l \in \mathbb{N}$ with the following property:
   - Every sequence $S \in \mathcal{F}(G)$ of length $|S| \ge l$ has a short zero-sum subsequence.
3. We denote by $\mathsf{s}(G)$ the smallest integer $l \in \mathbb{N}$ with the following property:
   - Every sequence $S \in \mathcal{F}(G)$ of length $|S| \ge l$ has a zero-sum subsequence T of length $|T| = n$.

   The invariant $\mathsf{s}(G)$ will be called the *Erdős–Ginzburg–Ziv constant* (EGZ constant for short).
4. We denote by $\mathsf{s}_{n\mathbb{N}}(G)$ the smallest integer $l \in \mathbb{N}$ with the following property:
   - Every sequence $S \in \mathcal{F}(G)$ of length $|S| \ge l$ has a nontrivial zero-sum subsequence T of length $|T| \equiv 0 \mod n$.

The investigation of these invariants has a long tradition in combinatorial number theory as well as in finite geometry. Indeed, the Erdős–Ginzburg–Ziv theorem, first proved in 1961 (see [32]) and stating that $\mathsf{s}(C_n) = 2n - 1$ (see Corollary 4.2.8), is considered as a starting point in zero-sum theory. As already pointed out by H. Harborth [94], $\mathsf{s}(C_n^r)$ is the smallest integer $l \in \mathbb{N}$ such that every set of l lattice point in an r-dimensional Euclidean space contains n elements which have a centroid with integral coordinates. Moreover, if φ is the maximal size of a cap in $\mathrm{AG}(r, 3)$, then $\mathsf{s}(C_3^r) = 2\varphi + 1$ (see [28, Section 5] for the connection with finite geometry).

The invariant $\eta(G)$ is a crucial tool in the inductive method which roughly works as follows: for the investigation of a given sequence $S \in \mathcal{F}(G)$ proceed in the following three steps:

- Find a suitable subgroup $K \subset G$ and consider the natural epimorphism $\varphi\colon G \to G/K$.
- Consider a factorization $S = S_0 S_1 \cdot \ldots \cdot S_k$ such that $|S_i|$ is small and $\varphi(S_i) \in \mathcal{B}(G/K)$ for all $i \in [1, k]$.
- Investigate the sequences $T = \sigma(S_1) \cdot \ldots \cdot \sigma(S_k) \in \mathcal{F}(K)$ and $S_0 T \in \mathcal{F}(G)$. Clearly, if S is zero-sum free, then $S_0 T$ is zero-sum free too.

The inductive method was already used successfully by J.E. Olson and P. van Emde Boas in the 1960s, and then it was more and more refined by W. Gao and many other authors. After having done the necessary preparations in Lemmas 4.2.4

and 4.2.5 we will demonstrate the power of this method in Theorems 4.2.10 and 4.2.12 (the polynomial method – see the article "The polynomial method in additive combinatorics" by G. Károlyi in Part III– and coverings by cosets – see [71, Chapter 5.6], [128, 102] – are further important methods, which however cannot be discussed here).

Our main result in this Section is Theorem 4.2.10 which gives, for groups G of rank $\mathsf{r}(G) \le 2$, the precise values of $\mathsf{d}(G)$, $\eta(G)$ and $\mathsf{s}(G)$. For $\mathsf{d}(G)$ this was shown independently by J.E. Olson and D. Kruyswijk in the late 1960s. The result on $\mathsf{s}(G)$ is based on C. Reiher's work [114]. The proof of Theorem 4.2.10, as presented here, follows the lines from [71, Theorem 5.8.3]. On our way we show the theorem of Erdős–Ginzburg–Ziv (Corollary 4.2.8; for some recent development in the flavor of Erdős–Ginzburg–Ziv see [40, 39, 82, 84, 86]).

Lemma 4.2.2.

1. *We have* $\mathsf{D}(G) \le \eta(G) \le \mathsf{s}(G) - \exp(G) + 1$.
2. *Let* $G = C_{n_1} \oplus \cdots \oplus C_{n_r}$ *with* $r = \mathsf{r}(G)$ *and* $1 < n_1 \mid \cdots \mid n_r$. *If* $r \ge 2$, *then* $\eta(G) \ge \mathsf{d}^*(G) + n_1$.

Proof. 1. The inequality $\mathsf{D}(G) \le \eta(G)$ follows by Lemma 2.1.2.3 and the very definition of $\eta(G)$. For the proof of the second inequality let $n = \exp(G)$, and consider a sequence $S \in \mathcal{F}(G)$ of length $|S| \ge \mathsf{s}(G) - n + 1$. We must prove that S has a short zero-sum subsequence. The sequence $T = 0^{n-1}S \in \mathcal{F}(G)$ satisfies $|T| \ge \mathsf{s}(G)$, and therefore there exists a zero-sum subsequence $T' = 0^k S'$ of T, where $k \in [0, n-1]$, $S' \mid S$ and $|T'| = |S'| + k = n$. Hence S' is a short zero-sum subsequence of S.

2. Let $r \ge 2$ and let $(e_1, \ldots, e_r)$ be a basis of G such that $\operatorname{ord}(e_i) = n_i$ for all $i \in [1, r]$,

$$e_0 = \sum_{i=1}^r e_i \quad \text{and} \quad S = e_0^{n_1 - 1} \prod_{i=1}^r e_i^{n_i - 1} \in \mathcal{F}(G).$$

We assert that S has no short zero-sum subsequence. Let

$$T = e_0^{n_0} \prod_{i=1}^r e_i^{n_i'}, \quad \text{where} \quad n_0 \in [0, n_1 - 1] \ \text{and} \ n_i' \in [0, n_i - 1] \ \text{for all} \ i \in [1, r],$$

be a nontrivial zero-sum subsequence of S. Then $n_0 \ge 1$ by Lemma 2.1.3. Since $0 = \sigma(T) = (n_1' + n_0)e_1 + \ldots + (n_r' + n_0)e_r$, it follows that $n_i' + n_0 \equiv 0 \bmod n_i$ for all $i \in [1, r]$, and $1 \le n_i' + n_0 \le 2n_i - 2$ implies $n_i' = n_i - n_0$ for all $i \in [1, r]$. Hence

$$|T| = n_0 + \sum_{i=1}^r (n_i - n_0) = n_r + \sum_{i=1}^{r-1} (n_i - n_0) > n_r = \exp(G),$$

and thus T is not a short zero-sum sequence of S over G. □

Lemma 4.2.3. *Let $S \in \mathcal{F}(G)$ and $n \ge 2$.*

1. *If $|S| \ge \mathsf{D}(G \oplus C_n)$ and $\mathsf{D}(G \oplus C_n) \le 3n-1$, then S has a zero-sum subsequence $T \in \mathcal{B}(G)$ of length $|T| \in \{n, 2n\}$.*
2. *Suppose that $\mathsf{D}(G) \le 2n-1$, $\mathsf{D}(G \oplus C_n) \le 3n-1$ and $|S| \ge \mathsf{D}(G \oplus C_n)$. Then S has a zero-sum subsequence $T \in \mathcal{B}(G)$ of length $|T| \in [1, n]$. In particular, if $n \le \exp(G)$, then $\eta(G) \le \mathsf{D}(G \oplus C_n)$.*
3. *If $\exp(G) = n$, then*

$$\mathsf{D}(G) + n - 1 \le \mathsf{s}_{n\mathbb{N}}(G) \le \min\{\mathsf{s}(G), \mathsf{D}(G \oplus C_n)\}.$$

Proof. Let $G \oplus C_n = G \oplus \langle e \rangle$ with $\operatorname{ord}(e) = n$, so that every $h \in G \oplus C_n$ has a unique representation $h = g + je$, where $g \in G$ and $j \in [0, n-1]$. We define $\varphi \colon G \to G \oplus C_n$ by $\varphi(g) = g + e$ for every $g \in G$.

1. Since $\varphi(S) \in \mathcal{F}(G \oplus C_n)$ and $|\varphi(S)| = |S| \ge \mathsf{D}(G \oplus C_n)$, S has a subsequence T with $1 \le |T| \le \mathsf{D}(G \oplus C_n) \le 3n-1$ such that $\varphi(T)$ has sum zero. Because $0 = \sigma\big(\varphi(T)\big) = \sigma(T) + |T|e \in G \oplus C_n$, we obtain that $\sigma(T) = 0$, $|T| \equiv 0 \mod n$, and $|T| \le 3n-1$ implies $|T| \in \{n, 2n\}$.

2. If $|S| \ge \mathsf{D}(G \oplus C_n)$, then by 1. there exists a zero-sum subsequence T of S such that $|T| \in \{n, 2n\}$. If $|T| \le n$, we are done. If $|T| = 2n$, then $|T| > \mathsf{D}(G)$ implies $T = T_1T_2$ for some zero-sum subsequences T_1, T_2 with $1 \le |T_1| \le |T_2|$, and T_1 is the desired subsequence of S.

3. If $S \in \mathcal{F}(G)$ is a zero-sum free sequence of length $|S| = \mathsf{D}(G) - 1$, then the sequence $0^{n-1}S$ has no zero-sum subsequence of length divisible by n. Thus $\mathsf{D}(G) + n - 2 = |0^{n-1} \cdot S| < \mathsf{s}_{n\mathbb{N}}(G)$. By definition we have $\mathsf{s}_{n\mathbb{N}}(G) \le \mathsf{s}(G)$.

In order to verify that $\mathsf{s}_{n\mathbb{N}}(G) \le \mathsf{D}(G \oplus C_n)$, let $S = \prod_{i=1}^{l} g_i \in \mathcal{F}(G)$ with $l = \mathsf{D}(G \oplus C_n)$. Then the sequence $\prod_{i=1}^{l}(g_i + e) \in \mathcal{F}(G \oplus C_n)$ has a zero-sum subsequence T of length $|T| \equiv 0 \mod n$, and whence the same is true for S. □

Lemma 4.2.4. *Let $\varphi \colon G \to \overline{G}$ be a group homomorphism and let $k \in \mathbb{N}$.*

1. *If $S \in \mathcal{F}(G)$ and $|S| \ge (k-1)\exp(\overline{G}) + \mathsf{s}(\overline{G})$, then S admits a product decomposition $S = S_1 \cdot \ldots \cdot S_k S'$, where $S_1, \ldots, S_k, S' \in \mathcal{F}(G)$ and, for every $i \in [1, k]$, $\varphi(S_i)$ has sum zero and length $|S_i| = \exp(\overline{G})$.*
2. *If $S \in \mathcal{F}(G)$ and $|S| \ge (k-1)\exp(\overline{G}) + \eta(\overline{G})$, then S admits a product decomposition $S = S_1 \cdot \ldots \cdot S_k S'$, where $S_1, \ldots, S_k, S' \in \mathcal{F}(G)$ and, for every $i \in [1, k]$, $\varphi(S_i)$ has sum zero and length $|S_i| \in [1, \exp(\overline{G})]$.*

Proof. 1. Suppose that for some $j \in [0, k-1]$ we have found a product decomposition $S = S_1 \cdot \ldots \cdot S_j S'$ where $S_1, \ldots, S_j, S' \in \mathcal{F}(G)$ and, for every $i \in [1, j]$, $\varphi(S_i)$ has sum zero and length $|S_i| = \exp(\overline{G})$. Then

$$|\varphi(S')| = |S'| = |S| - j\exp(\overline{G}) \ge \big(k-1-j\big)\exp(\overline{G}) + \mathsf{s}(\overline{G}) \ge \mathsf{s}(\overline{G}),$$

and therefore S' has a subsequence S_{j+1} such that $\varphi(S_{j+1})$ has sum zero and length $|S_{j+1}| = \exp(\overline{G})$. Now the assertion follows by induction on j.

2. This is proved in precisely the same way as 1. □

Lemma 4.2.5. *Let* $K \subset G$ *be a subgroup.*

1. *If* $S \in \mathcal{F}(G)$ *and* $|S| \ge (\mathsf{s}(K)-1)\exp(G/K) + \mathsf{s}(G/K)$, *then* S *has a zero-sum subsequence* T *of length* $|T| = \exp(K)\exp(G/K)$. *In particular, if* $\exp(G) = \exp(K)\exp(G/K)$, *then*
$$\mathsf{s}(G) \le (\mathsf{s}(K)-1)\exp(G/K) + \mathsf{s}(G/K)\,.$$
2. *If* $S \in \mathcal{F}(G)$ *and* $|S| \ge (\eta(K)-1)\exp(G/K) + \eta(G/K)$, *then* S *has a zero-sum subsequence* T *of length* $1 \le |T| \le \exp(K)\exp(G/K)$. *In particular, if* $\exp(G) = \exp(K)\exp(G/K)$, *then*
$$\eta(G) \le (\eta(K)-1)\exp(G/K) + \eta(G/K)\,.$$
3. $\mathsf{d}(G) \le \mathsf{d}(K)\exp(G/K) + \max\{\mathsf{d}(G/K), \eta(G/K) - \exp(G/K) - 1\}$.

Proof. Let $\varphi\colon G \to G/K$ denote the canonical epimorphism. If $K = \{0\}$, then all assertions are obvious. Suppose that $K \ne \{0\}$.

1. Let $S \in \mathcal{F}(G)$ be a sequence with $|S| \ge (\mathsf{s}(K)-1)\exp(G/K) + \mathsf{s}(G/K)$. By Lemma 4.2.4.1, S has a product decomposition $S = S_1 \cdot \ldots \cdot S_{\mathsf{s}(K)}S'$, where $S_1, \ldots, S_{\mathsf{s}(K)}, S' \in \mathcal{F}(G)$ and, for every $i \in [1, \mathsf{s}(K)]$, $\varphi(S_i)$ has sum zero and length $|S_i| = \exp(G/K)$. Then the sequence $\sigma(S_1) \cdot \ldots \cdot \sigma(S_{\mathsf{s}(K)}) \in \mathcal{F}(K)$ has a zero-sum subsequence V of length $|V| = \exp(K)$, say
$$V = \prod_{i \in I} \sigma(S_i)\,, \quad \text{where} \quad I \subset [1, \mathsf{s}(K)] \quad \text{and} \quad |I| = \exp(K)\,.$$
Thus the sequence
$$T = \prod_{i \in I} S_i$$
is a zero-sum subsequence of S of length $|T| = |I|\exp(G/K) = \exp(K)\exp(G/K)$.

2. This is proved in precisely the same way as 1.

3. Let $S \in \mathcal{F}(G)$ be a sequence of length
$$|S| > \mathsf{d}(K)\exp(G/K) + \max\{\mathsf{d}(G/K), \eta(G/K) - \exp(G/K) - 1\}\,.$$
We must prove that S is not zero-sum free. Since $|S| \ge (\mathsf{d}(K)-1)\exp(G/K) + \eta(G/K)$, Lemma 4.2.4.2 provides us with a product decomposition $S = S_1 \cdot \ldots \cdot S_{\mathsf{d}(K)}S'$, where $S_1, \ldots, S_{\mathsf{d}(K)}, S' \in \mathcal{F}(G)$ and, for every $i \in [1, \mathsf{d}(K)]$, $\varphi(S_i)$

has sum zero and length $|S_i| \in [1, \exp(G/K)]$. Now we obtain $|S'| \ge |S| - \exp(G/K)\mathsf{d}(K) > \mathsf{d}(G/K)$, and therefore S' has a nontrivial subsequence S_0 such that $\varphi(S_0)$ has sum zero. Hence $V = \sigma(S_0)\sigma(S_1) \cdot \ldots \cdot \sigma(S_{\mathsf{d}(K)}) \in \mathcal{F}(K)$, and $|V| > \mathsf{d}(K)$ implies that V is not zero-sum free. Hence $T = S_0 S_1 \cdot \ldots \cdot S_{\mathsf{d}(K)}$ is a subsequence of S which is not zero-sum free. □

The following result was found independently by several authors. Its history is described in [92].

Proposition 4.2.6. *Let* $S \in \mathcal{F}(G)$ *be a sequence of length* $|S| \ge |G|$ *and* $k' = \max\{\mathrm{ord}(g) \mid g \in \mathrm{supp}(S)\}$. *Then* S *has a nontrivial zero-sum subsequence* T *of length* $|T| \le \min\{\mathsf{h}(S),\, k'\}$.

Proof. We set $k = \mathsf{h}(S)$. If $k' \le k$, let $g \in G$ be such that $\mathsf{v}_g(S) = k$. Then $T = g^{\mathrm{ord}(g)}$ has the desired property. Hence it is sufficient to prove that S has a zero-sum subsequence of T of length $|T| \in [1, k]$. If $0 \in \mathrm{supp}(S)$, we set $T = 0$. Thus suppose that $0 \notin \mathrm{supp}(S)$. There exists a decomposition $S = S_1 \cdot \ldots \cdot S_k$, where $S_1, \ldots, S_k \in \mathcal{F}(G)$ are squarefree. For $i \in [1, k]$, let $B_i = \mathrm{supp}(S_i)$, $A_i = B_i \cup \{0\}$, and assume that S has no zero-sum subsequence as required. This implies that $r_{A_1,\ldots,A_k}(0) = 1$, and thus Corollary 4.1.3 implies that

$$|A_1 + \ldots + A_k| \ge |A_1| + \ldots + |A_k| - k + 1 = |B_1| + \ldots + |B_k| + 1 = |G| + 1\,,$$

a contradiction. □

Theorem 4.2.7 (Gao). *We have* $\eta(G) \le |G|$ *and* $\mathsf{s}(G) \le |G| + \exp(G) - 1$.

Proof. The first inequality is an immediate consequence of Proposition 4.2.6. In order to verify the upper bound for $\mathsf{s}(G)$, we set $n = \exp(G)$ and we must prove that every sequence $S \in \mathcal{F}(G)$ of length $|S| \ge |G| + n - 1$ has a zero-sum subsequence of length n. Thus assume that

$$S = g_1^{k_1} \cdot \ldots \cdot g_l^{k_l} \in \mathcal{F}(G)\,,$$

where $|S| \ge |G| + n - 1$, $k = k_1 \ge \cdots \ge k_l \ge 1$ and $g_1, \ldots, g_l \in G$ are distinct. If $k \ge n$, then g_1^n is a zero-sum subsequence of length n. Therefore we assume that $k \le n - 1$ and $l \ge 2$, and we consider the sequence

$$U = (g_2 - g_1)^{k_2} \cdot \ldots \cdot (g_l - g_1)^{k_l} \in \mathcal{F}(G)\,.$$

It is sufficient to prove that U has a zero-sum subsequence V such that $n - k \le |V| \le n$. Indeed, if $V = (g_2 - g_1)^{k_2'} \cdot \ldots \cdot (g_l - g_1)^{k_l'}$ is such a zero-sum subsequence, where $k_i' \in [0, k_i]$ for all $i \in [2, l]$ and $n - k \le k_2' + \ldots + k_l' \le n$, then $0 \le n - (k_2' + \ldots + k_l') \le k$, and the sequence

$$T = g_1^{n - (k_2' + \ldots + k_l')} g_2^{k_2'} \cdot \ldots \cdot g_l^{k_l'}$$

is a zero-sum subsequence of S of length n.

Since $|U| = |S| - k \geq |G| + n - 1 - k \geq |G|$ and $\eta(G) \leq |G|$, it follows that U has a short zero-sum subsequence. Let V be a short zero-sum subsequence of U of maximal length and assume, contrary to our requirement, that $|V| \leq n-k-1$. If $U = VV'$, then $|V'| = |U| - |V| \geq (|G| + n - 1 - k) - (n - k - 1) = |G|$, and by Proposition 4.2.6 it follows that V' has a zero-sum subsequence V'' of length

$$1 \leq |V''| \leq \max\{\mathsf{v}_g(V') \mid g \in G\} \leq \max\{\mathsf{v}_g(U) \mid g \in G\} \leq k\,.$$

Then VV'' is a zero-sum subsequence of U of length

$$|V| < |VV''| = |V| + |V''| \leq (n - k - 1) + k = n - 1,$$

a contradiction to the maximality of $|V|$. □

In the same spirits (using Proposition 4.2.6 and Theorem 4.2.7) W. Gao proved that $|G| + \mathsf{d}(G)$ is the smallest integer $l \in \mathbb{N}$ such that every sequence $T \in \mathcal{F}(G)$ of length $|T| \geq l$ has a zero-sum subsequence of length $|G|$ (see [42] and [71, Proposition 5.7.9]). There is a weighted generalization of this theorem by Y. ould Hamidoune [92] (for more of this flavor see [66, 1, 84, 78, 85]). For cyclic groups Gao's Theorem 4.2.7 reduces to the classical result of Erdős–Ginzburg–Ziv.

Corollary 4.2.8 (Erdős–Ginzburg–Ziv). *For every $n \in \mathbb{N}$ we have*

$$\eta(C_n) = n \quad \textit{and} \quad \mathsf{s}(C_n) = 2n - 1\,.$$

Proof. By Lemma 4.2.2 and Theorem 4.2.7, we obtain

$$n = \mathsf{D}(C_n) \leq \eta(C_n) \leq |C_n| = n,$$

and thus $\eta(C_n) = n$. Again by Lemma 4.2.2 and Theorem 4.2.7 we get

$$2n - 1 = \eta(C_n) + \exp(C_n) - 1 \leq \mathsf{s}(C_n) \leq |C_n| + \exp(C_n) - 1 = 2n - 1,$$

and thus $\mathsf{s}(C_n) = 2n - 1$. □

Proposition 4.2.9 (Reiher). *For every prime $p \in \mathbb{P}$ we have $\mathsf{s}(C_p \oplus C_p) \leq 4p - 3$.*

Proof. The original proof by C. Reiher (see [114]) is based on the theorem of Chevalley–Warning (see [130, Theorem 9.24]). A proof using group algebras may be found in [71, Proposition 5.8.1], and for a generalization see [125]. □

Theorem 4.2.10. *Let $G = C_{n_1} \oplus C_{n_2}$ with $1 \leq n_1 \mid n_2$. Then*

$$\mathsf{s}(G) = 2n_1 + 2n_2 - 3\,, \quad \eta(G) = 2n_1 + n_2 - 2 \quad \textit{and} \quad \mathsf{d}(G) = n_1 + n_2 - 2 = \mathsf{d}^*(G)\,.$$

Proof. By Corollaries 4.2.8 and 2.1.4, the result holds for $n_1 = 1$. Suppose that $n_1 > 1$ and note that $\exp(G) = n_2$. By Lemma 2.1.3 we have $\mathsf{d}^*(G) \leq \mathsf{d}(G)$. Now Lemma 4.2.2 implies that

$$\eta(G) \geq 2n_1 + n_2 - 2 \quad \text{and} \quad \mathsf{s}(G) \geq \eta(G) + n_2 - 1 \geq 2n_1 + 2n_2 - 3\,.$$

Thus it remains to show that $\mathsf{s}(G) \le 2n_1 + 2n_2 - 3$ and $\mathsf{d}(G) \le n_1 + n_2 - 2$.

We use induction on $\exp(G)$. If $p \in \mathbb{P}$ and $G = C_p \oplus C_p$, then $\mathsf{d}(G) = 2p - 2$ by Theorem 2.2.6, and Proposition 4.2.9 implies $\mathsf{s}(G) \le 4p - 3$.

Assume now that $p \in \mathbb{P}$, $p \,|\, n_1$, $p < n_2$ and set $m_i = p^{-1} n_i$ for $i \in \{1, 2\}$. Then the assertions are true for the groups $pG \cong C_{m_1} \oplus C_{m_2}$ and $G/pG \cong C_p \oplus C_p$. By Lemma 4.2.5.1 we obtain

$$\mathsf{s}(G) \le \big(\mathsf{s}(pG) - 1\big)p + \mathsf{s}(G/pG) \le (2m_1 + 2m_2 - 4)p + (4p - 3) = 2n_1 + 2n_2 - 3\,,$$

and Lemma 4.2.5.3 implies

$$\begin{aligned} \mathsf{d}(G) \le \ & \mathsf{d}(pG)p + \max\{\mathsf{d}(G/pG),\, \eta(G/pG) - p - 1\} \\ = \ & (m_1 + m_2 - 2)p + \max\{2p - 2,\, (3p - 2) - p - 1\} = n_1 + n_2 - 2. \end{aligned}$$ □

The next corollary was first proved in [52, Theorem 6.7].

Corollary 4.2.11. *Let* $\exp(G) = n$. *If* G *is either a* p*-group or* $\mathsf{r}(G) \le 2$, *then* $\mathsf{s}_{n\mathbb{N}}(G) = \mathsf{d}(G) + n$.

Proof. If G is a p-group, then the assertion follows from Theorem 2.2.6 and from Lemma 4.2.3.3.

Suppose that $G = C_{n_1} \oplus C_{n_2}$ with $1 \le n_1 \mid n_2$. Then Lemma 4.2.3.3 implies that $\mathsf{d}(G) + n_2 \le \mathsf{s}_{n\mathbb{N}}(G)$ whence it remains to prove that $\mathsf{s}_{n\mathbb{N}}(G) \le \mathsf{d}(G) + n_2 = n_1 + 2n_2 - 2$. If $n_1 = 1$, this follows from Theorem 4.2.10 and again from Lemma 4.2.3.3. Suppose that $n_1 > 1$, and let $S \in \mathcal{F}(G)$ be a sequence of length $|S| = n_1 + 2n_2 - 2$. We have to show that S has a zero-sum subsequence of length n_2 or $2n_2$.

Let $K = G \oplus C_{n_2} = G \oplus \langle e \rangle$ with $\operatorname{ord}(e) = n_2$, so that every $h \in G \oplus C_{n_2}$ has a unique representation $h = g + je$, where $g \in G$ and $j \in [0, n_2 - 1]$. We define $\psi \colon G \to K$ by $\psi(g) = g + e$ for every $g \in G$. Thus it suffices to show that $\psi(S)$ has a nontrivial zero-sum subsequence. We distinguish two cases.

Case 1: $n_1 = n_2$.

We set $n = n_1$ and proceed by induction on n. If n is prime, then G is a p-group and the assertion holds. Suppose that n is composite, p a prime divisor of n and $\varphi \colon K \to K$ the multiplication by p. Then $pG \cong C_{n/p} \oplus C_{n/p}$ and $\operatorname{Ker}(\varphi) \cong C_p^3$. Since by Theorem 4.2.10, $\mathsf{s}(pG) = 4(n/p) - 3$ and $|S| = 3n - 2 \ge (3p - 4)(n/p) + 4n/p - 3$, S admits a product decomposition $S = S_1 \cdot \ldots \cdot S_{3p-3} S'$ such that, for all $i \in [1, 3p - 3]$, $\varphi(S_i)$ has sum zero and length $|S_i| = n/p$ (see Lemma 4.2.4.1). Then $|S'| = 3n/p - 2 = \mathsf{s}_{(n/p)\mathbb{N}}(C_{n/p} \oplus C_{n/p})$, and thus S' has a subsequence S_{3p-2} such that $\varphi(S_{3p-2})$ has sum zero and length $|S_{3p-2}| \in \{n/p, 2n/p\}$. This implies that

$$\prod_{i=1}^{3p-2} \sigma\big(\psi(S_i)\big) \in \mathcal{F}\big(\operatorname{Ker}(\varphi)\big)\,.$$

Since $\mathsf{D}(\mathrm{Ker}(\varphi)) = 3p-2$, there exists a nonempty subset $I \subset [1, 3p-2]$ such that

$$\sum_{i \in I} \sigma(\psi(S_i)) = 0 \quad \text{whence} \quad \prod_{i \in I} \psi(S_i)$$

is a nontrivial zero-sum subsequence of $\psi(S)$.

Case 2: $n_2 > n_1$.

Let $m = n_1^{-1} n_2$ and let $\varphi\colon K = C_{n_1} \oplus C_{n_2}^2 \to C_{n_1} \oplus mC_{n_2}^2$ be a map which is the identity on the first component and the multiplication by m on the second and on the third component whence $\mathrm{Ker}(\varphi) \cong C_m \oplus C_m$ and $\varphi(G) \cong C_{n_1} \oplus C_{n_1}$. Since $\mathsf{s}(C_{n_1} \oplus C_{n_1}) = 4n_1 - 3$ and $|S| = n_1 + 2n_2 - 2 \ge (2m-3)n_1 + (4n_1 - 3)$, S admits a product decomposition $S = S_1 \cdot \ldots \cdot S_{2m-2} S'$, where for all $i \in [1, 2m-2]$, $\varphi(S_i)$ has sum zero and length $|S_i| = n_1$. Then $|S'| = 3n_1 - 2$, and since by *Case* 1, $\mathsf{s}_{n_1\mathbb{N}}(C_{n_1} \oplus C_{n_1}) = 3n_1 - 2$, the sequence S' has a subsequence S_{2m-1} such that $\varphi(S_{2m-1})$ has sum zero and length $|S_{2m-1}| \in \{n_1, 2n_1\}$. This implies that

$$\prod_{i=1}^{2m-1} \sigma(\psi(S_i)) \in \mathcal{F}(\mathrm{Ker}(\varphi)) \,.$$

Since $\mathsf{D}(\mathrm{Ker}(\varphi)) = 2m - 1$, there exists a nonempty subset $I \subset [1, 2m-1]$ such that

$$\sum_{i \in I} \sigma(\psi(S_i)) = 0 \quad \text{whence} \quad \prod_{i \in I} \psi(S_i)$$

is a nontrivial zero-sum subsequence of $\psi(S)$. □

We briefly discuss the state of the art concerning groups of higher rank (more detailed information can be found in [52]). A conjecture by W. Gao states that we always have $\eta(G) = \mathsf{s}(G) - \exp(G) + 1$. This was recently proved for p-groups G, where p is odd and $\mathsf{D}(G) = 2\exp(G) - 1$ (see [125]). C. Elsholtz et al. showed that, for all odd $n \ge 3$,

$$\eta(C_n^3) \ge 8n - 7, \quad \mathsf{s}(C_n^3) \ge 9n - 8, \quad \eta(C_n^4) \ge 19n - 18 \quad \text{and} \quad \mathsf{s}(C_n^4) \ge 20n - 19 \,,$$

and it is conjectured that all bounds are sharp (see [29, 28, 27, 113] for recent results).

It is conjectured that, if $\mathsf{r}(G) = 3$ or $G = C_n^r$ with $n, r \ge 3$, then $\mathsf{d}(G) = \mathsf{d}^*(G)$. On the other hand, for every $r \ge 4$ there are infinitely many groups G of rank $\mathsf{r}(G) = r$ such that $\mathsf{d}(G) > \mathsf{d}^*(G)$ (see [75] and [49, Theorem 3.3]). We end with a result (see [15]) providing more groups G with $\mathsf{d}(G) = \mathsf{d}^*(G)$ and whose proof demonstrates once more the power of the inductive method (for recent results see [6, 4]).

Theorem 4.2.12. *Let* $G = K \oplus C_{km}$ *where* $k, m \in \mathbb{N}$ *and* $K \subset G$ *is a subgroup with* $\exp(K) | m$. *If* $\mathsf{d}(K \oplus C_m) = \mathsf{d}(K) + m - 1$ *and* $\eta(K \oplus C_m) \le \mathsf{d}(K) + 2m$, *then* $\mathsf{d}(G) = \mathsf{d}(K) + km - 1$.

Proof. Clearly, we have $\mathsf{d}(G) \ge \mathsf{d}(K)+\mathsf{d}(C_{km}) = \mathsf{d}(K)+km-1$. Now let $S \in \mathcal{F}(G)$ be a sequence of length $|S| = \mathsf{d}(K)+km$. We have to show that S has a nontrivial zero-sum subsequence.

We consider the map $\varphi\colon G \to G$ which maps an element $g = h + a$, where $h \in K$ and $a \in C_{km}$, to $h + ka$ for all $g \in G$. Then $\operatorname{Ker}(\varphi) \cong C_k$ and $\varphi(G) \cong K \oplus C_m$. Since

$$|\varphi(S)| = |S| = (k-2)m + \big(\mathsf{d}(K) + 2m\big) \quad \text{and} \quad \eta(K \oplus C_m) \le \mathsf{d}(K) + 2m\,,$$

Lemma 4.2.4 provides us with a product decomposition

$$S = S_1 \cdot \ldots \cdot S_{k-1} S',$$

where $S_1, \ldots, S_{k-1}, S' \in \mathcal{F}(G)$ and, for every $i \in [1, k]$, $\varphi(S_i)$ has sum zero and length $|S_i| \in [1, \exp(K \oplus C_m)] = [1, m]$. Thus we get

$$|S'| = |S| - \sum_{i=1}^{k-1} |S_i| \ge |S| - (k-1)m = \mathsf{d}(K) + m = \mathsf{D}(K \oplus C_m)\,,$$

and hence S' has a subsequence S_k such that $\varphi(S_k)$ has sum zero. Thus

$$\prod_{i=1}^{k} \sigma(S_i) \in \mathcal{F}\big(\operatorname{Ker}(\varphi)\big)\,,$$

and there is a nonempty subset $I \subset [1, k]$ such that $\prod_{i \in I} \sigma(S_i)$ has sum zero. Hence $\prod_{i \in I} S_i$ is a nontrivial zero-sum subsequence of S. □

Corollary 4.2.13. *Let $G = K \oplus C_{km}$ where $k, m \in \mathbb{N}$, $p \in \mathbb{P}$ a prime, m a power of p and $K \subset G$ is a p-subgroup with $\mathsf{d}(K) \le m - 1$. Then $\mathsf{d}(G) = \mathsf{d}^*(G)$.*

Proof. Since $K \oplus C_m$ is a p-group, Theorem 2.2.6 implies that

$$\mathsf{d}(K \oplus C_m) = \mathsf{d}^*(K \oplus C_m) = \mathsf{d}^*(K) + m - 1 = \mathsf{d}(K) + m - 1\,.$$

Since $\exp(K)$ is a p-power and $\exp(K) - 1 \le \mathsf{d}(K) \le m - 1$, it follows that $\exp(K)$ divides m. By Lemma 4.2.3 we infer that

$$\eta(K \oplus C_m) \le \mathsf{d}(K \oplus C_m^2) + 1 = \mathsf{d}(K) + 2m - 1\,.$$

Thus all assumptions of Theorem 4.2.12 are satisfied and we obtain that

$$\mathsf{d}(G) = \mathsf{d}(K) + km - 1 = \mathsf{d}^*(K) + km - 1 = \mathsf{d}^*(G)\,. \qquad \square$$

Chapter 5

Inverse zero-sum problems and arithmetical consequences

The investigation of inverse problems has a long tradition in combinatorial number theory (see [107, 37]), and more recently it has been promoted by applications in the theory of non-unique factorizations. In this chapter we discuss the inverse problems associated with the invariants $\mathsf{D}(G)$, $\eta(G)$ and $\mathsf{s}(G)$. More precisely, we investigate the structure of sequences of length $\mathsf{D}(G) - 1$ ($\eta(G) - 1$ or $\mathsf{s}(G) - 1$, respectively) that do not have a zero-sum subsequence (of the required length). Recent results on the structure of $\Sigma(S)$ for (long) zero-sum free sequences may be found in [9, 58, 127, 132, 60].

We start with cyclic groups, then we deal with groups of the form $G = C_n^r$, and finally we outline some consequences in factorization theory.

5.1 Cyclic groups

Clearly, we can rephrase Corollary 2.1.4.1 as follows: Let G be cyclic of order $n \geq 2$ and $S \in \mathcal{F}(G)$ a sequence of length $\mathsf{D}(G) - 1 = \eta(G) - 1$. Then S has no (short) zero-sum subsequence if and only if $S = g^{n-1}$ for some $g \in G$ with $\mathrm{ord}(g) = n$.

We present a strong structural result on long zero-sum free sequences recently achieved by S. Savchev and Fang Chen [116]. Among others, their result settles a problem on the index of zero-sum sequences (studied in [16, 46, 112, 18], see Corollary 5.1.9) and provides information on multiplicities of elements, a topic studied by many authors before (Corollary 5.1.10).

Definition 5.1.1.

1. Let $g \in G$ be a non-zero element with $\mathrm{ord}(g) = n < \infty$. For a sequence

$$S = (n_1 g) \cdot \ldots \cdot (n_l g), \quad \text{where } l \in \mathbb{N}_0 \quad \text{and} \quad n_1, \ldots, n_l \in [1, n],$$

we define

$$\|S\|_g = \frac{n_1 + \ldots + n_l}{n}.$$

Obviously, S has sum zero if and only if $\|S\|_g \in \mathbb{N}_0$.

2. Let $S \in \mathcal{F}(G)$ be a sequence for which $\langle \operatorname{supp}(S) \rangle \subset G$ is cyclic. Then we call

$$\operatorname{ind}(S) = \min\{\|S\|_g \mid g \in G \text{ with } \langle \operatorname{supp}(S) \rangle = \langle g \rangle\} \in \mathbb{Q}_{\ge 0}$$

the *index of* S.

3. If G is cyclic, then let $\mathsf{l}(G)$ denote the smallest integer $l \in \mathbb{N}$ such that every minimal zero-sum sequence $S \in \mathcal{F}(G)$ of length $|S| \ge l$ satisfies $\operatorname{ind}(S) = 1$.

Lemma 5.1.2. *Let G be cyclic and $S \in \mathcal{F}(G)$. Then*

$$\begin{aligned} \operatorname{ind}(S) &= \min\{\|S\|_g \mid g \in G \textit{ with } \operatorname{supp}(S) \subset \langle g \rangle\} \\ &= \min\{\|S\|_g \mid g \in G \textit{ with } G = \langle g \rangle\}. \end{aligned}$$

Proof. We set $|G| = n$,

$$I_1 = \min\{\|S\|_g \mid g \in G \text{ with } \operatorname{supp}(S) \subset \langle g \rangle\} \quad \text{and}$$

$$I_2 = \min\{\|S\|_g \mid g \in G \text{ with } G = \langle g \rangle\}.$$

Let $|S| = l$, $g \in G$ with $\operatorname{ord}(g) = m$ and $S = (a_1 g) \cdot \ldots \cdot (a_l g)$ with $a_1, \ldots, a_l \in [1, m]$ such that $\|S\|_g = I_1$. First we verify that $I_1 = I_2$ and then we show that $I_1 = \operatorname{ind}(S)$.

1. Obviously, we have $I_1 \le I_2$, and it remains to verify the reverse inequality. There is an element $h \in G$ with $\langle h \rangle = G$ and $\frac{n}{m} h = g$. Thus we obtain

$$S = (a_1 \frac{n}{m} h) \cdot \ldots \cdot (a_l \frac{n}{m} h) \quad \text{with} \quad a_1 \frac{n}{m}, \ldots, a_l \frac{n}{m} \in [1, n],$$

$$\|S\|_h = \frac{\frac{n}{m} a_1 + \ldots + \frac{n}{m} a_l}{n} = \frac{a_1 + \ldots + a_l}{m} = \|S\|_g,$$

and hence $I_2 \le \|S\|_h = \|S\|_g = I_1$.

2. Obviously, we have $I_1 \le \operatorname{ind}(S)$, and it remains to verify the reverse inequality. We set $\langle a_1 g, \ldots, a_l g \rangle = K$ and pick an $a \in [1, m]$ with $a \mid m$ and $K = \langle a g \rangle$. Then $\operatorname{ord}(ag) = a^{-1} m$. For every $i \in [1, l]$ we have $a_i g \in \langle a g \rangle = \{ag, 2ag, \ldots, (a^{-1} m) ag\}$ and hence $a_i = a a_i'$ with $a_i' \in [1, a^{-1} m]$. Thus we obtain

$$S = (a_1' a g) \cdot \ldots \cdot (a_l' a g),$$

$$\|S\|_g = \frac{a(a_1' + \ldots + a_l')}{m} = \frac{a_1' + \ldots + a_l'}{a^{-1} m} = \|S\|_{ag},$$

and hence $I_1 = \|S\|_g = \|S\|_{ag} \ge \operatorname{ind}(S)$. □

Definition 5.1.3. A sequence $S \in \mathcal{F}(G)$ is called

- *smooth* if $S = (n_1 g) \cdot \ldots \cdot (n_l g)$, where $l = |S| \in \mathbb{N}$, $g \in G$, $1 = n_1 \le \cdots \le n_l$, $n = n_1 + \ldots + n_l < \operatorname{ord}(g)$ and $\Sigma(S) = \{g, 2g, \ldots, ng\}$ (in this case we say more precisely that S is g-smooth).
- *a splittable atom* if $S = (g_1 + g_2)T$ for some $g_1, g_2 \in G$ and $T \in \mathcal{F}(G)$ such that $S \in \mathcal{A}(G)$ and $g_1 g_2 T \in \mathcal{A}(G)$.

Lemma 5.1.4. *Let $g \in G$ and $k, l, n_1, \ldots, n_l \in \mathbb{N}$ such that $l \ge k/2$ and $n = n_1 + \ldots + n_l < k \le \operatorname{ord}(g)$. If $1 \le n_1 \le \cdots \le n_l$ and $S = (n_1 g) \cdot \ldots \cdot (n_l g)$, then $\Sigma(S) = \{g, 2g, \ldots, ng\}$, and S is g-smooth.*

Proof. We start with the following assertion.

A. For every $i \in [0, l-1]$ we have $n_{i+1} \le 1 + n_1 + \ldots + n_i$, and in particular we have $n_1 = 1$.

Proof of **A**. Assume to the contrary that there is an $i \in [0, l-1]$ such that $n_{i+1} \ge 2 + n_1 + \ldots + n_i$. Then for all $j \in [i+1, l]$ we have $n_j \ge n_{i+1} \ge 2 + i$ and therefore

$$k > n_1 + \ldots + n_l \ge i + (l-i)(2+i) = 2l + i(l-i-1) \ge k\,,$$

a contradiction.

For every $i \in [1, l]$ we set $S_i = (n_1 g) \cdot \ldots \cdot (n_i g)$ and assert that $\Sigma(S_i) = \{jg \mid j \in [1, n_1 + \ldots + n_i]\}$. We proceed by induction on i. For $i = 1$, this holds since $n_1 = 1$. If $i \in [1, l-1]$, then **A** and the induction hypothesis imply that

$$\begin{aligned}\Sigma(S_{i+1}) &= \Sigma(S_i) \cup \Big(n_{i+1} g + \big(\Sigma(S_i) \cup \{0\}\big)\Big) \\ &= \{jg \mid j \in [1, n_1 + \ldots + n_i]\} \cup \{jg \mid j \in [n_{i+1}, n_1 + \ldots + n_i + n_{i+1}]\} \\ &= \{jg \mid j \in [1, n_1 + \ldots + n_{i+1}]\}\,.\end{aligned}$$

□

Lemma 5.1.5. *Let $S \in \mathcal{F}(G)$ and $b \in G$ such that $Sb \in \mathcal{A}^*(G)$ and $|\Sigma(Sb)| = |\Sigma(S)| + 1$.*

1. *We have $\Sigma(S) = \{b, 2b, \ldots, sb\} \cup P_1 \cup \cdots \cup P_m$, where $s \in [1, \operatorname{ord}(b) - 2]$, $m \in \mathbb{N}_0$ and $P_1, \ldots, P_m \in G/\langle b \rangle$ are distinct cosets different from $\langle b \rangle$. Moreover, $sb = \sigma(S)$ and $|\Sigma(Sb^j)| = |\Sigma(Sb^{j-1})| + 1$ for all $j \in [1, \operatorname{ord}(b) - s]$.*
2. *If $c \in G$ is such that $Sc \in \mathcal{A}^*(G)$ and $|\Sigma(Sc)| = |\Sigma(S)| + 1$, then $c = b$.*

Proof. 1. Since $Sb \in \mathcal{A}^*(G)$, $|\Sigma(Sb)| = |\Sigma(S)| + 1$ and $\sigma(S) + b \in \Sigma(Sb)$, it follows that $\Sigma(Sb) \setminus \Sigma(S) = \{\sigma(S) + b\}$, and since $b \in \Sigma(Sb)$ and $\sigma(S) \ne 0$, we obtain that $b \in \Sigma(S)$. Let $s \in \mathbb{N}$ be maximal such that $\{b, 2b, \ldots, sb\} \subset \Sigma(S)$. Then $(s+1)b \in \Sigma(Sb)$, hence $(s+1)b = \sigma(S) + b$ and $sb = \sigma(S)$. Moreover, $Sb \in \mathcal{A}^*(G)$ implies that $s + 1 < \operatorname{ord}(b)$.

In order to prove that $\Sigma(S) \setminus \{b, 2b, \ldots, sb\}$ is the union of proper cosets of $\langle b \rangle$ in G, we have to verify that $c \in \Sigma(S) \setminus \{b, 2b, \ldots, sb\}$ implies that $c + b \in \Sigma(S) \setminus \{b, 2b, \ldots, sb\}$. Indeed, if $c \in \Sigma(S) \setminus \{b, 2b, \ldots, sb\}$, then $c + b \in \Sigma(Sb)$. If $c + b \notin \Sigma(S) \setminus \{b, 2b, \ldots, sb\}$, then $c + b = tb$ for some $t \in [1, s+1]$, and thus $c = (t-1)b \in \{0, b, 2b, \ldots, sb\}$, a contradiction. Hence it follows that $c + b \in \Sigma(S) \setminus \{b, 2b, \ldots, sb\}$.

If $j \in [1, \mathrm{ord}(b) - s]$, then

$$\Sigma(Sb^j) = \{b, 2b, \ldots, (s+j)b\} \cup P_1 \cup \cdots \cup P_m = \Sigma(Sb^{j-1}) \cup \{(s+j)b\},$$

and since $s + j \le \mathrm{ord}(b)$, we obtain $|\Sigma(Sb^j)| = |\Sigma(Sb^{j-1})| + 1$.

2. Let $c \in G$ be such that $Sc \in \mathcal{A}^*(G)$ and $|\Sigma(Sc)| = |\Sigma(S)| + 1$. Then $R = \{c, b+c, 2b+c, \ldots, sb+c\} \subset \Sigma(Sc)$, and since $|R| \ge 2$ we have $R \cap \Sigma(S) \ne \emptyset$. Suppose there is some $j \in [1, m]$ such that $R \cap P_j \ne \emptyset$. Then $R \subset P_j$, but $sb + c = \sigma(S) + c \in \Sigma(Sc) \setminus \Sigma(S)$ and thus $sb + c \notin P_j$, a contradiction. Hence it follows that $R \cap \{b, 2b, \ldots, sb\} \ne \emptyset$, say $rb + c = tb$ for some $r \in [0, s]$ and $t \in [1, s]$. Thus we obtain $c = (t-r)b$ and $R = \{(t - r + j)b \mid j \in [0, s]\}$. If $t \le r$, this implies $0 \in R$, a contradiction. If $t \ge r + 2$, then we obtain $\{(s+1)b, (s+2)b\} \subset R \setminus \Sigma(S) \subset \Sigma(Sc) \setminus \Sigma(S)$, contradicting $|\Sigma(Sc)| = |\Sigma(S)| + 1$. Therefore we get $t = r + 1$ and $c = b$. □

Lemma 5.1.6. *Let G be cyclic of order $n \ge 3$, $S = S_1 S_2 \in \mathcal{A}^*(G)$ where $|S| = l \ge \frac{n+1}{2}$, $2 \le |S_1| = k + 1 \le l$ and $|\Sigma(S_1)| \ge 2k + 1$. Then there is a decomposition $S_2 = S_2' c$ with $S_2' \in \mathcal{F}(G)$ and $c \in G$ such that $|\Sigma(S)| = |\Sigma(S_1 S_2')| + 1$.*

Proof. Let $t \in \mathbb{N}_0$ be maximal such that there is a decomposition $S_2 = g_1 \cdot \ldots \cdot g_t T$, where $g_1, \ldots, g_t \in G$ and $T \in \mathcal{F}(G)$ are such that

$$|\Sigma(S_1 g_1 \cdot \ldots \cdot g_j)| \ge |\Sigma(S_1 g_1 \cdot \ldots \cdot g_{j-1})| + 2 \quad \text{for all} \quad j \in [1, t].$$

Then we have

$$n - 1 \ge |\Sigma(S)| \ge |\Sigma(S_1 g_1 \cdot \ldots \cdot g_t)| \ge |\Sigma(S_1)| + 2t \ge 2k + 1 + 2t.$$

Therefore it follows that $2t + 2 \le n - 2k$, whence $t + 1 \le n/2 - k < l - k$ and $|T| = |S| - |S_1| - t = l - (k+1) - t > 0$. Let $c \in \mathrm{supp}(T)$. Since $|\Sigma(S_1 g_1 \cdot \ldots \cdot g_t c)| = |\Sigma(S_1 g_1 \cdot \ldots \cdot g_t)| + 1$ (by the maximality of t), Lemma 5.1.5 implies that

$$\Sigma(S_1 g_1 \cdot \ldots \cdot g_t) = \{c, 2c, \ldots, sc\} \cup P_1 \cup \cdots \cup P_m,$$

where $s \in [1, \mathrm{ord}(c) - 2]$, $m \in \mathbb{N}_0$ and $P_1, \ldots, P_m \in G/\langle c \rangle$ are proper cosets. If $d \in \mathrm{supp}(c^{-1}T)$, then (again by the maximality of t) $|\Sigma(S_1 g_1 \cdot \ldots \cdot g_t d)| = |\Sigma(S_1 g_1 \cdot \ldots \cdot g_t)| + 1$, and Lemma 5.1.5 implies that $d = c$. Hence $T = c^j$ for some $j \in [1, \mathrm{ord}(c) - s]$ and $S_2' = g_1 \cdot \ldots \cdot g_t c^{j-1}$ fulfills our requirements. □

Lemma 5.1.7. *Let G be cyclic of order $n \ge 3$, $S \in \mathcal{A}^*(G)$ with $\Sigma(S) = G^\bullet$, $|S| = l \ge \frac{n+1}{2}$, $a \in \mathrm{supp}(S)$ and S_a a maximal a-smooth subsequence of S. If $S_a \ne S$, then $(-\sigma(S)) \mid S_a^{-1} S$.*

Proof. Let $S_a = (s_1a) \cdot \ldots \cdot (s_ka)$, where $k, s_1, \ldots, s_k \in \mathbb{N}$, $1 = s_1 \le s_2 \le \cdots \le s_k$, $s = s_1 + \ldots + s_k < \mathrm{ord}(a)$, and assume that $S_a \ne S$. Then $S = S_abS_2$, where $b \in G$ and $S_2 \in \mathcal{F}(G)$, and we have $\Sigma(S_ab) = P_1 \cup P_2$ where $P_1 = \Sigma(S_a) = \{a, 2a, \ldots, sa\}$ and $P_2 = b + \big(\Sigma(S_a) \cup \{0\}\big) = \{b, b+a, \ldots, b+sa\}$. We shall prove that

A. $P_1 \cap P_2 = \emptyset$.

Proof of **A**. The assertion is obvious if $b \notin \langle a \rangle$. Thus assume that $b = ta$ for some $t \in [1, \mathrm{ord}(a)-1]$. Then $\Sigma(S_ab) = \{a, 2a, \ldots, sa, ta, (t+1)a, \ldots, (t+s)a\}$. If $t \le s$, then $t + s \ge s+1$ and S_ab is a-smooth, contradicting the maximality of S_a. If $t + s \ge \mathrm{ord}(a)$, then $0 \in \Sigma(S_ab)$, a contradiction. Hence $s < t < \mathrm{ord}(a) - s$ and $P_1 \cap P_2 = \emptyset$.

We apply Lemma 5.1.6 with $S_1 = S_ab$. Note that $2 \le k+1 = |S_ab| \le |S| = l$ and

$$n - 1 \ge |\Sigma(S_ab)| = |P_1| + |P_2| = 2s + 1 \ge 2k + 1\,.$$

Thus we arrive at a decomposition $S_2 = S_2'c$ for some $S_2' \in \mathcal{F}(G)$ and $c \in G$ such that $|\Sigma(S)| = |\Sigma(S_1S_2')| + 1 = |G| - 1$. Since $\Sigma(S) = \Sigma(S_1S_2') \cup \{\sigma(S)\}$ and $-c \notin \Sigma(S_1S_2')$, it follows that $c = -\sigma(S)$, whence $-\sigma(S) \,|\, S_2 \,|\, S_a^{-1}S$. □

Theorem 5.1.8 (Savchev-Chen). *Let G be cyclic of order $n \ge 3$.*

1. *If $S \in \mathcal{A}^*(G)$ and $|S| \ge \frac{n+1}{2}$, then S is g-smooth for some $g \in G$ with $\mathrm{ord}(g) = n$.*
2. *Let $U \in \mathcal{A}(G)$ be of length $|U| \ge \lfloor \frac{n}{2} \rfloor + 2$. Then $\mathrm{ind}(U) = 1$, and if U is not splittable, then $U = g^n$ for some $g \in G$.*

Proof. We may suppose that $U = g_0S$ with $|S| = l \ge \frac{n+1}{2}$. First we show that S is g-smooth for some $g \in G$. Then we show $\mathrm{ind}(U) = 1$ and $\mathrm{ord}(g) = n$.

1. Pick a sequence $S' \in \mathcal{A}^*(G)$ such that $S \,|\, S'$ and $|S'|$ is maximal. Then $\Sigma(S') = G^\bullet$ and $m = |S'| \ge |S| \ge \frac{n+1}{2}$. Let $a \in \mathrm{supp}(S')$ and let S_a be a maximal a-smooth subsequence of S'. If $S_a \ne S'$ and $h = -\sigma(S')$, then $h \,|\, S_a^{-1}S'$ by Lemma 5.1.7. Now let S_h be a maximal h-smooth subsequence of S'. Then $h \nmid S_h^{-1}S'$, since otherwise hS_h would be a longer h-smooth subsequence of S'. Hence again Lemma 5.1.7 implies that $S_h = S'$.

In any case S' is smooth, and thus we may assume that $S' = (n_1g) \cdot \ldots \cdot (n_mg)$ and $S = (n_1g) \cdot \ldots \cdot (n_lg)$, where $g \in G$, $m \in \mathbb{N}$, $n_1, \ldots, n_m \in \mathbb{N}$, $l \le m \le n_1 + \ldots + n_m < \mathrm{ord}(g) \le n < 2l$ and $1 \le n_1 \le \cdots \le n_l$. Thus Lemma 5.1.4 (with $k = \mathrm{ord}(g)$) implies that S is g-smooth.

2. By 1., $S = (n_1g) \cdot \ldots \cdot (n_lg)$ is g-smooth,whence $n_1, \ldots, n_l \in \mathbb{N}$ and $s = n_1 + \ldots + n_l < \mathrm{ord}(g)$. Then $-g_0 = \sigma(S) = sg$ and $U = (n_0g)(n_1g) \cdot \ldots \cdot (n_lg)$ with $n_0 = \mathrm{ord}(g) - s \in \mathbb{N}$ and hence $n_0 + \ldots + n_l = \mathrm{ord}(g)$. Let $m \in \mathbb{N}$ with $\mathrm{ord}(g) = n/m$ and $g' \in G$ with $mg' = g$. Then $U = (mn_0g') \cdot \ldots \cdot (mn_lg')$ and $mn_0 + \ldots + mn_l = n$ whence $\mathrm{ind}(U) = 1$. Since $n = mn_0 + \ldots + mn_l \ge m(l+1) \ge m\frac{n+3}{2}$, it follows that $m = 1$ and $g = g'$.

If there is some $i \in [0, l]$ with $n_i \ge 2$, say $i = 0$, then $\big((n_0 - 1)g\big)g(n_1g) \cdot \ldots \cdot (n_lg) \in \mathcal{A}(G)$, and hence U is splittable. □

Corollary 5.1.9 was achieved independently by S. Savchev and F. Chen, and by P. Yuan (see [133, Theorem 3.1] and [116, Proposition 10]).

Corollary 5.1.9. *Let G be cyclic of order $n \ge 1$. If $n \in \{1,2,3,4,5,7\}$, then $\mathsf{l}(G) = 1$, and otherwise we have $\mathsf{l}(G) = \lfloor \frac{n}{2} \rfloor + 2$.*

Proof. If $n \le 4$, it can be seen immediately that $\mathsf{l}(G) = 1$. If $n \in \{5,7\}$, a longer (but straightforward) case distinction shows that again $\mathsf{l}(G) = 1$ holds. Suppose that $n \in \mathbb{N}_{\ge 6} \setminus \{7\}$, and let $g \in G$ with $\mathrm{ord}(g) = n$. Since the sequence

$$S = \begin{cases} g^{\frac{n-4}{2}}(\frac{n}{2}g)(\frac{n+2}{2}g)^2 & \text{if } n \text{ is even,} \\ g^{\frac{n-5}{2}}(\frac{n+3}{2}g)^2(\frac{n-1}{2}g) & \text{if } n \text{ is odd} \end{cases}$$

is a minimal zero-sum sequence with $\mathrm{ind}(S) = 2$, it follows that $\mathsf{l}(G) \ge \lfloor \frac{n}{2} \rfloor + 2$, and hence equality holds by Theorem 5.1.8.2. □

The second statement of Corollary 5.1.10 was first proved by J.D. Bovey, P. Erdős and I. Niven [10] and the fourth statement by A. Geroldinger and Y. ould Hamidoune [72]. Note that the bounds given in Corollary 5.1.10.4 are attained (see [71, Example 5.4.7]).

Corollary 5.1.10. *Let G be cyclic of order $n \ge 3$, $S \in \mathcal{F}(G)$ a zero-sum free sequence of length*

$$|S| \ge \frac{n+1}{2}.$$

1. *For all $g \in \mathrm{supp}(S)$ we have $\mathrm{ord}(g) \ge 3$.*
2. *There exists some $g \in \mathrm{supp}(S)$ with $\mathsf{v}_g(S) \ge 2|S| - n + 1$.*
3. *There exists some $g \in \mathrm{supp}(S)$ with $\mathsf{v}_g(S) \ge |S| - \frac{n-1}{3}$.*
4. *There exists some $g \in \mathrm{supp}(S)$ with $\mathrm{ord}(g) = n$ such that*

$$\mathsf{v}_g(S) \ge \frac{n+5}{6} \quad \text{if } n \text{ is odd} \quad \text{and} \quad \mathsf{v}_g(S) \ge 3 \quad \text{if } n \text{ is even}.$$

Proof. 1. The assertion is clear for odd n. Thus suppose that $n = 2m$ for some $m \ge 2$. By Theorem 5.1.8, $S = (n_1 g) \cdot \ldots \cdot (n_l g)$ for some $g \in G$ with $\mathrm{ord}(g) = n$, $1 = n_1 \le \cdots \le n_l$ and $\Sigma(S) = \{g, 2g, \ldots, sg\}$ where $s = n_1 + \ldots + n_l$. Assume to the contrary that S contains some element of order 2. Then there is an $i \in [1,l]$ with $n_i = m$ and hence $s = n_1 + \ldots + n_l \ge l - 1 + m \ge 2m$, a contradiction to S zero-sum free.

2. We write S in the form $S = S_1 \cdot \ldots \cdot S_k (gh)^l g^{m-l}$, where $k, m \in \mathbb{N}_0$, $l \in [0,m]$, $g \ne h$, $S_1, \ldots, S_k$ are squarefree, and $|S_i| = 3$ for all $i \in [1,k]$. Clearly, we have $|\Sigma(gh)| = 3$, and using 1. we obtain that $|\Sigma(S_i)| \ge 6$ for all $i \in [1,k]$. By Corollary 4.1.3 it follows that

$$n - 1 \ge |\Sigma(S)| \ge 6k + 3l + (m - l) \ge 6k + 2l + 2m - \mathsf{v}_g(S) = 2|S| - \mathsf{v}_g(S)$$

and therefore $\mathsf{v}_g(S) \ge 2|S| - n + 1$.

3. and 4. The sequence $S_1 = \big(-\sigma(S)\big)S$ is a minimal zero-sum sequence of length $|S_1| \ge (n+3)/2$. Thus Corollary 5.1.9 implies that $\operatorname{ind}(S_1) = 1$. Thus there is an element $h \in G$ with $\operatorname{ord}(h) = n$ such that

$$S_1 = (xh)h^u(2h)^v(x_1h) \cdot \ldots \cdot (x_th)\,,$$

where $x \in [1, n-1], xh = -\sigma(S)$, $u, v, t \in \mathbb{N}_0$, $x_1, \ldots, x_t \in [3, n-1]$ and $x + u + 2v + (x_1 + \ldots + x_t) = n$. Clearly, we have

$$|S| = u + v + t \quad \text{and} \quad u + 2v + 3t = n - r \text{ for some } r \in \mathbb{N}\,,$$

which implies that $2u + v = 3|S| - (n - r)$ and hence

$$\max\{u, v\} \ge |S| - \frac{n-r}{3} \ge |S| - \frac{n-1}{3}\,.$$

If n is odd, then $\operatorname{ord}(h) = \operatorname{ord}(2h) = n$ and

$$\max\{\mathsf{v}_h(S), \mathsf{v}_{2h}(S)\} = \max\{u, v\} \ge |S| - \frac{n-1}{3} \ge \frac{n+5}{6}\,.$$

If n is even, then $|S| \ge (n/2) + 1$ and

$$\mathsf{v}_h(S) = u = 2|S| - n + r + t \ge 2 + r + t \ge 3. \qquad \square$$

Without proof we cite a most recent result by W. Gao et. al. (see [61]).

Theorem 5.1.11. *Let G be cyclic of order $n \ge 3$. If $S \in \mathcal{F}(G)$ is a zero-sum free sequence of length*

$$|S| \ge \frac{6n+28}{19}\,, \quad \textit{then} \quad \mathsf{h}(S) \ge \frac{6|S| - n + 1}{17}\,.$$

Next we consider the inverse problem with respect to the $\mathsf{s}(G)$-invariant. The first result in this direction was achieved independently by B. Peterson and T. Yuster [109, Theorem 1] and by A. Bialostocki and P. Dierker [7, Lemma 4]. It runs as follows.

Proposition 5.1.12. *Let G be cyclic of order $n \ge 2$ and $S \in \mathcal{F}(G)$ a sequence of length $|S| = \mathsf{s}(G) - 1$. Then the following statements are equivalent:*

(a) *S has no zero-sum subsequence of length n.*

(b) *$S = (gh)^{n-1}$ where $g, h \in G$ with $\operatorname{ord}(g - h) = n$.*

Proposition 5.1.12 was the starting point for a huge variety of investigations (see [11, 34, 12, 43, 62, 8, 131, 63, 95, 65]). We present a recent result, achieved by S. Savchev and F. Chen in [117], which characterizes all sequences of length greater than or equal to $(3n-1)/2$ that have no zero-sum subsequence of length n (Theorem 5.1.16). This easily implies Proposition 5.1.12, and moreover the lower bound is, in a certain sense, best possible (see Remark 5.1.17). We start with the following result of W. Gao (see [41, Theorem 1]).

Proposition 5.1.13 (Gao). *Let $S \in \mathcal{F}(G)$ be a sequence of length $|S| = |G| + k$ with $k \in \mathbb{N}_0$, and suppose that for every $g \in G$ and every subsequence T of S of length $|T| = k+1$ the sequence $g + T$ has a zero-sum subsequence. Then*

$$\Sigma_{|G|}(S) = \bigcap_{g \in G} \Sigma(g + S)\,.$$

In particular, if $\mathsf{h}(S) = \mathsf{v}_0(S)$, then $\Sigma_{\ge |G|}(S) = \Sigma_{|G|}(S)$.

Proof. We set $|G| = n$. Let $S = a^h T$ with $h = \mathsf{h}(S)$ and $|S| = n + k$. Without restriction we may suppose that $a = 0$. Obviously, we have

$$\Sigma_n(S) \subset \bigcap_{g \in G} \Sigma(g + S) \subset \Sigma(S)\,.$$

Hence, for the main statement, it suffices to show that $\Sigma(S) \subset \Sigma_n(S)$. But this inclusion clearly implies the "in particular" statement too. We pick $b \in \Sigma(S)$ and distinguish two cases.

Case 1: $h \ge n$.

If $b = 0$, then $0^n \mid S$ and $0 = \sigma(0^n) \subset \Sigma_n(S)$. Suppose that $b \ne 0$. Since $b \in \Sigma(S)$, there is a subsequence T' of T with $b = \sigma(T')$. Since $\mathsf{D}(G) \le n$, we may assume that $|T'| \le n$. This implies that $T' 0^{n - |T'|} \mid S$ and $b = \sigma(T' 0^{n-|T'|}) \subset \Sigma_n(S)$.

Case 2: $h \le n - 1$.

If $b \ne 0$, then $b \in \Sigma(T)$. If $b = 0$, then $n + k - h \ge k + 1$, and thus the assumption implies that $b \in \Sigma(T)$. Therefore in both cases we have $b \in \Sigma(T)$, and we assert that there is a subsequence U of T with $b = \sigma(U)$ and $|U| \in [n - h, n]$. If this holds, then $U 0^{n-|U|} \mid S$ and $b = \sigma(U 0^{n-|U|}) \in \Sigma_n(S)$.

Let U be a subsequence of T of maximal length such that $b = \sigma(U)$. By the maximality of $|U|$ it follows that $0 \notin \Sigma(U^{-1}T)$ and hence $|U^{-1}T| \le k$ by assumption. This implies that $|U| \ge |T| - k = n - h$. If $|U| \le n$, then we are done. Suppose that $|U| > n$. By Proposition 4.2.6, U admits a product decomposition $U = U_1 \cdot \ldots \cdot U_\varphi U'$ where all $U_i \in \mathcal{B}(G)$ are of length $|U_i| \le h$ for $i \in [1, \varphi]$ and $U' \in \mathcal{F}(G)$ with $n - h \le |U'| \le n$. Since $b = \sigma(U) = \sigma(U')$, the sequence U' has the required property. □

Although Proposition 5.1.12 is a straightforward consequence of the main result 5.1.16, we give a simple independent proof based on Proposition 5.1.13.

Proof of Proposition 5.1.12. (a) ⇒ (b). Let $S \in \mathcal{F}(G)$ be a sequence of length $2n - 2$ which has no zero-sum subsequence of length n, that is, $0 \notin \Sigma_n(S)$. For every $g \in G$ we have $|g + S| = 2n - 2 \ge n$, hence $0 \in \Sigma(g + S)$ and consequently $\Sigma_n(S) \ne \bigcap_{g \in G} \Sigma(g + S)$. By Proposition 5.1.13, there is some $g \in G$ and some subsequence T of S such that $|T| = n - 1$ and $g + T$ is zero-sum free. Then Corollary 2.1.4 implies that $g + T = a^{n-1}$ and thus $T = (a - g)^{n-1}$. Let $c = a - g$

and $S = c^{n-1}U$ for some sequence $U \in \mathcal{F}(G)$. Then $-c + S = 0^{n-1}(-c+U)$. Since $-c + S$ has no zero-sum subsequence of length n, it follows that $-c+U$ is zero-sum free whence $-c + U = d^{n-1}$ for some $d \in G$ with $\operatorname{ord}(d) = n$. Thus it follows that $S = c^{n-1}(c+d)^{n-1}$.

(b) $\Rightarrow$ (a). Since $\operatorname{ord}(g-h) = n$, the shifted sequence $-h+S = 0^{n-1}(g-h)^{n-1}$ has no zero-sum subsequence of length n, and hence S has no zero-sum subsequence of length n. $\square$

We continue with a simple observation which will be needed frequently in what follows. Let $g \in G$ with $\operatorname{ord}(g) = n$ and $S = (n_1 g) \cdot \ldots \cdot (n_l g)$ where $l \in \mathbb{N}_0$ and $n_1, \ldots, n_l \in [1, n]$. Then $n\|S\|_g = n_1 + \ldots + n_l$,

$$g - S = \big((n - n_1 + 1)g\big) \cdot \ldots \cdot \big((n - n_l + 1)g\big)$$

and

$$n\|g - S\|_g = \sum_{i=1}^{l}(n - n_i + 1) = \sum_{\substack{i=1 \\ n_i \neq n}}^{l} (n - n_i) + |S| \,.$$

Proposition 5.1.14. *Let G be cyclic of order $n \ge 3$, $k \in [1, n-1]$, $g \in G$ with $\operatorname{ord}(g) = n$ and $S, S_1, S_2 \in \mathcal{F}(G)$ such that $S = S_1S_2$, $|S| = n + k - 1$, $\|S_1\|_g < 1$ and $\|g - S_2\|_g < 1$.*

1. *S has no zero-sum subsequence of length n.*
2. *$k \le |S_1| < n$, $k \le |S_2| < n$ and $b - a \ge k$ where $a, b \in [1, n]$, $(ag) \,|\, S_1$ and $(bg) \,|\, S_2$. In particular, $\gcd(S_1, S_2) = 1$.*
3. (i) *We have*

$$\mathsf{v}_g(S) + \mathsf{v}_0(S) \ge 2k, \ \max\{\mathsf{v}_g(S), \mathsf{v}_0(S)\} \ge k \quad \text{and}$$
$$\min\{\mathsf{v}_g(S), \mathsf{v}_0(S)\} \ge 2k - n + 1 \,.$$

   (ii) *The following statements are equivalent:*
   (a) *$\mathsf{v}_g(S) + \mathsf{v}_0(S) = 2k$.*
   (b) *$S_1 = g^{2p-n+1}(2g)^{n-1-p}$ and $S_2 = 0^{2q-n+1}(-g)^{n-1-q}$ where $p, q \in [(n-1)/2, n-1]$ and $p + q = n + k - 1$.*

   (iii) *The following statements are equivalent:*
   (a) *$\max\{\mathsf{v}_g(S), \mathsf{v}_0(S)\} = k$.*
   (b) *$n + k$ is odd, $S_1 = g^k(2g)^{(n-k-1)/2}$ and $S_2 = 0^k(-g)^{(n-k-1)/2}$.*
4. *If $k \ge \frac{n-1}{2}$, then $\mathsf{h}(S) = \max\{\mathsf{v}_g(S), \mathsf{v}_0(S)\}$.*

Proof. 1. Let $S' = S_1'S_2'$ be a zero-sum subsequence of S where $S_1' \,|\, S_1$ and $S_2' \,|\, S_2$. We set

$$S_1' = (a_1 g) \cdot \ldots \cdot (a_r g) \quad \text{and} \quad S_2' = (b_1 g) \cdot \ldots \cdot (b_s g),$$

where $r, s \in \mathbb{N}_0$ and $a_1, \ldots, a_r, b_1, \ldots, b_s \in [1, n]$. Since S' has sum zero, we infer that $\sum_{i=1}^r a_i \equiv \sum_{j=1}^s (n - b_j) \mod n$. Since $0 \le \sum_{i=1}^r a_i \le n\|S_1\|_g < n$ and

$$0 \le \sum_{j=1}^s (n - b_j) \le n\|g - S_2\|_g - |S_2| < n - |S_2| \le n\,,$$

we infer that $r \le \sum_{i=1}^r a_i = \sum_{j=1}^s (n - b_j) < n - |S_2|$ and hence $|S'| = |S_1' S_2'| \le r + |S_2| < n$.

2. Since $|S_1 S_2| = n + k - 1$, $|S_1| \le n\|S_1\|_g < n$ and $|S_2| = |g - S_2| \le n\|g - S_2\|_g < n$, we obtain the first two inequalities. We set

$$M = \max\{a \in [1, n] \mid (ag) \mid S_1\} + \max\{n - b + 1 \mid b \in [1, n], (bg) \mid S_2\}\,.$$

Then

$$2(n-1) \ge n\|S_1\|_g + n\|g - S_2\|_g \ge M + (|S_1| - 1) + (|S_2| - 1) \ge M + n + k - 3\,,$$

and hence $M \le n - k + 1$. If $a, b \in [1, n]$ such that $(ag) \mid S_1$ and $(bg) \mid S_2$, then $a + n - b + 1 \le M \le n - k + 1$ and hence $b - a \ge k$.

3.(i) Since $\|S_1\|_g < 1$, we get $0 \nmid S_1$ and thus $\mathsf{v}_0(S) = \mathsf{v}_0(S_2)$. Similarly, we get $\mathsf{v}_g(S) = \mathsf{v}_g(S_1)$. We have

$$n - 1 \ge n\|S_1\|_g \ge \mathsf{v}_g(S_1) + 2(|S_1| - \mathsf{v}_g(S_1)) = 2|S_1| - \mathsf{v}_g(S)$$

and

$$n - 1 \ge n\|g - S_2\|_g \ge \mathsf{v}_0(S_2) + 2(|S_2| - \mathsf{v}_0(S_2)) = 2|S_2| - \mathsf{v}_0(S)\,.$$

Adding these two inequalities we obtain

$$2(n-1) \ge 2|S| - (\mathsf{v}_g(S) + \mathsf{v}_0(S)) = 2(n + k - 1) - (\mathsf{v}_g(S) + \mathsf{v}_0(S)), \qquad (*)$$

and hence $\mathsf{v}_g(S) + \mathsf{v}_0(S) \ge 2k$. Using 2., we get $k \le \max\{\mathsf{v}_g(S), \mathsf{v}_0(S)\} \le n - 1$ and thus $\min\{\mathsf{v}_g(S), \mathsf{v}_0(S)\} \ge 2k - n + 1$.

3.(ii) The implication (b) $\Rightarrow$ (a) is obvious, and hence it suffices to show that (a) $\Rightarrow$ (b).

The equality $\mathsf{v}_g(S) + \mathsf{v}_0(S) = 2k$ holds if and only if $n - 1 = 2|S_1| - \mathsf{v}_g(S)$ and $n - 1 = 2|S_2| - \mathsf{v}_0(S)$. These conditions imply that $S_1 = g^{\mathsf{v}_g(S_1)}(2g)^{\mathsf{v}_{2g}(S_1)}$, $S_2 = 0^{\mathsf{v}_0(S_2)}(-g)^{\mathsf{v}_{-g}(S_2)}$, $\mathsf{v}_g(S_1) = 2|S_1| - n + 1 \ge 0$ and $\mathsf{v}_0(S_2) = 2|S_2| - n + 1 \ge 0$. Using 2. we infer that $|S_i| \in [n-1)/2, n-1]$ for $i \in \{1, 2\}$. Now setting $p = |S_1|$ and $q = |S_2|$ we obtain the assertion.

3.(iii) Again it suffices to show that (a) $\Rightarrow$ (b). Suppose that $\max\{\mathsf{v}_g(S), \mathsf{v}_0(S)\} = k$. Since $\mathsf{v}_g(S) + \mathsf{v}_0(S) \ge 2k$, it follows that $\mathsf{v}_g(S) = \mathsf{v}_0(S) = k$. Thus we have again equality in $(*)$ and in the two previous inequalities, which implies the assertion.

4. Since $\mathsf{v}_g(S) + \mathsf{v}_0(S) \ge 2k$ and $k \ge \frac{n-1}{2}$, it follows that

$$|g^{-\mathsf{v}_g(S)} 0^{-\mathsf{v}_0(S)} S| \le |S| - 2k = n - 1 - k \le k\,.$$

By 3. we have $\max\{\mathsf{v}_g(S), \mathsf{v}_0(S)\} \ge k$, and thus the assertion follows. □

Lemma 5.1.15. *Let* $g \in G$ *with* $\operatorname{ord}(g) = n \ge 2$ *and* $S \in \mathcal{F}(\langle g \rangle)$ *with* $2|S| > n\|S\|_g$.

1. $\mathsf{v}_g(S) \ge 2|S| - n\|S\|_g$.
2. *For every* $x \in [2|S| - n\|S\|_g, n\|S\|_g]$ *there is a subsequence* S' *of* S *with* $|S'| \ge 2|S| - n\|S\|_g$ *and* $\sigma(S') = xg$.

Proof. 1. We set $S = g^{\mathsf{v}_g(S)} (a_1 g) \cdot \ldots \cdot (a_k g)$ where $k \in \mathbb{N}_0$ and $a_1, \ldots, a_k \in [2, n]$. Since $2(\mathsf{v}_g(S) + k) = 2|S| > n\|S\|_g = \mathsf{v}_g(S) + a_1 + \ldots + a_k$, it follows that

$$0 < 2|S| - n\|S\|_g = 2|S| - (\mathsf{v}_g(S) + a_1 + \ldots + a_k) = \mathsf{v}_g(S) - \sum_{i=1}^{k} (a_i - 2) \le \mathsf{v}_g(S)\,.$$

2. We start with the following assertion.

A. Let $k \in \mathbb{N}$ and $1 = a_0, a_1, \ldots, a_k \in \mathbb{N}$ such that $a_1 + \ldots + a_i \le 2i$ for all $i \in [1, k]$. Then

$$\Big\{ \sum_{i \in I} a_i \mid \emptyset \ne I \subset [0, k] \Big\} = \Big[1, \sum_{\nu=0}^{k} a_\nu \Big]\,.$$

Proof of **A**. We proceed by induction on k. If $k = 1$, then $a_1 \le 2$, and hence the assertion follows. Suppose that $k \ge 2$ and that the assertion holds for $k - 1$. Then

$$\begin{aligned} \Big\{ \sum_{i \in I} a_i \mid \emptyset \ne I \subset [0, k] \Big\} &= \Big\{ \sum_{i \in I} a_i \mid \emptyset \ne I \subset [0, k-1] \Big\} \cup \{a_k\} \\ &\qquad \cup \Big\{ a_k + \sum_{i \in I} a_i \mid \emptyset \ne I \subset [0, k-1] \Big\} \\ &= \Big[1, \sum_{\nu=0}^{k-1} a_\nu \Big] \cup \{a_k\} \cup \Big[1 + a_k, \sum_{\nu=0}^{k} a_\nu \Big]\,. \end{aligned}$$

Since $2 \sum_{\nu=0}^{k-1} a_\nu \ge 2k \ge a_1 + \ldots + a_k$, we infer that $a_k \le 1 + \sum_{\nu=0}^{k-1} a_\nu$, and thus **A** follows.

We set $m = 2|S| - n\|S\|_g$ and $S = g^m (a_1 g) \cdot \ldots \cdot (a_{|S|-m} g)$ where $1 \le a_1 \le \cdots \le a_{|S|-m} \le n$. Then $a_1 + \ldots + a_{|S|-m} = n\|S\|_g - m = 2(|S| - m)$ and hence

$$a_1 \le \frac{a_1 + a_2}{2} \le \frac{a_1 + a_2 + a_3}{3} \le \cdots \le \frac{a_1 + \ldots + a_{|S|-m}}{|S| - m} = 2\,,$$

which implies that $a_1 + \ldots + a_i \le 2i$ for all $i \in [1, |S| - m]$. Pick $x \in [2|S| - n\|S\|_g, n\|S\|_g] = [m, 2|S| - m]$ and set $y = x - (m-1) \in [1, 2(|S| - m) + 1]$. Then **A** implies that the sequence $g(a_1 g) \cdot \ldots \cdot (a_{|S|-m} g)$ has a subsequence S'' with $\sigma(S'') = yg$. Then $S' = g^{m-1} S''$ is a subsequence of S with $|S'| \ge m$ such that $\sigma(S') = xg$. □

Let $\exp(G) = n$ and $S \in \mathcal{F}(G)$. Obviously, S has no zero-sum subsequence of length n if and only if $-g + S$ has no zero-sum subsequence of length n for any $g \in G$. Thus the investigation of sequences, that have no zero-sum subsequences of length n, can be restricted to those sequences T with $\mathsf{h}(T) = \mathsf{v}_0(T)$.

Theorem 5.1.16 (Savchev–Chen). *Let G be cyclic of order $n \ge 3$ and $S \in \mathcal{F}(G)$ a sequence of length $|S| \ge \frac{3n-1}{2}$. Then the following statements are equivalent:*

(a) *S has no zero-sum subsequence of length n and $\mathsf{h}(S) = \mathsf{v}_0(S)$.*

(b) *$S = S_1 S_2$ where $S_1, S_2 \in \mathcal{F}(G)$ with $\|S_1\|_g < 1$ and $\|g - S_2\|_g < 1$ for some $g \in G$ with* $\operatorname{ord}(g) = n$.

Proof. (a) ⇒ (b). We set $S = 0^{\mathsf{h}(S)} S'$ with $S' \in \mathcal{F}(G)$. Then (a) and the "in particular" statement of Proposition 5.1.13 imply that for every zero-sum subsequence T of S we have $|T| < n$. In particular, $\mathsf{v}_0(S) < n$.

Let T be a zero-sum subsequence of S' of maximal length. Thus $U = T^{-1} S'$ is zero-sum free, $|T| + \mathsf{h}(S) < n$ and $|U| \ge \frac{3n-1}{2} - (n-1) = \frac{n+1}{2}$. By Theorem 5.1.8.1, there is some $g \in G$ with $\operatorname{ord}(g) = n$ such that U is g-smooth. We set

$$T = g^{\mathsf{v}_g(T)} (b_1 g) \cdot \ldots \cdot (b_q g),$$

where $q \in \mathbb{N}_0$ and $b_1, \ldots, b_q \in [2, n-1]$. We continue with the following two assertions.

A1. Let $I \subset [1, q]$ such that $\varphi = n - \sum_{i \in I} (n - b_i) \in [2, n\|U\|_g]$. Then

$$|I| \ge \begin{cases} 2|U| - n\|U\|_g & \text{if } \ \varphi \in [2|U| - n\|U\|_g, n\|U\|_g], \\ \varphi & \text{if } \ \varphi \in [2, 2|U| - n\|U\|_g - 1], \end{cases}$$

and $b_i > n\|U\|_g$ for all $i \in [1, q]$.

A2. $n\|U\|_g + \sum_{j=1}^{q} (n - b_j) < n$.

Suppose that **A2** holds. We set

$$S_1 = g^{\mathsf{v}_g(T)} U \quad \text{and} \quad S_2 = 0^{\mathsf{v}_0(S)} (b_1 g) \cdot \ldots \cdot (b_q g),$$

and then clearly $S = S_1 S_2$. Since T has sum zero, we get $\mathsf{v}_g(T) \equiv \sum_{j=1}^{q} (n - b_j) \mod n$. Because S has no zero-sum subsequence of length n, we have $\mathsf{v}_g(T) \in [0, n-1]$, and by **A2** we have $\sum_{j=1}^{q} (n - b_j) < n$. Thus $\mathsf{v}_g(T) = \sum_{j=1}^{q} (n - b_j)$, and again by **A2** we infer that $n\|S_1\|_g = \mathsf{v}_g(T) + n\|U\|_g < n$. Furthermore, it follows that

$$n\|g - S_2\|_g = \sum_{j=1}^{q}(n - b_j) + \big(\mathsf{v}_0(S) + q\big) = \mathsf{v}_g(T) + \mathsf{h}(S) + q = |T| + \mathsf{h}(S) < n\,.$$

Proof of **A1**. Note that $2|U| - n\|U\|_g \ge 2|U| - (n-1) \ge 2$.

Let $\varphi \in [2|U| - n\|U\|_g, n\|U\|_g] \subset [2, n-1]$. By Lemma 5.1.15.2 there is a subsequence U' of U with $|U'| \ge 2|U| - n\|U\|_g$ and $\sigma(U') = \varphi g = \big(\sum_{i\in I} b_i\big)g$. This implies that $|I| \ge |U'| \ge 2|U| - n\|U\|_g$, because $|I| < |U'|$ and replacing $\prod_{i\in I}(b_i g)$ by U' would yield a zero-sum subsequence T' of S with $|T'| > |T|$, a contradiction to the maximality of $|T|$.

Let $\varphi \in [2, 2|U| - n\|U\|_g - 1]$. By Lemma 5.1.15.1 there is a subsequence $U' = g^\varphi$ of U with $\sigma(U') = \varphi g = \big(\sum_{i\in I} b_i\big)g$. As above it follows that $|I| \ge |U'| = \varphi$.

Let $i \in [1, q]$ and assume to the contrary that $b_i \le n\|U\|_g$. Then $2 \le b_i = n - (n - b_i) \le n\|U\|_g$, which yields a contradiction to the lower bound of $|I|$ for $I = \{i\}$.

Proof of **A2**. If $q = 0$, then the assertion follows because U is g-smooth. Assume to the contrary that $q \ge 1$ and $n\|U\|_g + \sum_{j=1}^{q}(n - b_j) \ge n$. Thus there is some minimal $I \subset [1, q]$, say $I = [1, m]$, such that $\varphi = n - \sum_{j=1}^{m}(n - b_j) \le n\|U\|_g$, and hence $\varphi + (n - b_j) > n\|U\|_g$ for all $j \in [1, m]$. Since $b_m > n\|U\|_g$ by **A1**, we get

$$\varphi > n\|U\|_g - (n - b_m) > n\|U\|_g - (n - n\|U\|_g) = 2n\|U\|_g - n \ge 2|U| - n \ge 1\,.$$

If $\varphi \in [2|U| - n\|U\|_g, n\|U\|_g]$, then **A1** implies $m \ge 2|U| - n\|U\|_g$ and hence

$$\begin{aligned} n\|U\|_g + 1 &\le n - \sum_{j=1}^{m-1}(n - b_j) \\ &\le n - (m-1) \le n - (2|U| - n\|U\|_g - 1) = (n - 2|U|) + n\|U\|_g + 1\,, \end{aligned}$$

a contradiction to $2|U| \ge n + 1$.

Suppose that $\varphi \in [2, 2|U| - n\|U\|_g - 1]$. Then **A1** implies $m \ge \varphi$, and since $n - b_j \ge n\|U\|_g + 1 - \varphi > 0$ for all $j \in [1, m]$, we infer that

$$\begin{aligned} n &= \varphi + \sum_{j=1}^{m}(n - b_j) \ge \varphi + m\big(n\|U\|_g + 1 - \varphi\big) \\ &\ge \varphi + \varphi\big(n\|U\|_g + 1 - \varphi\big) = \varphi\big(n\|U\|_g + 2 - \varphi\big)\,. \end{aligned}$$

Consider the quadratic function $f\colon \mathbb{R} \to \mathbb{R}$, defined by $f(x) = x^2 - (n\|U\|_g + 2)x + n$ for all $x \in \mathbb{R}$, and observe that the above inequality states that $f(\varphi) \ge 0$. However, the maximum value of f in the interval between 2 and $2|U| - n\|U\|_g - 1$ equals $f(2) = n - 2n\|U\|_g \le n - 2|U| < 0$, a contradiction.

(b) $\Rightarrow$ (a). This follows from Proposition 5.1.14. $\square$

Remarks 5.1.17. Let G be cyclic of order $n \ge 3$.

1. Let $S \in \mathcal{F}(G)$ be a sequence of length $n+k-1$, with $k \in [(n+1)/2, n-1]$ that has no zero-sum subsequence of length n, and suppose that $\mathsf{h}(S) = \mathsf{v}_0(S)$. Then Theorem 5.1.16 and Proposition 5.1.14 give structural information on S. In particular, we get $\mathsf{v}_g(S) + \mathsf{v}_0(S) \ge 2k$. Thus if $k = n-1$, we obtain the description of S given in Proposition 5.1.12.

2. We give an example of a sequence S of length $|S| = \lfloor (3n-2)/2 \rfloor$, that has no zero-sum subsequence of length n, but that does not satisfy the structural description given in Theorem 5.1.16.(b). For an element $h \in G$ with $\operatorname{ord}(h) = n$, where $n \ge 9$ for odd n and $n \ge 6$ for even n, we set

$$S = \begin{cases} 0^{n-1}(2h)^{\frac{n}{2}-1}(3h) & \text{if } n \text{ is even,} \\ 0^{n-1}(2h)^{\frac{n-5}{2}}(3h)^2 & \text{if } n \text{ is odd.} \end{cases}$$

Obviously, S has the asserted length and no zero-sum subsequence of length n. Assume to the contrary that there exists an element $g \in G$ with $\operatorname{ord}(g) = n$ such that the condition in Theorem 5.1.16.(b) holds. Then Proposition 5.1.14.4 implies that $n-1 = \mathsf{h}(S) = \max\{\mathsf{v}_g(S), \mathsf{v}_0(S)\}$, and hence $S_2 = 0^{n-1}$. Then $S_1 = S^{-1}S$ and $\|S_1\|_g \ge 1$, a contradiction.

5.2 Groups of higher rank

We focus on groups of the form $G = C_n^r$, with $n, r \in \mathbb{N}$ and $n \ge 2$, and we start our discussion with the following two properties.

Property C. Every sequence S over G of length $|S| = \eta(G) - 1$, that has no zero-sum subsequence of length in $[1, n]$, has the form $S = T^{n-1}$ for some sequence T over G.

Property D. Every sequence S over G of length $|S| = \mathsf{s}(G) - 1$, that has no zero-sum subsequence of length n, has the form $S = T^{n-1}$ for some sequence T over G.

For groups of rank two, Property **C** was first considered by P. van Emde Boas and Property **D** by W. Gao (see [30, 45], [44, Lemma 4.7]).

If $r = 1$, then G has Property **D** by Proposition 5.1.12. It follows from the very definition that C_2^r satisfies Property **D**, and a straightforward argument shows that C_3^r satisfies Property **D** (see [28, Lemma 2.3.3] and the subsequent discussion). In [52, Conjecture 7.2] it is conjectured that every group $G = C_n^r$, where $r \in \mathbb{N}$ and $n \in \mathbb{N}_{\ge 2}$, has Property **D** (see [129, 57]). Groups of rank two will be considered in some detail below.

Following [47, Proposition 2.7] and [64], we work out the relationship between Property **C** and Property **D**. We need the technical Property **D1** for an arbitrary group G with $\exp(G) = n$.

Property D1. Every sequence S over G of length $|S| = \mathsf{s}(G) - 1$ that has no zero-sum subsequence of length n satisfies $\mathsf{h}(S) \ge \lfloor \frac{n-1}{2} \rfloor$.

Lemma 5.2.1. *Let* $\exp(G) = n$ *and* $S \in \mathcal{F}(G)$.

1. *If* $g \in G$ *with* $\mathsf{v}_g(S) \ge \lfloor \frac{n-1}{2} \rfloor$ *and* S *has no zero-sum subsequence of length* n, *then* S *has a subsequence* T *of length* $|T| \ge |S| - n + 1$ *such that* $-g + T$ *has no short zero-sum subsequence.*

2. *If* $i \in [1, (n+2)/2]$ *and* $|S| = \eta(G) + i - 1$, *then* S *has a zero-sum subsequence* T *of length* $|T| \in [i, n]$.

3. *If* $|S| = \eta(G) + n - 1$ *and* $\mathsf{h}(S) \ge \lfloor \frac{n-1}{2} \rfloor$, *then* S *has a zero-sum subsequence of length* n.

4. *If* G *has Property* **D1**, *then* $\mathsf{s}(G) = \eta(G) + n - 1$.

Proof. 1. Without restriction we may suppose that $g = 0$, say $S = 0^v R$ with $v \ge \lfloor \frac{n-1}{2} \rfloor$ and $R \in \mathcal{F}(G)$. Let U be a zero-sum subsequence of R such that $|U| \le n$ is maximal. Then, by the maximality of $|U|$, either $|U| > n/2$ or $U^{-1}R$ has no short zero-sum subsequence. Since S has no zero-sum subsequence of length n, it follows that $|U| + v < n$. Thus $|U| \le n/2$ and $T = U^{-1}R$ has no short zero-sum subsequence. Since $|S| = |T| + |U| + v$ and $|U| + v < n$, we obtain that $|T| > |S| - n$.

2. We proceed by induction on i. For $i = 1$, the assertion is clear. Now suppose the assertion holds for $i \in [1, n/2]$, and we have to show it for $i+1$. Then S has a zero-sum subsequence T_1 of length $|T_1| \in [i, n]$. If $|T_1| \ge i+1$, then we are done. Otherwise, $|T_1| = i$ and $|T_1^{-1}S| \ge \eta(G)$. Thus $T_1^{-1}S$ has a short zero-sum subsequence T_2. If $|T_2| \ge i+1$, then we are done. Otherwise, $|T_2| \in [1, i]$ and T_1T_2 is a zero-sum subsequence of S of length $1 + i \le |T_1T_2| \le 2i \le n$.

3. We may suppose that $S = 0^v R$ with $v \ge \lfloor \frac{n-1}{2} \rfloor$ and $R \in \mathcal{F}(G)$. If $v \ge n$, then we are done. Suppose that $v \le n - 1$. Since $|R| = \eta(G) + n - 1 - v$ and $1 \le n - v \le (n+2)/2$, 2. implies that R has a zero-sum subsequence T of length $|T| \in [n - v, n]$. Thus $0^{n-|T|}|T|$ is a zero-sum subsequence of S of length n.

4. Lemma 4.2.2.1 implies that $\mathsf{s}(G) \ge \eta(G) + n - 1$. Let $S \in \mathcal{F}(G)$ be a sequence of length $|S| = \mathsf{s}(G) - 1$ that has no zero-sum subsequence of length n. By Property **D1** we have $\mathsf{h}(S) \ge \lfloor \frac{n-1}{2} \rfloor$, and hence 3. implies that $\mathsf{s}(G) - 1 = |S| \le \eta(G) + n - 2$. □

Proposition 5.2.2. *Let* $G = C_n^r$ *with* $r, n \ge 2$. *Then the following statements are equivalent*:

(a) *G has Property* **D**.

(b) *G has Properties* **C** *and* **D1**.

Proof. (a) $\Rightarrow$ (b). By definition, G has Property **D1**. To show that G satisfies Property **C** as well, let $S \in \mathcal{F}(G)$ be a sequence of length $\eta(G) - 1$ which has no short zero-sum subsequence. We consider the sequence

$$T = 0^{n-1} S .$$

If T has a zero-sum subsequence T' of length $|T'| = n$, then $T' = 0^k S'$ with $k' \in [0, n-1]$ whence S' is a short zero-sum subsequence of S. Since Property **D** holds, Lemma 5.2.1.4 implies that $|T| = \eta(G) - 1 + (n-1) = \mathsf{s}(G) - 1$. Therefore Property **D** implies that S has the required form.

(b) $\Rightarrow$ (a). Let $S \in \mathcal{F}(G)$ be a sequence of length $|S| = \mathsf{s}(G) - 1$ that has no zero-sum subsequence of length n. By Property **D1**, S may be written in the form $S = g^{\mathsf{h}(S)} S'$ where $g \in$G, $S' \in \mathcal{F}(G)$ and $\mathsf{h}(S) \ge \lfloor \frac{n-1}{2} \rfloor$. By Lemma 5.2.1.1, S has a subsequence T of length $|T| \ge |S| - n + 1 \ge \eta(G) - 1$ such that $-g + T$ has no short zero-sum subsequence. Clearly, we may suppose that $|T| = \eta(G) - 1$. Since G has Property **C**, it follows that there are $a_1, \ldots, a_k \in G$ such that $-g + T = (a_1 \cdot \ldots \cdot a_k)^{n-1}$. Since $0 \notin \{a_1, \ldots, a_k\}$, it follows that $T \mid S'$. This implies that $\mathsf{h}(S) = n - 1$ and hence $S = \big(g(g+a_1) \cdot \ldots \cdot (g + a_k)\big)^{n-1}$. $\square$

Suppose that Property **D** holds. Then, by definition, there exists some $c(G) \in \mathbb{N}$ such that $\mathsf{s}(G) = c(G)(n-1) + 1$. For $r = 1$ we have $\mathsf{c}(G) = 2$ and for $r = 2$ we have $c(G) = 4$ (see Theorem 4.2.10). In case of higher ranks, bounds for $c(G)$ are given by N. Alon and M. Dubiner [2] and then in [100, 29, 28, 27].

Property **C** and Property **D** are both multiplicative, provided that the $c(\cdot)$-invariants of all involved groups coincide (see [56, Theorem 3.2]).

Theorem 5.2.3. *Let* $G = C_{mn}^r$ *with* $m, n, r \in \mathbb{N}$.

1. *If both* C_m^r *and* C_n^r *have Property* **D** *and*

$$\frac{\mathsf{s}(C_m^r) - 1}{m-1} = \frac{\mathsf{s}(C_n^r) - 1}{n-1} = \frac{\mathsf{s}(C_{mn}^r) - 1}{mn-1},$$

then G *has Property* **D**.

2. *If both* C_m^r *and* C_n^r *have Property* **C** *and*

$$\frac{\eta(C_m^r) - 1}{m-1} = \frac{\eta(C_n^r) - 1}{n-1} = \frac{\eta(C_{mn}^r) - 1}{mn-1},$$

then G *has Property* **C**.

From now on we restrict our discussion on groups of rank two. For $G = C_n \oplus C_n$ with $n \ge 2$ the following property was first addressed in [49].

Property B. Every minimal zero-sum sequence S over G of length $|S| = \mathsf{D}(G) = 2n - 1$ contains some element with multiplicity $n - 1$.

It is conjectured that every group of the above form satisfies Property **B**. Several equivalent conditions to Property **B** may be found in [71, Section 5.8]. Proposition 5.2.6 shows that if G satisfies one of the Properties **B**, **C** or **D**, then the structure of the extremal sequences is completely determined. In order to work this out we need some preparations.

Lemma 5.2.4. *Let* $G = C_n \oplus C_n$ *with* $n \geq 2$ *and* $S \in \mathcal{F}(G)$. *If* $|S| = 3n - 3$ *and* S *has no short zero-sum subsequence, then* S *has a minimal zero-sum subsequence* T *of length* $|T| = 2n - 1$.

Proof. Suppose that $|S| = 3n - 3$ and that S has no short zero-sum subsequence, and set $W = 0S \in \mathcal{F}(G)$. Then Corollary 4.2.11 implies that W has a zero-sum subsequence U of length $|U| \in \{n, 2n\}$, and by our assumption on S we get $|U| = 2n$. If U is a subsequence of S, then $\mathsf{D}(G) = 2n - 1$ implies that $U = U_1 U_2$, where both U_1 and U_2 are nontrivial zero-sum sequences. Therefore, either U_1 or U_2 is a short zero-sum subsequence of S, a contradiction. Therefore, $U = 0T$ with $|T| = 2n - 1$, and since S has no short zero-sum subsequence, it follows that T is a minimal zero-sum sequence. □

Theorem 5.2.5. *Let* $G = C_n \oplus C_n$ *with* $n \geq 2$ *and let* $S \in \mathcal{A}(G)$ *be a minimal zero-sum sequence of length* $|S| = 2n - 1$.

1. *For every* $g \in \operatorname{supp}(S)$ *we have* $\operatorname{ord}(g) = n$.
2. *If* $|\operatorname{supp}(S)| = 3$, *then* S *contains some element with multiplicity* $n - 1$.
3. *If* G *has Property* **B**, *then* G *has Property* **C**.

Proof. 1. See [71, Theorem 5.8.4]. The proof is done by the inductive method.

2. See [101, Theorem 1]. The proof uses the theory of continued fractions.

3. By Corollary 4.2.11 this follows from [50, Theorem 6.2]. □

Proposition 5.2.6. *Let* $G = C_n \oplus C_n$ *with* $n \geq 2$ *and let* $S \in \mathcal{F}(G)$.

1. *If* S *has length* $\mathsf{D}(G)$, *then the following statements are equivalent*:
   (a) S *is a minimal zero-sum sequence and contains some element with multiplicity* $n - 1$.
   (b) *There exists a basis* (e_1, e_2) *of* G *and integers* $x_1, \ldots, x_n \in [0, n-1]$ *with* $x_1 + \ldots + x_n \equiv 1 \mod n$ *such that*
   $$S = e_1^{n-1} \prod_{\nu=1}^{n} (x_\nu e_1 + e_2) .$$
2. *If* S *has length* $\eta(G) - 1$, *then the following statements are equivalent*:
   (a) $S = T^{n-1}$ *for some* $T \in \mathcal{F}(G)$ *and* S *has no short zero-sum subsequence.*

(b) *There exists a basis* (e_1, e_2) *of* G *and some* $x \in [1, n-1]$ *with* $\gcd(x, n) = 1$ *such that*

$$S = \big(e_1 e_2 (x e_1 + e_2)\big)^{n-1}.$$

3. *If* S *has length* $\mathsf{s}(G) - 1$*, then the following statements are equivalent:*

(a) $S = T^{n-1}$ *for some* $T \in \mathcal{F}(G)$ *and* S *has no zero-sum subsequence of length* n*.*

(b) *For every* $g \in \operatorname{supp}(S)$ *there exists a basis* (e_1, e_2) *of* G *and some* $x \in [1, n-1]$ *with* $\gcd(x, n) = 1$ *such that*

$$-g + S = \big(0 e_1 e_2 (x e_1 + e_2)\big)^{n-1}.$$

Proof. In all three items, the implications (b) $\Rightarrow$ (a) are obvious. Thus it remains to verify the converse.

1. Let $S = {e_1}^{n-1} g_1 \cdot \ldots \cdot g_n$, where $g_1, \ldots, g_n \in G$. Then $\operatorname{ord}(e_1) = n$, and there exists some $\widetilde{e} \in G$ such that $(e_1, \widetilde{e})$ is a basis of G. For $i \in [1, n]$ let x_i, $y_i \in [0, n-1]$ be such that $g_i = x_i e_1 + y_i \widetilde{e}$. Since

$$\sigma(S) = (n - 1 + x_1 + \ldots + x_n) e_1 + (y_1 + \ldots + y_n)\widetilde{e} = 0,$$

we obtain that $x_1 + \ldots + x_n \equiv 1 \mod n$, and that $B = (y_1 \widetilde{e}) \cdot \ldots \cdot (y_n \widetilde{e})$ has sum zero. We assert that B is even a minimal zero-sum sequence. Indeed, otherwise there exists some $\emptyset \ne I \subsetneq [1, n]$ such that

$$\sum_{i \in I} y_i \equiv 0 \mod n.$$

If $k \in [0, n-1]$ is such that

$$\sum_{i \in I} x_i \equiv n - k \mod n, \quad \text{then} \quad {e_1}^k \prod_{i \in I} (x_i e_1 + y_i \widetilde{e})$$

has sum zero, a contradiction to $S \in \mathcal{A}(G)$. Now Corollary 2.1.4.1 implies that $y_1 = \cdots = y_n = y$, where $\gcd(y, n) = 1$, and we set $e_2 = y\widetilde{e}$ to complete the proof.

2. Let $S = T^{n-1}$ be as in 2.(a). By Lemma 5.2.4, S has a minimal zero-sum subsequence U of length $|U| = 2n - 1$. Since $3 = |\operatorname{supp}(S)| = |\operatorname{supp}(U)|$, Theorem 5.2.5.2 and item 1. imply that U has a form as given in 1.(b). Thus there exists a basis (e_1, e_2) of G such that $T = e_1 (x_1 e_1 + e_2)(x_2 e_1 + e_2)$ with $0 \le x_1 < x_2 \le n - 1$. Clearly, $(f_1, f_2) = (e_1, x_1 e_1 + e_2)$ is a basis of G and hence $T = f_1 f_2 (x f_1 + f_2)$ with $x \in [1, n-1]$. Assume to the contrary that $\gcd(x, n) = a > 1$ and set $n' = n/a$. Then $(x f_1 + f_2)^{n'} f_2^{n-n'}$ is a short zero-sum subsequence of S, a contradiction.

3. Let $S = T^{n-1}$ be as in 3.(a) and let $g \in \operatorname{supp}(S)$. Then $T = gU$ for some $U \in \mathcal{F}(G)$. Since $-g + U^{n-1}$ has no short zero-sum subsequence, 2. implies that $-g + S$ has the required form. □

The argument in Proposition 5.2.6.2 stems from [121]. Moreover, in that paper Wolfgang A. Schmid established a characterization of the structure of all minimal zero-sum sequences over $C_{n_1} \oplus C_{n_2}$, where $1 < n_1 \mid n_2$, of length $\mathsf{D}(C_{n_1} \oplus C_{n_2}) = n_1 + n_2 - 1$, under the hypothesis that $C_{n_1} \oplus C_{n_1}$ has Property **B**. Analogous results were derived for Properties **C** and **D**.

The next theorem reduces the question whether Property **B** holds for all groups under discussion to groups $C_p \oplus C_p$ where p is a prime (see [54]). Property **B** is verified for some small primes in [50, 5].

Theorem 5.2.7. *Let* $G = C_n \oplus C_n$ *with* $n \ge 2$. *If for every prime divisor* p *of* n *the group* $C_p \oplus C_p$ *has Property* **B**, *then* G *has Property* **B**.

Theorem 5.2.8. *Let* $G = C_p \oplus C_p$ *for some odd prime* p *and let* $S \in \mathcal{F}(G)$.

1. *If* S *is a minimal zero-sum sequence of length* $|S| = \mathsf{D}(G)$, *then* $|\mathrm{supp}(S)| \in [3, p]$.
2. *If* S *is zero-sum free of length* $\mathsf{D}(G) - 1$, *then each two distinct elements of* $\mathrm{supp}(S)$ *are independent.*
3. *If* S *is zero-sum free of length* $\mathsf{D}(G) - 1$, $\varepsilon > 0$ *and* p *sufficiently large, then* S *contains some element* g *with multiplicity* $\mathsf{v}_g(S) > p^{1/4-\varepsilon}$.

Proof. 1. See [71, Proposition 5.8.5]. Note that for every $j \in [3, p]$ there is an $S_j \in \mathcal{A}(G)$ of length $|S_j| = \mathsf{D}(G)$ and with $|\mathrm{supp}(S_j)| = j$.

2. See [71, Corollary 5.6.9]. The proof is based on a covering result.

3. See [56, Theorem 4.1]. The proof is based on a Theorem of J.A. Dias da Silva and Y. ould Hamidoune [21] which runs as follows: If $A \in \mathcal{F}(G)$ is a squarefree sequence and $k \in [1, |A|]$, then

$$|\Sigma_k(A)| \ge \min\{p, k(|A| - k) + 1\}. \qquad \square$$

5.3 Arithmetical consequences

Some simple arithmetical consequences of Property **B** can be found in [71, Chapter 6]. In this Section we restrict our attention to cyclic groups and start with a result first proved in [48] and based on Theorem 5.1.8.

Theorem 5.3.1. *Let* H *be a Krull monoid with cyclic class group* G *of order* $n \ge 2$ *such that every class contains a prime. Then for every* $k \in \mathbb{N}$ *we have* $\rho_{2k+1}(H) = kn + 1$.

Proof. By Theorem 1.3.4, it suffices to consider $\mathcal{B}(G)$. Assume to the contrary that there is some $k \in \mathbb{N}$ such that $\rho_{2k+1}(G) \ge kn + 2$. Let $k \in \mathbb{N}$ be minimal with this property whence $\rho_{2k-1}(G) = (k-1)n + 1$. Thus there are $B \in \mathcal{B}(G)$

and $U_1, \ldots, U_{2k+1}, V_1, \ldots, V_\rho \in \mathcal{A}(G)$ with $\rho = \rho_{2k+1}(G)$, $|U_1| \ge \cdots \ge |U_{2k+1}|$, $|V_1| \le \cdots \le |V_\rho|$ and

$$B = U_1 \cdot \ldots \cdot U_{2k+1} = V_1 \cdot \ldots \cdot V_\rho \,. \tag{$*$}$$

We may suppose that $|B|$ is maximal such that $(*)$ holds. Since $\rho_{2k}(G) = kn$, it follows that $0 \nmid B$ whence $|U_{2k+1}| \ge 2$.

Let $\ell \in [0, \rho]$ such that $|V_\ell| = 2$ and $3 \le |V_{\ell+1}|$, and assume that $(*)$ is a representation with maximal ℓ. Then $h(-h) \nmid V_{\ell+1} \cdot \ldots \cdot V_\rho$ for any $h \in G$. Indeed, assume to the contrary that there are $h \in G$ and distinct $i, j \in [\ell+1, \rho]$ such that $h \,|\, V_i$ and $-h \,|\, V_j$. Then $V_i V_j = h(-h)V'$ with $V' \in \mathcal{B}(G)$, and by the maximality of ρ it follows that $V' \in \mathcal{A}(G)$. But this contradicts the maximal choice of ℓ.

Next we prove that $\ell \ge 1$. If $\ell = 0$, then $3\rho \le |B| \le (2k+1)n$ implies $\rho \le \frac{n}{3}(2k+1) < kn + 1$, a contradiction.

By the maximality of $|B|$ it follows that, for all $i \in [1, 2k+1]$, U_i is not splittable, and we assert that $|U_{2k}| \ge \lfloor \frac{n}{2} \rfloor + 2$. Indeed, assume to the contrary that $|U_{2k}| \le \lfloor \frac{n}{2} \rfloor + 1$. Then

$$2\rho \le |B| \le \Big((2k-1)n + 2(\frac{n}{2} + 1)\Big) = 2kn + 2\,,$$

and hence $\rho \le kn + 1$, a contradiction. Thus Theorem 5.1.8 implies that $U_i = g_i^n$ for all $i \in [1, 2k]$.

Assume to the contrary that there are $i, j \in [1, 2k+1]$ such that $U_i = g^n$ and $U_j = (-g)^n$. Then $\ell \ge n$, and after renumbering, if necessary, we may suppose that $V_1 = \cdots = V_n = (-g)g$. Since $(U_i U_j)^{-1} B = V_{n+1} \cdot \ldots \cdot V_\rho$, it follows that $(k-1)n + 1 = \rho_{2k-1}(G) \ge \rho - n \ge (k-1)n + 2$, a contradiction. Thus there are no two U_i, U_j of such a form and hence $\ell \le |U_{2k+1}|$.

If $|U_{2k+1}| \ge \lfloor \frac{n}{2} \rfloor + 2$, then Theorem 5.1.8 implies that $U_{2k+1} = g_{2k+1}^n$ for some $g_{2k+1} \in G$. Since $\ell \ge 1$, it follows that $g_{2k+1} \in \{-g_1, \ldots, -g_{2k}\}$, a contradiction. Thus

$$|U_{2k+1}| \le \lfloor \frac{n}{2} \rfloor + 1\,,$$

and therefore we obtain that

$$\begin{aligned} \rho &\le \ell + \frac{|B| - 2\ell}{3} = \frac{|B| + \ell}{3} = \frac{2kn + |U_{2k+1}| + \ell}{3} \\ &\le \frac{2kn + 2|U_{2k+1}|}{3} \le \frac{2kn + n + 2}{3} \le kn + \frac{2}{3}\,, \end{aligned}$$

a contradiction. □

Corollary 5.3.2. *Let H be a Krull monoid with cyclic class group G of order $n \ge 2$ such that every class contains a prime. Then for every $k \in \mathbb{N}$ and every $l \in \mathbb{N}_0$ we have $\mathcal{V}_k(H) = [\lambda_k(H), \rho_k(H)]$,*

$$\rho_{2k+j}(H) = kn + j \ \textit{for}\ j \in [0,1] \quad \textit{and} \quad \lambda_{ln+j}(H) = \begin{cases} 2l + j & \textit{for}\ j \in [0,1], \\ 2l + 2 & \textit{for}\ j \in [2, n-1], \end{cases}$$

provided that $ln + j \geq 1$.

Proof. As in the proof of Theorem 3.1.3, it suffices to consider the block monoid $\mathcal{B}(G)$. If $n = 2$, then $\mathcal{B}(G)$ is half-factorial whence for all $k \in \mathbb{N}$ we have $\lambda_k(G) = k = \rho_k(G)$. Suppose that $n \geq 3$, and let $k \in \mathbb{N}$. By Theorem 3.1.3 we obtain that $\mathcal{V}_k(H) = [\lambda_k(G), \rho_k(G)]$. The assertion on $\rho_{2k+j}(G)$ follows from Theorem 5.3.1, and the assertion on $\lambda_{ln+j}(G)$ follows from Corollary 3.1.4. □

Corollary 5.3.3. *Let G be either cyclic or an elementary 2-group with Davenport constant $\mathsf{D}(G) = n \geq 4$. If G' is a finite abelian group with $\mathcal{L}(G) = \mathcal{L}(G')$, then $G \cong G'$.*

Proof. Suppose that $\mathcal{L}(G) = \mathcal{L}(G')$. Then Proposition 3.3.1 implies that $\Delta(G) = \Delta(G')$ and $\rho_k(G) = \rho_k(G')$ for all $k \in \mathbb{N}$. By Corollary 2.3.6 it follows that G' is either cyclic or an elementary 2-group. Corollary 3.1.5 and Theorem 5.3.1 imply that

$$\rho_3(C_n) = n + 1 < \lfloor \frac{3n}{2} \rfloor = \rho_3(C_2^{n-1}),$$

and thus we obtain that $G \cong G'$. □

Bibliography

[1] S.D. Adhikari and P. Rath, *Davenport constants with weights and some related questions*, Integers **6** (2006), Paper A30, 6 p.

[2] N. Alon and M. Dubiner, *A lattice point problem and additive number theory*, Combinatorica **15** (1995), 301–309.

[3] E. Balandraud, *Une variante de la méthode isopérimétrique de Hamidoune, appliquée au théorème de Kneser*, Ann. Inst. Fourier **58** (2008), 915–943.

[4] G. Bhowmik, I. Halupczok, and J.-C. Schlage-Puchta, *Inductive methods and zero-sum free sequences*, manuscript.

[5] ———, *The structure of maximal zero-sum free sequences*, manuscript.

[6] G. Bhowmik and J.-C. Schlage-Puchta, *Davenport's constant for groups of the form $\mathbb{Z}_3 \oplus \mathbb{Z}_3 \oplus \mathbb{Z}_{3d}$*, Additive Combinatorics, CRM Proceedings and Lecture Notes, vol. 43, Amer. Math. Soc., 2008.

[7] A. Bialostocki and P. Dierker, *On the Erdős-Ginzburg-Ziv theorem and the Ramsey numbers for stars and matchings*, Discrete Math. **110** (1992), 1–8.

[8] A. Bialostocki, P. Dierker, D. Grynkiewicz, and M. Lotspeich, *On some developments of the Erdős-Ginzburg-Ziv Theorem II*, Acta Arith. **110** (2003), 173–184.

[9] B. Bollobás and I. Leader, *The number of k-sums modulo k*, J. Number Theory **78** (1999), 27–35.

[10] J.D. Bovey, P. Erdős, and I. Niven, *Conditions for zero sum modulo n*, Canad. Math. Bull. **18** (1975), 27–29.

[11] Y. Caro, *Zero-sum Ramsey numbers–stars*, Discrete Math. **104** (1992), 1–6.

[12] ———, *Zero-sum problems – a survey*, Discrete Math. **152** (1996), 93–113.

[13] S.T. Chapman (ed.), *Arithmetical Properties of Commutative Rings and Monoids*, Lecture Notes in Pure and Appl. Math., vol. 241, Chapman & Hall/CRC, 2005.

[14] S.T. Chapman and J. Coykendall, *Half-factorial domains, a survey*, Non-Noetherian Commutative Ring Theory, Mathematics and Its Applications, vol. 520, Kluwer Academic Publishers, 2000, pp. 97–115.

[15] S.T. Chapman, M. Freeze, W. Gao, and W.W. Smith, *On Davenport's constant of finite abelian groups*, Far East J. Math. Sci. **5** (2002), 47–54.

[16] S.T. Chapman, M. Freeze, and W.W. Smith, *Minimal zero sequences and the strong Davenport constant*, Discrete Math. **203** (1999), 271–277.

[17] S.T. Chapman and W.W. Smith, *Factorization in Dedekind domains with finite class group*, Israel J. Math. **71** (1990), 65–95.

[18] ______, *A characterization of minimal zero-sequences of index one in finite cyclic groups*, Integers **5(1)** (2005), Paper A27, 5 p.

[19] X. Chen and P. Yuan, *A note on Kneser's theorem*, JP J. Algebra Number Theory Appl. **6** (2006), 77–83.

[20] J. Coykendall, *Extensions of half-factorial domains: a survey*, Arithmetical Properties of Commutative Rings and Monoids, Lecture Notes in Pure and Appl. Math., vol. 241, Chapman & Hall/CRC, 2005, pp. 46–70.

[21] J.A. Dias da Silva and Y. ould Hamidoune, *Cyclic spaces for Grassmann derivatives and additive theory*, Bull. London Math. Soc. **26** (1994), 140–146.

[22] J.M. Deshouillers and G.A. Freiman, *A step beyond Kneser's theorem for abelian finite groups*, Proc. London Math. Soc. **86** (2003), 1–28.

[23] J.M. Deshouillers, B. Landreau, and A.A. Yudin, *Structure Theory of Set Addition*, vol. 258, Astérisque, 1999.

[24] M. DeVos, *A short proof of Kneser's addition theorem for abelian groups*, manuscript.

[25] M. DeVos, L. Goddyn, and B. Mohar, *A generalization of Kneser's addition theorem*, Adv. Math., to appear.

[26] L. Diracca, *On a generalization of the exchange property to modules with semilocal endomorphism rings*, J. Algebra **313** (2007), 972–987.

[27] Y. Edel, *Sequences in abelian groups G of odd order without zero-sum subsequences of length* $\exp(G)$, Des. Codes Cryptography **47** (2008), 125–134.

[28] Y. Edel, C. Elsholtz, A. Geroldinger, S. Kubertin, and L. Rackham, *Zero-sum problems in finite abelian groups and affine caps*, Quarterly. J. Math., Oxford II. Ser. **58** (2007), 159–186.

[29] C. Elsholtz, *Lower bounds for multidimensional zero sums*, Combinatorica **24** (2004), 351–358.

[30] P. van Emde Boas, *A combinatorial problem on finite abelian groups II*, Reports ZW-1969-007, Math. Centre, Amsterdam, 1969.

[31] P. van Emde Boas and D. Kruyswijk, *A combinatorial problem on finite abelian groups*, Reports ZW-1967-009, Math. Centre, Amsterdam, 1967.

[32] P. Erdős, A. Ginzburg, and A. Ziv, *Theorem in the additive number theory*, Bull. Research Council Israel **10** (1961), 41–43.

[33] B.W. Finklea, T. Moore, V. Ponomarenko, and Z.J. Turner, *Invariant polynomials and minimal zero sequences*, Involve **1** (2008), 159–165.

[34] C. Flores and O. Ordaz, *On the Erdős-Ginzburg-Ziv theorem*, Discrete Math. **152** (1996), 321–324.

[35] M. Freeze and A. Geroldinger, *Unions of sets of lengths*, Funct. Approx., Comment. Math. **39** (2008), 149–162.

[36] G.A. Freiman, *Foundations of a Structural Theory of Set Addition*, Transl. Math. Monogr., vol. 37, Amer. Math. Soc., 1973.

[37] ______, *Structure theory of set addition*, Structure Theory of Set Addition, vol. 258, Astérisque, 1999, pp. 1–33.

[38] G.A. Freiman and A. Geroldinger, *An addition theorem and its arithmetical application*, J. Number Theory **85** (2000), 59–73.

[39] L. Gallardo, G. Grekos, L. Habsieger, F. Hennecart, B. Landreau, and A. Plagne, *Restricted addition in $\mathbb{Z}/n\mathbb{Z}$ and an application to the Erdős-Ginzburg-Ziv problem*, J. London Math. Soc. **65** (2002), 513–523.

[40] L. Gallardo, G. Grekos, and J. Pihko, *On a variant of the Erdős-Ginzburg-Ziv problem*, Acta Arith. **89** (1999), 331–336.

[41] W. Gao, *Addition theorems for finite abelian groups*, J. Number Theory **53** (1995), 241–246.

[42] ______, *A combinatorial problem on finite abelian groups*, J. Number Theory **58** (1995), 100–103.

[43] ______, *An addition theorem for finite cyclic groups*, Discrete Math. **163** (1997), 257- 265.

[44] ______, *On Davenport's constant of finite abelian groups with rank three*, Discrete Math. **222** (2000), 111–124.

[45] ______, *Two zero sum problems and multiple properties*, J. Number Theory **81** (2000), 254- 265.

[46] ______, *Zero sums in finite cyclic groups*, Integers **0** (2000), Paper A14, 9 p.

[47] ———, *On zero sum subsequences of restricted size II*, Discrete Math. **271** (2003), 51–59.

[48] W. Gao and A. Geroldinger, *On products of k atoms*, Monatsh. Math. **156** (2009), 141–157.

[49] ———, *On long minimal zero sequences in finite abelian groups*, Period. Math. Hungar. **38** (1999), 179–211.

[50] ———, *On zero-sum sequences in $\mathbb{Z}/n\mathbb{Z} \oplus \mathbb{Z}/n\mathbb{Z}$*, Integers **3** (2003), Paper A08, 45 p.

[51] ———, *On a property of minimal zero-sum sequences and restricted sumsets*, Bull. London Math. Soc. **37** (2005), 321–334.

[52] ———, *Zero-sum problems in finite abelian groups: a survey*, Expo. Math. **24** (2006), 337–369.

[53] ———, *On the number of subsequences with given sum of sequences over finite abelian p-groups*, Rocky Mountain J. Math. **37** (2007), 1541–1550.

[54] W. Gao, A. Geroldinger, and D.J. Grynkiewicz, *Inverse zero-sum problems III*, submitted.

[55] W. Gao, A. Geroldinger, and F. Halter-Koch, *Group algebras of finite abelian groups and their applications to combinatorial problems*, Rocky Mt. J. Math., to appear.

[56] W. Gao, A. Geroldinger, and W.A. Schmid, *Inverse zero-sum problems*, Acta Arith. **128** (2007), 245–279.

[57] W. Gao, Q.H. Hou, W.A. Schmid, and R. Thangadurai, *On short zero-sum subsequences II*, Integers **7** (2007), Paper A21, 22 p.

[58] W. Gao and I. Leader, *Sums and k-sums in abelian groups of order k*, J. Number Theory **120** (2006), 26–32.

[59] W. Gao and Y. Li, *Remarks on group rings and the Davenport constant*, Ars Combin., to appear.

[60] W. Gao, Y. Li, J. Peng, and F. Sun, *On subsequence sums of a zero-sum free sequence II*, Electron. J. Comb. **15** (2008), Research paper 117.

[61] ———, *Subsums of a zero-sum free subset of an abelian group*, Electron. J. Comb. **15** (2008), Research paper 116.

[62] W. Gao and Y. ould Hamidoune, *Zero sums in abelian groups*, Combin. Probab. Comput. **7** (1998), 261–263.

[63] W. Gao, A. Panigrahi, and R. Thangadurai, *On the structure of p-zero-sum free sequences and its application to a variant of Erdős-Ginzburg-Ziv theorem*, Proc. Indian Acad. Sci. Math. Sci. **115** (2005), 67–77.

[64] W. Gao and R. Thangadurai, *On the structure of sequences with forbidden zero-sum subsequences*, Colloq. Math. **98** (2003), 213–222.

[65] W. Gao, R. Thangadurai, and J. Zhuang, *Addition theorems on the cyclic groups of order* p^l, Discrete Math. **308** (2008), 2030–2033.

[66] W. Gao and J. Zhuang, *Sequences not containing long zero-sum subsequences*, European J. Combin. **27** (2006), 777–787.

[67] A. Geroldinger, *Über nicht-eindeutige Zerlegungen in irreduzible Elemente*, Math. Z. **197** (1988), 505–529.

[68] ______, *Chains of factorizations in weakly Krull domains*, Colloq. Math. **72** (1997), 53–81.

[69] A. Geroldinger and D.J. Grynkiewicz, *On the arithmetic of Krull monoids with finite Davenport constant*, J. Algebra **321** (2009), 1256–1284.

[70] A. Geroldinger and F. Halter-Koch, *Non-unique factorizations: a survey*, Multiplicative Ideal Theory in Commutative Algebra (J.W. Brewer, S. Glaz, W. Heinzer, and B. Olberding, eds.), Springer, 2006, pp. 217–226.

[71] ______, *Non-Unique Factorizations. Algebraic, Combinatorial and Analytic Theory*, Pure Appl. Math., vol. 278, Chapman & Hall/CRC, 2006.

[72] A. Geroldinger and Y. ould Hamidoune, *Zero-sumfree sequences in cyclic groups and some arithmetical application*, J. Théor. Nombres Bordeaux **14** (2002), 221–239.

[73] A. Geroldinger and W. Hassler, *Arithmetic of Mori domains and monoids*, J. Algebra **319** (2008), 3419–3463.

[74] ______, *Local tameness of v-noetherian monoids*, J. Pure Appl. Algebra **212** (2008), 1509–1524.

[75] A. Geroldinger and R. Schneider, *On Davenport's constant*, J. Combin. Theory Ser. A **61** (1992), 147–152.

[76] B. Girard, *Inverse zero-sum problems and algebraic invariants*, Acta Arith. **135** (2008), 231–246.

[77] ______, *A new upper bound for the cross number of finite abelian groups*, Israel J. Math., to appear.

[78] S. Griffiths, *The Erdős-Ginzburg-Ziv theorem with units*, Discrete Math. **308** (2008), 5473–5484.

[79] P.A. Grillet, *Commutative Semigroups*, Kluwer Academic Publishers, 2001.

[80] D.J. Grynkiewicz, *On extending Pollard's theorem for t-representable sums*, Israel J. Math., to appear.

[81] ______, *A step beyond Kemperman's structure theorem*, Mathematika, to appear.

[82] ———, *On an extension of the Erdős-Ginzburg-Ziv Theorem to hypergraphs*, European J. Combin. **26** (2005), 1154–1176.

[83] ———, *Quasi-periodic decompositions and the Kemperman structure theorem*, European J. Combin. **26** (2005), 559–575.

[84] ———, *A weighted Erdős-Ginzburg-Ziv Theorem*, Combinatorica **26** (2006), 445–453.

[85] D.J. Grynkiewicz, E. Marchan, and O. Ordaz, *Representation of finite abelian group elements by subsequence sums*, J. Théor. Nombres Bordx., to appear.

[86] D.J. Grynkiewicz, O. Ordaz, M.T. Varela, and F. Villarroel, *On Erdős-Ginzburg-Ziv inverse theorems*, Acta Arith. **129** (2007), 307–318.

[87] F. Halter-Koch, *Non-unique factorizations of algebraic integers*, Funct. Approx. Comment. Math. **39** (2008), 49–60.

[88] ———, *Ideal Systems. An Introduction to Multiplicative Ideal Theory*, Marcel Dekker, 1998.

[89] Y. ould Hamidoune, *The global isoperimetric methodology applied to Kneser's theorem*, manuscript.

[90] ———, *Hyper-atoms and the Kemperman's critical pair theory*, manuscript.

[91] ———, *Some additive applications of the isoperimetric approach*, Ann. Inst. Fourier **58** (2008), 2007–2036.

[92] ———, *A weighted generalization of Gao's $n+D-1$ theorem*, Comb. Probab. Comput. **17** (2008), 793–798.

[93] Y. ould Hamidoune, O. Serra, and G. Zémor, *On some subgroup chains related to Kneser's theorem*, J. Théor. Nombres Bordx. **20** (2008), 125–130.

[94] H. Harborth, *Ein Extremalproblem für Gitterpunkte*, J. Reine Angew. Math. **262** (1973), 356–360.

[95] F. Hennecart, *La fonction de Brakemeier dans le problème d'Erdős-Ginzburg-Ziv*, Acta Arith. **117** (2005), 35–50.

[96] F. Kainrath, *Elasticity of finitely generated domains*, Houston J. Math. **31** (2005), 43–64.

[97] ———, *On local half-factorial orders*, Arithmetical Properties of Commutative Rings and Monoids, Lecture Notes in Pure and Appl. Math., vol. 241, Chapman & Hall/CRC, 2005, pp. 316–324.

[98] J.H.B. Kemperman, *On small sumsets in an abelian group*, Acta Math. **103** (1960), 63–88.

[99] H. Kraft and C. Procesi, *Classical invariant theory, a primer*, http://www.math.unibas.ch/~kraft/, 2000.

[100] S. Kubertin, *Nullsummen in* $\mathbb{Z}_p^d$, Master's thesis, Technical University Clausthal, 2002.

[101] G. Lettl and W.A. Schmid, *Minimal zero-sum sequences in* $C_n \oplus C_n$, European J. Combin. **28** (2007), 742–753.

[102] G. Lettl and Zhi-Wei Sun, *On covers of abelian groups by cosets*, Acta Arith. **131** (2008), 341–350.

[103] H.B. Mann, *Additive group theory – a progress report*, Bull. Amer. Math. Soc. (N.S.) **79** (1973), 1069–1075.

[104] ______, *Addition Theorems: The Addition Theorems of Group Theory and Number Theory*, R.E. Krieger, 1976.

[105] H.B. Mann and J.E. Olson, *Sums of sets in the elementary abelian group of type* (p, p), J. Combin. Theory Ser. A **2** (1967), 275–284.

[106] C.P. Milies and S.K. Sehgal, *An Introduction to Group Rings*, Kluwer Academic Publishers, 2002.

[107] M.B. Nathanson, *Additive Number Theory: Inverse Problems and the Geometry of Sumsets*, Springer, 1996.

[108] J.E. Olson, *A combinatorial problem on finite abelian groups I*, J. Number Theory **1** (1969), 8–10.

[109] B. Peterson and T. Yuster, *A generalization of an addition theorem for solvable groups*, Canad. J. Math. **36** (1984), 529–536.

[110] A. Plagne and W.A. Schmid, *On large half-factorial sets in elementary p-groups: maximal cardinality and structural characterization*, Israel J. Math. **145** (2005), 285–310.

[111] ______, *On the maximal cardinality of half-factorial sets in cyclic groups*, Math. Ann. **333** (2005), 759–785.

[112] V. Ponomarenko, *Minimal zero sequences of finite cyclic groups*, Integers **4** (2004), Paper A24, 6p.

[113] A. Potechin, *Maximal caps in* $AG(6, 3)$, Des. Codes Cryptogr. **46** (2008), 243–259.

[114] C. Reiher, *On Kemnitz' conjecture concerning lattice points in the plane*, Ramanujan J. **13** (2007), 333–337.

[115] K. Rogers, *A combinatorial problem in abelian groups*, Proc. Cambridge Philos. Soc. **59** (1963), 559–562.

[116] S. Savchev and F. Chen, *Long zero-free sequences in finite cyclic groups*, Discrete Math. **307** (2007), 2671–2679.

[117] ______, *Long n-zero-free sequences in finite cyclic groups*, Discrete Math. **308** (2008), 1–8.

[118] P. Scherk, *Distinct elements in a set of sums*, Amer. Math. Monthly **62** (1955), 46–47.

[119] W.A. Schmid, *Arithmetical characterization of class groups of the form* $\mathbb{Z}/n\mathbb{Z} \oplus \mathbb{Z}/n\mathbb{Z}$ *via the system of sets of lengths*, Abh. Math. Univ. Semin. Hamburg, to appear.

[120] ______, *Characterization of class groups of Krull monoids via their systems of sets of lengths: a status report*, HRI Conference Proceedings 2009, to appear.

[121] ______, *Inverse zero-sum problems II*, submitted.

[122] ______, *A realization theorem for sets of lengths*, J. Number Theory, to appear.

[123] ______, *Differences in sets of lengths of Krull monoids with finite class group*, J. Théor. Nombres Bordeaux **17** (2005), 323–345.

[124] ______, *Half-factorial sets in finite abelian groups: a survey*, Grazer Math. Ber. **348** (2005), 41–64.

[125] W.A. Schmid and J.J. Zhuang, *On short zero-sum subsequences over p-groups*, Ars Combin., to appear.

[126] O. Serra, *An isoperimetric method for the small sumset problem*, Surveys in Combinatorics 2005, London Math. Soc. Lecture Note Ser., vol. 327, Cambridge University Press, 2005, pp. 119–152.

[127] Fang Sun, *On subsequence sums of a zero-sumfree sequence*, Electron. J. Combin. **14** (2007), Paper R52, 9 p.

[128] Zhi-Wei Sun, *Zero-sum problems for abelian p-groups and covers of the integers by residue classes*, Israel J. Math., to appear.

[129] B. Sury and R. Thangadurai, *Gao's conjecture on zero-sum sequences*, Proc. Indian Acad. Sci. Math. Sci. **112** (2002), 399–414.

[130] T. Tao and V.H. Vu, *Additive Combinatorics*, Cambridge University Press, 2006.

[131] C. Wang, *Note on a variant of the Erdős-Ginzburg-Ziv Theorem*, Acta Arith. **108** (2003), 53–59.

[132] P. Yuan, *Subsequence sums of a zero-sumfree sequence*, Eur. J. Comb. **30** (2009), 439–446.

[133] ______, *On the index of minimal zero-sum sequences over finite cyclic groups*, J. Combin. Theory, Ser. A **114** (2007), 1545–1551.

Part II

Sumsets and Structure

Imre Z. Ruzsa

Introduction

This is a somewhat extended version of my course given in the DocCourse in Barcelona, Spring 2008. Two students, Itziar Bardají and Lluís Vena helped me to prepare the final version, and here I wish to express my sincere thanks to them.

The course is devoted to certain aspects of combinatorial number theory, or additive combinatorics as it is now often called. This change of terminology reflects a shift in the emphasis of problems investigated. First it was mainly infinite sequences and finite sets of integers; this naturally led to sets of residues, then sets in finite groups, and also sets of lattice points, then sets in general commutative groups. (We shall now and then mention results that do not need commutativity, but will not pursue this aim forcefully.)

In classical additive number theory we start with a given set, say of primes, and try to understand how an integer can be expressed as a sum of elements of this set. In combinatorial (or structural, or inverse) theory we do the opposite: Given an additive assumption about a set, say that it has few or many sums, we try to understand its structure. (This is not always explicit in the formulation; we can equivalently say "if a set has few sums, it has property A" or "if a set has property non-A, it has many sums"; we will not attempt uniformity here.)

Historically, combinatorial additive theory grew out of the classical one. Though a few isolated results existed before, the turning point is Schnirelmann's approach to the Goldbach problem. Goldbach's conjecture asserts that any integer greater than 3 can be expressed as a sum of two or three primes, depending on parity. Schnirelmann proved the weaker result that there is a bound k so that every integer is a sum of at most k primes, or in other words, the primes form an additive basis. To this end he established that (very loosely speaking; exact formulations will be given in Chapter 3) integers that can be written as a sum of two primes have positive density; and every set having positive density is a basis. For the Goldbach problem Schnirelmann's approach was soon superseded by Vinogradov's trigonometric sum method; however, it kindled the interest in addition of general sets.

In the first chapter we consider questions of the following kind. Suppose we know the cardinality of a (finite) set and we know also the number of sums of pairs. What can we say about the number of differences, or of sums of triples? The understanding of such cardinality problems is of paramount importance for understanding the structure.

First we explain the most important tool, Plünnecke's inequality, then two further inequalities independent of it. These will be applied to study the connection between $|A|$, $|A+B|$ and $|A+kB|$, with particular emphasis on the case $B=A$.

In the second chapter we prove Freiman's structure theorem on the structure of sets with small doubling. In addition to Plünnecke's inequality, we introduce some basics of Fourier analysis and the important notion of Bohr sets which play a significant role in the proof.

In the third chapter we tell results connecting geometrical position or position within a group and cardinality of sumsets. These include Kneser's theorem and a theorem by Freiman relating the cardinality of the sumset with its diameter. The second part of the chapter deals with the sumset problem in higher dimensions and discusses its connections with geometrical position. The chapter ends with Hovanskii's theorem showing that the growth of the n-fold sum of a finite set is polynomial in an abelian semigroup.

In the fourth chapter we give some results about density. We review the results of Schnirelmann and Mann which provide lower bound for the density of a sumset in terms of the density of the summands. We also discuss the Erdős result which shows that addition of an additive base, even with zero density, increases the density of the sumset. We give some examples and Plünnecke's improvement of this result.

In the fifth chapter we explore some connections with topology and measure. While measure is a more sophisticated concept than cardinality of finite sets, the results are often simpler and sometimes also easier to prove. Although there are several topologies on the set of integers as a discrete set, we will mainly discuss the connection of the Bohr topology to additive properties.

The course contains exercises and problems; the difference between them is that an exercise is what I can solve. Some exercises are called a *prexercise*; this means that the solution will be included in the text, just I feel it may benefit the reader to meditate on it at that point before (or better, instead) reading the proof.

Notation

Let A and B be sets in a (mostly commutative) group. We will call the group operation addition and use additive notation. The *sumset* of these sets is

$$A + B = \{a + b : a \in A,\, b \in B\}.$$

Similarly

$$A - B = \{a - b : a \in A,\, b \in B\} = \{a + (-b)\}.$$

For repeated addition we write

$$kA = A + \cdots + A,\ k \text{ times};$$

in particular, $1A = A$, $0A = \{0\}$.

The set kA is typically different from

$$k \cdot A = \{ka : a \in A\};$$

this will appear rarely.[1]

We will use $\mathbb{Z}_q = \mathbb{Z}/(q\mathbb{Z})$ to denote the set of residue classes modulo q.

[1] This should cause no difficulty in a country where ortography distinguishes between -ll– and –l·l–.

Chapter 1

Cardinality inequalities

1.1 Introduction

Let A, B be sets in a group, $|A| = m$, $|B| = n$. The cardinality of $A + B$ can be anywhere between $\max(m, n)$ and mn. Our aim is to understand the connection between this size and the structure of these sets.

Exercise 1. If $A, B \subset \mathbb{Z}$, $|A| = m$, $|B| = n$, prove that

$$|A + B| \geq m + n - 1$$

and describe the cases of equality.

Exercise 2. Given three positive integers m, n, s such that $m + n - 1 \leq s \leq mn$, find sets $A, B \subset \mathbb{Z}$ such that $|A| = m$, $|B| = n$, $|A + B| = s$.

In this chapter we present some inequalities of the following kind: If a sumset, say $A + B$, is small (in various senses), then so are some other sums. The most frequently applied one of them sounds as follows.

Theorem 1.1.1. *Let A, B be finite sets in a commutative group and write $|A| = m$, $|A + B| = \alpha m$. For arbitrary non-negative integers k, l we have*

$$|kB - lB| \leq \alpha^{k+l} m.$$

Observe that there is no a priori assumption on the size of B; however, with such assumptions sometimes the conclusion can be strengthened.

We end this introduction by mentioning some basic ideas.

(i) **Direct product.** Assume $A_1, A_2, \ldots, A_k$ are subsets of a group G with cardinalities of sumsets

$$|A_{i_1} + A_{i_2} + \cdots + A_{i_m}| = N(i_1, \ldots, i_m).$$

Let $A'_1, \dots, A'_k$ be another collection of sets in another group G' with corresponding values $N'(\dots)$. If we form the direct products

$$B_i = A_i \times A'_i = \{(a, b) : a \in A, b \in B\} \subset G \times G',$$

then we have

$$|B_{i_1} + B_{i_2} + \dots + B_{i_m}| = N(i_1, \dots, i_m) N'(i_1, \dots, i_m).$$

This explains the multiplicative nature of many of the results – when a quantity is estimated in terms of others, this is mostly in the form of a product of powers. This method can often be used to build large examples starting from a single one. It will be used now and then in the opposite way: We apply a result for a power of a small set to get better results for the small set (see Section 1.6).

(ii) **Projection.** If we start from sets of integers, the above construction gives us sets of integral vectors. This is, however, not an essential difference. If we have sets $A_i \subset \mathbb{Z}^d$ and a *finite* number of sum-cardinalities are prescribed, then we can construct sets of integers that behave the same way. Indeed, the linear map

$$(x_1, \dots, x_d) \rightarrow x_1 + m x_2 + \dots + m^{d-1} x_d$$

will not add any new coincidence between sums if m is large enough.

This observation will be used without any further mentioning. If we construct a set in $\mathbb{Z}^d$ with certain properties, we shall tacitly realize that a set of integers can also be constructed if necessary; a set in several dimensions often exhibits the structure more clearly.

Exercise 3. Extend Exercise 1 to sets in $\mathbb{Z}^d$.

On the other hand, if we know that a set is proper d-dimensional, this may yield further results.

Exercise 4. Improve Exercise 1 for sets in $\mathbb{Z}^2$ that do not lie on a single line.

(iii) **Torsion.** The above consideration shows that from our point of view the structure of $\mathbb{Z}^k$ is not richer than that of $\mathbb{Z}$. We can add that no torsion-free group produces anything new either. Indeed, let G be a torsion-free group and take a finite subset (the union of all finite sets which we want to add). This generates a subgroup G'; and, as a finitely generated torsion-free group, G' is isomorphic to $\mathbb{Z}^k$ for some k.

Exercise 5. Extend Exercise 1 to sets in any commutative torsion-free group.

Exercise 6. Extend Exercise 1 to sets in any non-commutative torsion-free group.

1.2 Plünnecke's method

Plünnecke [41] developed a graph-theoretic method to estimate the density of sumsets $A + B$, where A has a positive density and B is a basis. I published a simplified version of his proof [49, 50]. Other accounts (of my version) were published by Malouf [34] and Nathanson [35]. In what follows we adopt Malouf's terminology.

Plünnecke observed that the cardinality properties of the sets A, $A + B$, $A + 2B$, ..., are well reflected by the following directed graph. We take $h + 1$ copies of the group where these sets are situated, and we build a graph on these sets as vertices by connecting an $x \in A + jB$ to a $y \in A + (j + 1)B$ if $y = x + b$ with some $b \in B$. We call this graph the *addition graph.* These graphs have certain properties which follow from the commutativity of addition, and hence Plünnecke called them *commutative*; we shall retain this terminology.

We consider directed graphs $\mathcal{G} = (V, E)$, where V is the set of vertices and E is that of the edges. If there is an edge from x to y, then we also write $x \to y$. A graph is *semicommutative* if for every collection $(x; y; z_1, z_2, \dots, z_k)$ of distinct vertices such that $x \to y$ and $y \to z_i$, there are distinct vertices $y_1, \dots, y_k$ such that $x \to y_i$ and $y_i \to z_i$. $\mathcal{G}$ is *commutative* if both $\mathcal{G}$ and the graph $\hat{\mathcal{G}}$ obtained by reversing the direction of every edge of $\mathcal{G}$ are semicommutative.

Our graphs will be of a special kind we call *layered.* By an *h-layered* graph we mean a graph with a fixed partition of the set of vertices

$$V = V_0 \cup V_1 \cup \dots \cup V_h$$

into $h + 1$ disjoint sets (layers) such that every edge goes from some V_{i-1} into V_i.

Exercise 7. If there is no isolated point, then this partition is unique.

To avoid the separate formulation of certain degenerate cases we do not exclude isolated points.

For $X, Y \subset V$, we define the *image* of X in Y as

$$\operatorname{im}(X, Y) = \{y \in Y : \text{ there is a directed path from some } x \in X \text{ to } y\}.$$

The *magnification ratio* is defined by

$$\mu(X, Y) = \min\left\{\frac{|\operatorname{im}(Z, Y)|}{|Z|} : Z \subset X, Z \neq \emptyset\right\}.$$

For a layered graph we write

$$\mu_j(\mathcal{G}) = \mu(V_0, V_j).$$

Now Plünnecke's main result can be stated as follows.

Theorem 1.2.1 (Plünnecke [41]). *In a commutative layered graph $\mu_j^{1/j}$ is decreasing.*

That is, for $j < h$ we have $\mu_h \le \mu_j^{h/j}$. An obvious (and typically the only available) upper estimate for μ_j is $|V_j|/|V_0|$. This yields the following corollary (in fact, an equivalent assertion).

Theorem 1.2.2. *Let $j < h$ be integers and $\mathcal{G}$ a commutative layered graph on the layers $V_0, \ldots, V_h$. Write $|V_0| = m$, $|V_j| = s$. There is an $X \subset V_0$, $X \neq \emptyset$, such that*

$$|\operatorname{im}(X, V_h)| \le (s/m)^{h/j}|X|.$$

Exercise 8. Deduce Theorem 1.2.1 from Theorem 1.2.2.

These fundamental results will be proved in the next three sections. Now we mention some important corollaries.

An application of the above theorem to the addition graph yields the following result.

Theorem 1.2.3. *Let $j < h$ be integers, A, B sets in a commutative group and write $|A| = m$, $|A + jB| = \alpha m$. There is an $X \subset A$, $X \neq \emptyset$, such that*

$$|X + hB| \le \alpha^{h/j}|X|.$$

It is not true in general that a proper choice for X is A itself. $|A+hB|$ can be much larger, it can be greater than $m^{1+C(h)}$, even if $\alpha < 2$. X has to be selected carefully. For more details on this phenomenon see Section 1.10.

Since $|X + hB| \ge |hB|$ and $|X| \le m$, we get the following immediate consequence.

Corollary 1.2.4. *Let $j < h$ be integers, A, B sets in a commutative group and write $|A| = m$, $|A + jB| = \alpha m$. We have*

$$|hB| \le \alpha^{h/j} m.$$

This is less general than Theorem 1.1.1, which will be proved in Section 1.8.

In the torsion-free case, using $|X + hB| \ge |X| + |hB| - 1$ instead (see Exercises 1, 5, 6) we obtain the following result, which is stronger for α near to 1 (and gives the correct order of magnitude).

Corollary 1.2.5. *Let $j < h$ be integers, A, B sets in a torsion-free commutative group and write $|A| = m$, $|A + jB| = \alpha m$. We have*

$$|hB| \le \left(\alpha^{h/j} - 1\right) m + 1.$$

Exercise 9. Let A, B be finite sets (in any commutative group), $|A| = n$, $|A+B| = \lambda n$. Show that there is a set T such that $|T| \le \lambda$ and $B \subset T + (A - A)$.

Exercise 10. Let A be a finite set (in any commutative group), $|A| = n$, $|2A| = \lambda n$. From Plünnecke's theorem we know that $|kA| \le \lambda^k n$. For fixed λ this is an exponential function of k. Find a bound of the form $f(k, \lambda)n$, where $f(k, \lambda)$ is, for fixed λ, a polynomial of k.

Prexercise. Let A be a finite set (in any commutative group). Prove that $|kA|$ is, for $k > k_0$, actually equal to some polynomial of k (Hovanskii's theorem). (The polynomial and the value of k_0 depend on the set A.)

Exercise 11. Let $A \subset \mathbb{Z}$. Show that

$$k|(k+1)A| \geq (k+1)|kA| - 1.$$

The commutativity of the addition graph requires two assumptions: one is the commutativity of addition, the other is that the same set B is added repeatedly. Still, an application to different summands and non-commutative operation is possible; we will consider this in Sections 1.6 and 1.11.

Besides the complete addition graph we used above, a more general graph may be useful. Given three sets A, B, C we build on them the *restricted addition graph* as follows. The layers will be $V_0 = A$, $V_1 = (A+B) \setminus C$, $V_j = (A+jB) \setminus (C+(j-1)B)$ for $j > 1$. (We can omit this distinction by defining $0B = \{0\}$.) Again, there is an edge from an $x \in V_j$ to a $y \in V_{j+1}$ if $y = x + b$ with some $b \in B$. The case $C = \emptyset$ returns the complete addition graph. An important case is $C = A$, where in each stage we get the "new sums".

Lemma 1.2.6. *The restricted addition graph is commutative.*

Proof. Consider a typical path of length 2, $x \to y \to z$ with $x \in V_{j-1}$, $y \in V_j$, $z \in V_{j+1}$. This means $y = x + b$, $z = y + b'$ with $b, b' \in B$. We claim that $x \to x+b' \to x+b'+b = z$ is also a path in our graph. To see this we only need to check $x+b' \in V_j$, that is, $x+b' \in A+jB$ and $x+b' \notin C+(j-1)B$. The first follows from $x \in V_{j-1}$, and the negation of the second would imply $z = x+b'+b \in C+jB$, which would contradict $z \in V_{j+1}$.

We apply this substitution to a collection $x \to y \to z_i$ to find distinct y_i with $x \to y_i \to z_i$, and to a collection $x_i \to y \to z$ to find $x_i \to y_i \to z$; this is what we need to establish commutativity. □

By applying Plünnecke's Theorem 1.2.1 to this graph we obtain the following.

Theorem 1.2.7. *Let $j < h$ be integers, A, B, C sets in a commutative group and write $|A| = m$, $|(A+jB) \setminus (C+(j-1)B)| = \alpha m$. There is an $X \subset A$, $X \neq \emptyset$, such that*

$$|(X+hB) \setminus (C+(h-1)B)| \leq \alpha^{h/j}|X|.$$

1.3 Magnification and disjoint paths

In this section we prove the following result.

Theorem 1.3.1. *Let $\mathcal{G}$ be a commutative layered graph with layers $V_0, \ldots, V_h$, $|V_0| = m$. If $\mu_h \geq 1$, then there are m (vertex)-disjoint paths from V_0 to V_h.*

The *outdegree* and *indegree* of a vertex x will be denoted by

$$d^+(x) = d^+(x, \mathcal{G}) = |\{y : x \to y\}|,$$

$$d^-(x) = d^-(x, \mathcal{G}) = |\{y : y \to x\}|.$$

Lemma 1.3.2. *In a commutative graph if $x \to y$, then we have*

$$d^+(x) \ge d^+(y), \tag{1.3.1}$$

$$d^-(x) \le d^-(y). \tag{1.3.2}$$

This is an immediate consequence of the definition of commutativity; we formulate it as a lemma to emphasize its importance.

Definition 1.3.3. Given a graph $\mathcal{G} = (V, E)$ and two sets $X, Y \subset V$ of vertices, the *channel* between them is the graph $\overline{\mathcal{G}}(X, Y) = (\overline{V}, \overline{E})$ defined as follows. We take all directed paths starting in X and ending in Y, put all the vertices on these paths (including the endpoints) into $\overline{V}$ and connect two vertices if they are connected in $\mathcal{G}$. (It is easily seen that in a layered graph this is the same as putting all the vertices on the above-mentioned paths into $\overline{E}$.)

Lemma 1.3.4. *If $\mathcal{G}$ is commutative, so is every channel $\overline{\mathcal{G}}(X, Y)$.*

This is again an immediate consequence of the definition.

We see that the inequalities of Lemma 1.3.2 hold for every channel in a commutative graph, and this is the only property we will use.

Exercise 12. Suppose that a directed graph has the property that inequalities (1.3.1) and (1.3.2) hold for every channel in it. Is it necessarily commutative?

Proof of Theorem 1.3.1. Let r be the maximal number of disjoint paths from V_0 to V_h. By Menger's theorem (see, e.g., Ore [38, Ch. 12], or almost any book on graph theory), we know that there is a separating set S of cardinality r, that is, a set with the property that every path contains a vertex from S.

For a vertex $x \in V_i$ we say that i is its *index*, and we denote it as $i = \operatorname{ind} x$.

From the separating sets of cardinality r we select one for which

$$\sum_{s \in S} \operatorname{ind} s \tag{1.3.3}$$

is minimal. We are going to show that

$$S \subset V_0 \cup V_h. \tag{1.3.4}$$

Let $\Omega_1, \dots, \Omega_r$ be r disjoint paths from V_0 to V_h. S has one element on each Ω_i, say s_i. Assume that (1.3.4) fails, and for some j, $1 \le j \le h-1$, we have

$$|S \cap V_j| = q > 0.$$

We may assume that

$$S \cap V_j = \{s_1, \ldots, s_q\}.$$

For $1 \le i \le q$, let x_i be the predecessor and y_i the successor of s_i on Ω_i, so that

$$\Omega_i = (\ldots, x_i, s_i, y_i, \ldots).$$

The set

$$S' = \{x_1, \ldots, x_q, t_{q+1}, \ldots, t_r\}$$

cannot separate V_0 and V_h because of the minimality of the index sum (1.3.3). Consequently, there is a path Γ from V_0 to V_h that avoids S'. It cannot avoid S, so it contains some vertex from $s_1, \ldots, s_q$, say s_1. The predecessor x of s_1 on Γ is a vertex

$$x \notin \{x_1, \ldots, x_q\}.$$

Write

$$M = \{s_1, \ldots, s_q\},\ M^+ = \{y_1, \ldots, y_q\},\ M^- = \{x_1, \ldots, x_q\},$$

$$M' = M^- \cup \{x\},\ \mathcal{G}' = \overline{\mathcal{G}}(M', M^+).$$

We claim that the set of vertices of $\mathcal{G}'$ is $M' \cup M \cup M^+$. To see this, suppose that there were a path Λ from x or from some x_i to some y_j that avoids M. In this case taking Γ or Ω_i from V_0 to x or x_i, then Λ to y_j, then Ω_j from y_j to V_h we would get a path from V_0 to V_h that avoids S, a contradiction.

Now we have the following chain of inequalities:

$$\begin{aligned}\sum_{i=1}^{q} d^+(x_i, \mathcal{G}') \ge \sum_{i=1}^{q} d^+(s_i, \mathcal{G}') &= \sum_{i=1}^{q} d^-(y_i, \mathcal{G}') \\ &\ge \sum_{i=1}^{q} d^-(s_i, \mathcal{G}') = \sum_{i=1}^{q} d^+(x_i, \mathcal{G}') + d^+(x, \mathcal{G}') \\ &> \sum_{i=1}^{q} d^+(s_i, \mathcal{G}'),\end{aligned}$$

a contradiction. Here the inequalities are applications of Lemma 1.3.2, the equalities express the fact that both sides are enumerations of the number of edges between M and M^+, and between M' and M, respectively. This contradiction proves (1.3.4). The separating property of S means that every upward path from $V_0 \setminus S$ must end in $V_h \cap S$. There are such paths (unless $S \supset V_0$, and in this case we are done), the assumption $\mu_h \ge 1$ means that the number of their possible endpoints is at least $|V_0 \setminus S|$, so we have

$$|V_h \cap S| \ge |V_0 \setminus S| = |V_0| - |V_0 \cap S|,$$

therefore

$$r = |S| = |V_h \cap S| + |V_0 \cap S| \ge |V_0|. \qquad \square$$

Corollary 1.3.5. *In a commutative graph if $\mu_h \geq 1$, then $\mu_j \geq 1$ for $1 \leq j \leq h$.*

Proof. Take a collection of m disjoint paths from V_0 to V_h. For any $X \subset V_0$ the paths that start from X cross V_j in $|X|$ different vertices that all belong to $\mathrm{im}(X, V_j)$. □

This is a particular case of Theorem 1.2.1 that will be used to deduce the general case in the next section.

1.4 Layered product

Definition 1.4.1. Let $\mathcal{G}' = (V', E')$ and $\mathcal{G}'' = (V'', E'')$ be h-layered graphs with layers V_i' and V_i'', respectievely Their *layered product* is the h-layered graph on the layers $V_i = V_i' \times V_i''$, and two vertices $(x', x'') \in V_i$ and $(y', y'') \in V_{i+1}$ are connected if both $x' \to y'$ and $x'' \to y''$. This graph will be denoted by $\mathcal{G} = \mathcal{G}'\mathcal{G}''$. For repeated products with identical factors the usual power notation $\mathcal{G}^n$ will be used.

Observe that this is a proper subgraph of the usual product of these graphs.

Lemma 1.4.2. *The layered product of commutative graphs is commutative as well.*

This is an immediate consequence of the definitions.

Lemma 1.4.3. *Magnification ratios are multiplicative: If $\mathcal{G}, \mathcal{G}', \mathcal{G}''$ are h-layered graphs with magnification ratios μ_i, μ_i', μ_i'', respectively, and $\mathcal{G} = \mathcal{G}'\mathcal{G}''$, then $\mu_i = \mu_i'\mu_i''$ for all i.*

Proof. The inequality $\mu_i \leq \mu_i'\mu_i''$ is obvious: If μ_i' is attained at a subset $Z' \subset V_0$ and μ_i'' at $Z'' \subset V_0''$, then $Z = Z' \times Z'' \subset V_0$ gives the upper bound.

To prove the reverse inequality first consider a special case: $h = 1$, $\mathcal{G}''$ consists of two copies W_0, W_1 of a set W, and from a vertex $w \in W_0$ there is a unique edge to the corresponding vertex in W_1, consequently $\mu_1'' = 1$.

Take a set

$$X \subset V_0 = V_0' \times W.$$

We have

$$X = \bigcup_{w \in W} X_w,$$

where X_w is the set of those elements of X whose second coordinate is w. We obtain

$$|\mathrm{im}(X, V_1)| = \sum |\mathrm{im}(X_w, V_1)| \geq \sum \mu_1' |X_w| = \mu_1' |X|$$

as desired.

Now consider the general case. We construct an auxiliary 3-layered graph $\mathcal{H}$ on the layers

$$U_0 = V_0' \times V_0'' = V_0, \; U_1 = V_j' \times V_0'', \; U_2 = V_j' \times V_j'' = V_j.$$

We connect an $(x', x'') \in U_0$ to $(y', x'') \in U_1$ (second coordinates equal) if there is a path from x' to y' in $\mathcal{G}'$, and we connect $(y', x'') \in U_1$ to $(y', y'') \in U_2$ (first coordinates equal) if there is a path from x'' to y'' in $\mathcal{G}''$. Clearly, from (x', x'') to (y', y'') in $\mathcal{H}$ if and only if there is one in $\mathcal{G}$, so

$$\mu_2(\mathcal{H}) = \mu_j(\mathcal{G}).$$

The subgraphs $\mathcal{H}_1$, spanned by $U_0 \cup U_1$, and $\mathcal{H}_2$, spanned by $U_1 \cup U_2$, fall into the particular case treated above, which means

$$\mu_1(\mathcal{H}_1) = \mu(U_0, U_1) \geq \mu_j(\mathcal{G}'), \ \mu_1(\mathcal{H}_2) = \mu(U_1, U_2) \geq \mu_j(\mathcal{G}'').$$

Finally we have

$$\mu_2(\mathcal{H}) = \mu(U_0, U_2) \geq \mu(U_0, U_1)\mu(U_1, U_2) \geq \mu_j(\mathcal{G}')\mu_j(\mathcal{G}'')$$

by the previous inequality and this completes the proof of the reverse inequality $\mu_i \geq \mu_i' \mu_i''$. □

1.5 The independent addition graph

We define the *independent addition graph* $\mathcal{I}_{nh}$ as follows. Take a set B (say, of integers), $|B| = n$, such that all h-fold sums $b_1 + \cdots + b_h$, $b_i \in B$, are different, unless they are rearrangements of each other, and $A = \{0\}$, and build the addition graph on them. Since $|V_0| = 1$, the j-th magnification ratio of this graph is clearly

$$\mu_j(\mathcal{I}_{nh}) = |V_j| = |jB|\,.$$

Exercise 13. Calculate $|jB|$ as a function of j and n.

Since the number of formal j-fold sums is n^j and a sum occurs at most $j!$ times, we have

$$\frac{n^j}{j!} \leq \mu_j(\mathcal{I}_{nh}) = |jB| \leq n^j. \tag{1.5.1}$$

We shall also use the inverse of this graph. Here we have

$$\mu_h(\hat{\mathcal{I}}_{nh}) = |hB|^{-1} \geq n^{-h},$$

and for $j < h$

$$\mu_j(\hat{\mathcal{I}}_{nh}) \leq \frac{|(h-j)B|}{|hB|} \leq h! n^{-j}.$$

Exercise 14. Find the exact value of $\mu_j(\hat{\mathcal{I}}_{nh})$.

These graphs will be used in the proof of Plünnecke's theorem.

Proof of Theorem 1.2.1. We want to prove that $\mu_h \le \mu_j^{h/j}$. We know that $\mu_j \ge 1$ whenever $\mu_h \ge 1$, and this settles the case $\mu_h = 1$.

Take now a graph $\mathcal{G}$ with $\mu_h < 1$. Consider the layered product $\mathcal{G}^* = \mathcal{G}^k \mathcal{I}_{nh}$. If we select k and n so that

$$\mu_h^k \frac{n^h}{h!} \ge 1,$$

then (with the natural notation) we have $\mu_h^* \ge 1$, hence $\mu_j^* \ge 1$, which then implies

$$\mu_j^k n^j \ge 1,$$

using the appropriate part of inequality (1.5.1).

To optimize this take

$$n = 1 + \left[\left(h! \mu_h^{-k} \right)^{1/h} \right] \le 2h!^{1/h} \mu_h^{-k/h} = c_h \mu_h^{-k/h}.$$

The previous inequality gives

$$\mu_j \ge n^{-j/k} \ge c_h^{-1j/k} \mu_h^{j/k} \to \mu_h^{j/h}$$

as $k \to \infty$.

Finally assume that $\mu_h > 1$. Consider the layered product $\mathcal{G}^* = \mathcal{G}^k \hat{\mathcal{I}}_{nh}$. If we select k and n so that

$$\mu_h^k n^{-h} \ge 1,$$

then similarly we get $\mu_h^* \ge 1$ and hence $\mu_j^* \ge 1$, which then implies

$$\mu_j^k h! n^{-j} \ge 1,$$

using inequality (1.5.1) again.

Our choice of n is now

$$n = \left[\mu_h^{k/h} \right],$$

and the previous inequality gives

$$\mu_j \ge h!^{-1/k} n^{j/k} \to \mu_h^{j/h}$$

as $k \to \infty$. □

1.6 Different summands

An application to different summands is less straightforward, however, the case $j = 1$ of Theorem 1.2.3 can be extended in this way as follows (see [49]).

Theorem 1.6.1. *Let A, $B_1, \ldots, B_h$ be sets in a commutative group G and write $|A| = m$, $|A + B_i| = \alpha_i m$. There is an $X \subset A$, $X \ne \emptyset$, such that*

$$|X + B_1 + \cdots + B_h| \le \alpha_1 \alpha_2 \cdots \alpha_h |X|. \tag{1.6.1}$$

Proof. Take auxiliary sets $T_1, \dots, T_h \subset G$ such that $|T_i| = n_i$ (which will be specified soon) and all the sums

$$y + t_1 + \dots + t_h,\ y \in A + B_1 + \dots + B_h,\ t_i \in T_i$$

are distinct. (This may be impossible in a finite group; in this case first embed it into an infinite one.) Now apply case $j = 1$ of Theorem 1.2.3 to the sets A and

$$B = \bigcup (B_i + T_i).$$

Observe that

$$|A + B| \le \sum |A + B_i + T_i| \le \sum |A + B_i|\,|T_i| = m \sum n_i \alpha_i.$$

We obtain the existence of a set $X \subset A$ such that

$$|X + hB| \le |X| \left(\sum n_i \alpha_i\right)^h.$$

On the other hand, $X + hB \supset X + B_1 + \dots + B_h + T_1 + \dots + T_h$, consequently we have

$$|X + hB| \ge |X + B_1 + \dots + B_h|\, n_1 \cdots n_h.$$

A comparison of these inequalities gives

$$|X + B_1 + \dots + B_h| \le \left(\sum n_i \alpha_i\right)^h (n_1 \cdots n_h)^{-1} |X|. \tag{1.6.2}$$

To make this quotient small we put $n_i = n/\alpha_i$ with suitable n; since the numbers α_i are rational, this can be achieved with integers. Then inequality (2.1.1) turns into

$$|X + B_1 + \dots + B_h| \le h^h \prod \alpha_i |X|, \tag{1.6.3}$$

which is worse than we claimed by a factor h^h.

To remove this factor we consider two 1-layered graphs. The first, say $\mathcal{G}$, is built on the sets A and $A + B_1 + \dots + B_h$ in the natural way. The other, say $\mathcal{G}'$, is built similarly from the direct powers

$$A^k = A \times \dots \times A, B_1^k, \dots, B_h^k$$

considered as sets in the k-th direct power of our initial group. Let μ and μ' be the magnification ratios of these graphs. The previous argument told us that $\mu \le h^h \prod \alpha_i$, and the same, when applied to the sets A^k, B_j^k instead, gives

$$\mu' \le h^h \left(\prod \alpha_j\right)^k.$$

Now observe that $\mathcal{G}'$ is isomorphic to $\mathcal{G}^k$, so $\mu' = \mu^k$ by Lemma 1.4.3. The above inequality thus reduces to

$$\mu \le h^{h/k} \prod \alpha_j.$$

with an arbitrary k. As $k \to \infty$, we obtain (1.6.1). □

The case of general j can also be extended (see the paper by Gyarmati, Matolcsi and Ruzsa, in preparation, [19]).

Theorem 1.6.2. *Let $j < h$ be integers, and let A, $B_1, \dots, B_h$ be finite sets in a commutative group G. Let $K = \{1, 2, \dots, h\}$, and for any $I \subset K$ put*

$$B_I = \sum_{i \in I} B_i,$$

$$|A| = m, \quad |A + B_I| = \alpha_I m.$$

Write

$$\beta = \left(\prod_{L \subset K, |L| = j} \alpha_L \right)^{(j-1)!(h-j)!/(h-1)!} . \tag{1.6.4}$$

There exists an $X \subset A$, $X \neq \emptyset$, such that

$$|X + B_K| \leq \beta |X|. \tag{1.6.5}$$

1.7 Plünnecke's inequality with a large subset

We show an extension of Theorem 1.2.2 with a bound on the size of the selected subset.

Theorem 1.7.1. *Let $j < h$ be integers, $\mathcal{G}$ a commutative layered graph on the layers $V_0, \dots, V_h$. Write $|V_0| = m$, $|V_j| = s$, $\gamma = h/j$. Let an integer k be given, $1 \leq k \leq m$. There is an $X \subset V_0$, $|X| \geq k$, such that*

$$|\operatorname{im}(X, V_h)| \leq \left(\frac{s}{m}\right)^\gamma + \left(\frac{s}{m-1}\right)^\gamma + \dots + \left(\frac{s}{m-k+1}\right)^\gamma + (|X| - k) \left(\frac{s}{m-k+1}\right)^\gamma . \tag{1.7.1}$$

Proof. We use induction on k. The case $k = 1$ is Theorem 1.2.2.

Assume we know it for k; we prove it for $k + 1$. The inductive assumption gives us a set X, $|X| \geq k$, with a bound on $|\operatorname{im}(X, V_h)|$ as given by (1.7.1). We want to find a set X' with $|X'| \geq k + 1$ and

$$|\operatorname{im}(X', V_h)| \leq \left(\frac{s}{m}\right)^\gamma + \left(\frac{s}{m-1}\right)^\gamma + \dots + \left(\frac{s}{m-k}\right)^\gamma + (|X'| - k - 1) \left(\frac{s}{m-k}\right)^\gamma . \tag{1.7.2}$$

If $|X| \geq k + 1$, we can put $X' = X$. If $|X| = k$, we apply Theorem 1.2.2 to the graph obtained from $\mathcal{G}$ by omitting the vertices in X. This yields a set $Y \subset V_0 \setminus X$ such that

$$|\operatorname{im}(Y, V_h)| \leq \left(\frac{s}{m-k}\right)^\gamma |Y|$$

and we put $X' = X \cup Y$. □

The following variant will be more comfortable for calculations.

Theorem 1.7.2. *Let $j < h$ be integers, $\mathcal{G}$ a commutative layered graph on the layers $V_0, \ldots, V_h$. Write $|V_0| = m$, $|V_j| = s$, $\gamma = h/j$. Let a real number t be given, $0 \le t < m$. There is an $X \subset V_0$, $|X| > t$, such that*

$$|\operatorname{im}(X, V_h)| \le \frac{s^\gamma}{\gamma}\left(\frac{1}{(m-t)^{\gamma-1}} - \frac{1}{m^{\gamma-1}}\right) + (|X| - t)\left(\frac{s}{m-t}\right)^\gamma. \qquad (1.7.3)$$

Proof. We apply Theorem 1.7.1 with $k = [t] + 1$. The right side of (1.7.3) can be written as $s^\gamma \int_0^{|X|} f(x)\,dx$, where $f(x) = (m-x)^{-\gamma}$ for $0 \le x \le t$, and $f(x) = (m-t)^{-\gamma}$ for $t < x \le |X|$. Since f is increasing, the integral is $\ge f(0) + f(1) + \cdots + f(|X| - 1)$. This exceeds the right side of (1.7.1) by a termwise comparation. ☐

We state the consequences of this result for the complete and restricted addition graphs.

Theorem 1.7.3. *Let $j < h$ be integers, A, B sets in a commutative group, and write $|A| = m$, $|A + jB| = s$, $\gamma = h/j$. Let a real number t be given, $0 \le t < m$. There is an $X \subset A$, $|X| > t$, such that*

$$|X + hB| \le \frac{s^\gamma}{\gamma}\left(\frac{1}{(m-t)^{\gamma-1}} - \frac{1}{m^{\gamma-1}}\right) + (|X| - t)\left(\frac{s}{m-t}\right)^\gamma.$$

Theorem 1.7.4. *Let $j < h$ be integers, A, B, C sets in a commutative group, and write $|A| = m$, $|(A + jB) \setminus (C + (j-1)B)| = s$, $\gamma = h/j$. Let a real number t be given, $0 \le t < m$. There is an $X \subset A$, $|X| > t$, such that*

$$|(X + hB) \setminus (C + (h-1)B)| \le \frac{s^\gamma}{\gamma}\left(\frac{1}{(m-t)^{\gamma-1}} - \frac{1}{m^{\gamma-1}}\right) + (|X| - t)\left(\frac{s}{m-t}\right)^\gamma.$$

We state separately the case $j = 1$, $h = 2$ which will be applied in what follows.

Corollary 1.7.5. *Let A, B be sets in a commutative group and write $|A| = m$, $|A + iB| = s$. Let a real number t be given, $0 \le t < m$. There is an $X \subset A$, $|X| > t$, such that*

$$|X + 2B| \le \frac{s^2}{(m-t)^2}\left(|X| - \frac{t(t+m)}{2m}\right).$$

Corollary 1.7.6. *Let A, B, C be sets in a commutative group and write $|A| = m$, $|(A + B) \setminus C| = s$. Let a real number t be given, $0 \le t < m$. There is an $X \subset A$, $|X| > t$ such that*

$$|(X + 2B) \setminus (C + B)| \le \frac{s^2}{(m-t)^2}\left(|X| - \frac{t(t+m)}{2m}\right).$$

Theorem 1.6.1 can also be modified to yield large subsets.

Theorem 1.7.7. *Let A, $B_1, \dots, B_h$ be sets in a commutative group G and write $|A| = m$, $|A + B_i| = \alpha_i m$. Let a real number t be given, $0 \le t < m$. There is an $X \subset A$, $X \ne \emptyset$, such that*

$$|X+B_1+\cdots+B_h| \le \alpha_1\alpha_2\cdots\alpha_h m^h \left(\frac{1}{h}\left(\frac{1}{(m-t)^{h-1}} - \frac{1}{m^{h-1}}\right) + \frac{(|X|-t)}{(m-t)^{h-1}}\right). \tag{1.7.4}$$

The proof follows that of Theorem 1.7.1 with the difference that in the inductive step we apply Theorem 1.6.1 to the sets $A \setminus X$, $B_1, \dots, B_h$. The available upper estimate for $|(A \setminus X) + B_i|$ is naturally $\alpha_i m$.

1.8 Sums and differences

With Plünnecke's method one can get various inequalities for cardinalities of sumsets, but it stops to work when differences are also involved (we shall give reasons why).

As far as we know the first inequality connecting sums and differences is due to Freiman and Pigaev [12]. They proved that

$$|A+A|^{3/4} \le |A-A| \le |A+A|^{4/3}. \tag{1.8.1}$$

We prove the following (see [45]).

Theorem 1.8.1. *Let A, Y, Z be finite sets in a (not necessarily commutative) group. We have*

$$|A||Y-Z| \le |A-Y||A-Z|. \tag{1.8.2}$$

Prexercise. Try to prove this inequality instead of reading the proof.

Exercise 15. Let

$$A = \left\{(x_1, \dots, x_d) \in \mathbb{Z}^d : x_i \ge 0, \sum x_i \le n\right\}.$$

a) Calculate $|A|$.

b) What are the elements of $A - A$?

c) What are the limits of $|A + A|/|A|$ and $|A - A|/|A|$ as $n \to \infty$ for fixed d?

Proof. We will map the pairs $(a, x), a \in A, x \in Y - Z$, into $(A - Y) \times (A - Z)$ in an injective way.

List the elements of Y somehow, say $y_1, \dots, y_k$. Now given a pair (a, x), from all possible representations of x in the form $x = y - z$, $y \in Y$, $z \in Z$, select the one for which $y = y_i$ with minimal i, and map this pair into $(a - y, a - z)$. Take another pair (a', x') with representation $x' = y' - z'$. If we have

$$a - y = a' - y',\ a - z = a' - z',$$

then subtracting these equations (carefully in the non-commutative case!) we get $y - z = y' - z'$. Since both representations are minimal in the above sense, we conclude that $y = y'$, then $z = z'$ and $a = a'$. □

Substituting $Y = -Y$ and $Z = -Z$, we obtain the following version:

$$|A||Y - Z| \le |Y + A||Z + A|. \tag{1.8.3}$$

Inequality (1.8.2) has the following interpretation (me cannot recall who made this observation). Define

$$\rho(X, Y) = \log \frac{|X - Y|}{\sqrt{|X|\,|Y|}}.$$

Then (1.8.2) can be written as

$$\rho(Y, Z) \le \rho(Y, A) + \rho(A, Z),$$

a triangle-inequality-like property. ρ is also symmetric. A marked difference from distances is that $\rho(X, X)$ is typically positive.

Exercise 16. Show that $\rho(X, Y) \ge 0$, and find the cases of equality.

Substituting $Y = Z = -A$ in (1.8.2) we obtain the following inequality.

Corollary 1.8.2. *If* $|A| = m$, $|2A| \le \alpha m$, *then* $|-A+A| \le \alpha^2 m$ *and* $|A-A| \le \alpha^2 m$.

The second inequality above follows from the first via replacing A by $-A$ and observing that $2(-A) = -(2A)$.

We can also substitute $Y = Z = -2A$ to obtain the following inequality.

Corollary 1.8.3. *If* $|A| = m$, $|3A| \le \alpha m$, *then* $|-2A + 2A| \le \alpha^2 m$.

The exponent 2 in Corollary 1.8.2 is best possible (though an improvement to something like $\alpha^2/(\log \alpha)^c$ is conceivable). This can be seen by considering the lattice points inside a d-dimensional simplex

$$\left\{(x_1, \dots, x_d) \in \mathbb{R}^d : x_i \ge 0, \sum x_i \le n\right\},$$

where $2^d \approx \alpha$. Denoting the volume of this simplex by v $(= n^d/d!)$, the number of lattice points is about v, the size of the sumset is about the volume of this simplex dilated by 2, that is, $2^d v$, while the size of the difference set is about the volume of the difference set of this simplex, which is easily calculated to be $\binom{2d}{d} v$. Note that for a convex set the volume of the sumset is always 2^d times the original; the volume of the difference set varies and, by a theorem of C. A. Rogers and G. G. Shephard [43], the simplex yields the maximum.

This example is analyzed in detail by Hennecart, Robert and Yudin [23]; they attribute the underlaying idea to Freiman and Pigaev's above-mentioned paper [12]. See also A. Granville's survey [16, Section 1.5].

One difference from the Plünnecke inequalities is the non-commutative nature of the above result. From Corollary 1.2.4 we obtain a similar implication: If $|A| = m$, $|A - A| \le \alpha m$, then $|2A| \le \alpha^2 m$. This fails in non-commutative groups; see Section 1.11.

Another difference is the following. To go to differences one would require the case "$h = -1$" of Theorem 1.2.3, which might be expected to sound as follows:

> "If $|A| = n$, $|A + B| = \alpha n$, then there is a nonempty $X \subset A$ such that $|X - B| \le \alpha' |X|$, with α' depending only on α."

This is, however, false; we have the following results by Gyarmati, Konyagin and Ruzsa [18].

Theorem 1.8.4. *Let $\alpha > 2$. Then for any $c < \frac{\sqrt{2}\log 2}{\sqrt{3}}$ and infinitely many m, there exist two sets A and B such that $|A| = m$, $|A + B| \le \alpha m$ and for any nonempty $X \subset A$, one has*

$$\frac{|X - B|}{|X|} \ge \exp\left(c\sqrt{(\log(\alpha/2))(\log m)(\log\log m)^{-1}}\right).$$

Theorem 1.8.5. *Let A and B be nonempty and finite subset of some abelian group such that $|A| = m$, $|A + B| \le \alpha m$. Then there exists some nonempty subset X of A such that*

$$\frac{|X - B|}{|X|} \le \alpha \exp\left(2\sqrt{(\log \alpha)(\log m)}\right). \tag{1.8.4}$$

Inequality (1.8.2) together with Plünnecke's can be used to deduce the basic Theorem 1.1.1, which is sufficient for most of the applications ([52], Lemma 3.3) and which we repeat below.

Theorem 1.8.6. *Let A, B be finite sets in a commutative group and write $|A| = m$, $|A + B| = \alpha m$. For arbitrary non-negative integers k, l we have*

$$|kB - lB| \le \alpha^{k+l} m. \tag{1.8.5}$$

Proof. By symmetry we may assume $k \le l$. Assume also $k \ge 1$, since the case $k = 0$ is contained in Corollary 1.2.4. An application of Theorem 1.2.3 with $j = 1, h = k$ gives us a set $X \subset A$ such that

$$|X + kB| \le \alpha^k |X|.$$

Another application with $j = k$, $h = l$ and X in the place of A gives a set $X' \subset X$ such that

$$|X' + lB| \le \alpha^l |X'|.$$

Now apply Theorem 1.8.2 to the sets $-X'$, kB and lB to obtain

$$|X'|\,|kB - lB| \le |X' + kB|\,|X' + lB| \le \alpha^{k+l} |X'|\,|X|.$$

Now we divide by $|X'|$ and use $|X| \le m$ to get inequality (1.8.5). □

The sum-sum analogue of Theorem 1.8.1 can be deduced from Plünnecke's inequality.

Theorem 1.8.7. *In any commutative group we have*

$$|A||Y+Z| \le |A+Y||A+Z|. \tag{1.8.6}$$

Proof. Indeed, applying Theorem 1.6.1 we get a set $X \subset A$ such that

$$|X+Y+Z| \le |X| \frac{|A+Y|}{|A|} \frac{|A+Z|}{|A|},$$

and to obtain (1.8.6) we just have to use $|X+Y+Z| \ge |Y+Z|$ and $|X| \le |A|$. □

Exercise 17. Prove Freiman and Pigaev's inequality (1.8.1).

1.9 Double and triple sums

We present an inequality which sometimes nicely complements Plünnecke's inequality.

Theorem 1.9.1. *Let X, Y, Z be finite sets in a commutative group. We have*

$$|X+Y+Z|^2 \le |X+Y||Y+Z||X+Z|. \tag{1.9.1}$$

This inequality may be extended to the non-commutative case as follows.

Theorem 1.9.2. *Let X, Y, Z be finite sets in a not necessarily commutative group. We have*

$$|X+Y+Z|^2 \le |X+Y||Y+Z| \max_{y \in Y} |X+y+Z|. \tag{1.9.2}$$

Proof. We use induction on $|Y|$. For $|Y| = 1$, (1.9.2) reduces to the obvious inequality

$$|X+y+Z| \le |X||Z|.$$

Assume now that we know (1.9.2) for smaller sets. Fix y as the element of Y which maximizes $|X+y+Z|$. Write

$$|X+y+Z| = m, Y \setminus \{y\} = Y',$$

$$|(X+Y+Z) \setminus (X+Y'+Z)| = a, |(X+Y) \setminus (X+Y')| = b, |(Y+Z) \setminus (Y'+Z)| = c.$$

With these notations, (1.9.2) can be rewritten as

$$(|X+Y'+Z|+a)^2 \le m(|X+Y'|+b)(|Y'+Z|+c). \tag{1.9.3}$$

We shall obtain (1.9.3) as the sum of the following three inequalities:

$$|X+Y'+Z|^2 \le m|X+Y'||Y'+Z|, \tag{1.9.4}$$

$$2a|X + Y' + Z| \le m(c|X + Y'| + b|Y' + Z|), \tag{1.9.5}$$

$$a^2 \le mbc. \tag{1.9.6}$$

Of these inequalities (1.9.4) follows from the induction hypothesis.

Clearly, every element of $(X+Y+Z)\setminus(X+Y'+Z)$ is of the form $x+y+z$ with $x \in X$, $z \in Z$, hence $a \le m$. We can map this set into the Cartesian product of $(X+Y)\setminus(X+Y')$ and $(Y+Z)\setminus(Y'+Z)$ by mapping a typical element $x+y+z$ into the pair $(x+y, y+z)$. This pair determines $x+y+z$ uniquely and clearly $x+y \notin X+Y'$ as otherwise we would have $x+y+z \in X+Y'+Z$; similarly $y+z \notin Y'+Z$. This mapping shows that $a \le bc$. The product of these inequalities gives (1.9.6).

By multiplying inequalities (1.9.4) and (1.9.6) and taking the square root we obtain

$$a|X + Y' + Z| \le m\sqrt{bc|X + Y'||Y' + Z|};$$

(1.9.5) now follows from the arithmetic-geometric mean inequality. □

In the commutative case this inequality can be extended to more than three sets as follows (see Gyarmati, Matolcsi and Ruzsa [20]).

Theorem 1.9.3. *Let $A_1, \dots, A_k$ be finite, nonempty sets in an arbitrary commutative semigroup. Put*

$$S = A_1 + \dots + A_k,$$

$$S_i = A_1 + \dots + A_{i-1} + A_{i+1} + \dots + A_k.$$

We have

$$|S| \le \left(\prod_{i=1}^{k} |S_i|\right)^{\frac{1}{k-1}}. \tag{1.9.7}$$

Curiously, one of the arguments relies on invertibility, the other on commutativity, so we do not have any result for non-commutative semigroups. Neither could we extend the above non-commutative argument for more than three summands, and hence the following question remains open.

Problem 1.9.4. Let $A_1, \dots, A_k$ be finite, nonempty sets in an arbitrary non-commutative group. Put

$$S = A_1 + \dots + A_k,$$

$$n_i = \max_{a \in A_i} |A_1 + \dots + A_{i-1} + a + A_{i+1} + \dots + A_k|.$$

Is it true that

$$|S| \le \left(\prod_{i=1}^{k} n_i\right)^{\frac{1}{k-1}} ? \tag{1.9.8}$$

We finish this section by a meditation on the sizes of $2A$ and $3A$.

Write $|A| = m$, $|2A| = n$. Corollary 1.2.4 implies $|3A| \le n^3/m^2$, and Theorem 1.9.1 implies $|3A| \le n^{3/2}$. The first is better for $n \le m^{4/3}$, the second for larger values. The two together describe the maximal possible value of $|3A|$ up to a constant.

Theorem 1.9.5. *Let m, n be positive integers satisfying $m \le n \le m^2$. There is a set A of integers such that $|A| \asymp m$, $|2A| \asymp n$ and*

$$|3A| \asymp \min\left(n^3/m^2, n^{3/2}\right).$$

Proof. We construct A in $\mathbb{Z}^3$. Take two integers k, l such that $k \le l \le k^3$ and put

$$A_1 = \{(x,y,z) : \ 0 \le x,y,z < k\},$$

$$A_2 = \{(x,0,0),(0,x,0),(0,0,x) : \ 0 \le x < l\},$$

and $A = A_1 \cup A_2$.

We have $m = |A| = k^3 + 3(l-k) \asymp k^3$, so the proper choice is $k \sim m^{1/3}$. Further $2A = 2A_1 \cup (A_1 + A_2) \cup 2A_2$. The cardinality of the parts is of order k^3, $k^2 l$ and l^2, respectively. The first is always smaller than the second, hence

$$n = |2A| \asymp \max(k^2 l, l^2);$$

the threshold of behavior is at $l = k^2$. Hence the proper choice of l is

$$l \sim \min(\sqrt{n}, n/m^{2/3})$$

and the claim follows from the fact that $|3A| \ge |3A_2| \ge l^3$. □

1.10 $A + B$ and $A + 2B$

In this section we consider the following problem. Let $|A| = m$, $|A+B| = \alpha m$. How large can $|A+2B|$ be? In the case $B = A$ the answer was given at the end of the last section. A similar bound can be found by Plünnecke's method if A and B are about the same size. Without any assumption on B, however, the situation changes.

An application of Theorem 1.9.1 immediately yields

$$|A+2B| \le |A+B|\sqrt{|2B|}; \tag{1.10.1}$$

this inequality was already proved differently in [58, Theorem 7.2]. To estimate $|2B|$ we can use Corollary 1.2.4 to obtain $|2B| \le \alpha^2 m$; combined with (1.10.1) we get

$$|A+2B| \le \alpha^2 m^{3/2}.$$

In [58, Theorem 7.1], examples are given (for every rational α and infinitely many m) such that

$$|A+2B| \geq \left(\frac{\alpha-1}{4}\right)^2 m^{3/2}. \tag{1.10.2}$$

These results describe the order of magnitude for fixed $\alpha > 1$ unless α is near to 1.

We now explore what happens for small values of α. In the extremal case $\alpha = 1$ clearly also $A + 2B = m$. The transition is somewhat less clear.

Theorem 1.10.1. *Let A, B be finite sets in a commutative group G, $|A| = m$, $|A+B| = \alpha m$, $1 < \alpha \leq 2$. We have*

$$|A+2B| \leq \alpha m + \frac{3}{2}(\alpha-1)m\sqrt{|2B|}, \tag{1.10.3}$$

consequently

$$|A+2B| \leq \alpha m + 3(\alpha-1)m^{3/2}; \tag{1.10.4}$$

if G is torsion-free, then

$$|A+2B| \leq \alpha m + 3(\alpha-1)^{3/2}m^{3/2}. \tag{1.10.5}$$

Proof. We apply Corollary 1.7.6 with the choice $C = A+b$, where b is an arbitrary element of B. The s in the hypothesis will be

$$s = |(A+B) \setminus (A+b)| = (\alpha-1)m,$$

and we obtain (for every $0 \leq t < m$) the existence of an $X \subset A$, $|X| > t$ such that

$$|(X+2B) \setminus (C+B)| \leq \frac{s^2}{(m-t)^2}\left(|X| - \frac{t(t+m)}{2m}\right).$$

Since $|C+B| = |A+B| = \alpha m$, this implies

$$|(X+2B)| \leq \alpha m + \frac{s^2}{(m-t)^2}\left(|X| - \frac{t(t+m)}{2m}\right).$$

For $A \setminus X$ we use an obvious estimate:

$$|(A \setminus X) + 2B| \leq |A \setminus X||2B| = (m - |X|)|2B|,$$

and sum the last two inequalities to get

$$|(A+2B)| \leq \alpha m + \left(\frac{s^2}{(m-t)^2} - |2B|\right)|X| + m|2B| - \frac{s^2}{(m-t)^2}\frac{t(t+m)}{2m}. \tag{1.10.6}$$

We choose t so that the coefficient of $|X|$ vanishes; that is,

$$\frac{s^2}{(m-t)^2} = |2B|. \tag{1.10.7}$$

Such a t exists in the interval $(0,m)$ as long as $|2B| \geq s^2/m^2 = (\alpha-1)^2$, which certainly holds under our assumption that $\alpha \leq 2$. (We do not really need this restriction; however, for $\alpha > 2$ this estimate is weaker than (1.10.1), due to the factor 3/2.) With this choice, (1.10.6) becomes

$$|(A+2B)| \leq \alpha m + |2B| \left(m - \frac{t(t+m)}{2m}\right) = \alpha m + |2B| \frac{(m-t)(2m+t)}{2m}. \quad (1.10.8)$$

We estimate $2m+t$ by $3m$, and we express $m-t$ by (1.10.7):

$$m - t = \frac{s}{\sqrt{|2B|}} = \frac{(\alpha-1)m}{\sqrt{|2B|}}.$$

After these substitutions (1.10.8) becomes (1.10.3).

To deduce (1.10.4) we use Corollary 1.2.4 and $\alpha \leq 2$.

To deduce (1.10.5) we use Corollary 1.2.5: In a torsion-free group

$$|2B| \leq 1 + (\alpha^2 - 1)m \leq 4(\alpha-1)m,$$

since $\alpha = |A+B|/m \geq 1 + 1/m$, and put this into (1.10.3). □

We remark that the summand αm in these estimates can actually be the main term, as α may be as small as $1 + O(1/m)$. In the general estimate (1.10.4), the threshold is $1+O(m^{-1/2})$, in the torsion-free estimate (1.10.5), it is $1+O(m^{-1/3})$.

Still there is a gap between the exponent 2 of $\alpha - 1$ in the example (1.10.2) and 3/2 in the upper estimate (1.10.5). We now show by an example that the exponent 1 of $\alpha - 1$ for general groups in (1.10.4) is exact. Take a group G which has two k-element subgroups H_1, H_2 such that $H_1 \cap H_2 = \{0\}$. Write $H = H_1 + H_2$ and let

$$A = H \cup \{a_1, \ldots, a_t\}, \quad B = H_1 \cup H_2,$$

where $a_1, \ldots, a_t$ lie in different nonzero cosets of H. Observe that $2B = H$. We have

$$m = |A| = k^2 + t,$$

$$\alpha m = |A+B| = k^2 + t(2k-1),$$

$$\alpha - 1 = \frac{2t(k-1)}{m},$$

$$|A+2B| = (t+1)k^2 = \alpha m + (\alpha-1)(k-1)m/2.$$

Since $k - 1 \sim \sqrt{m}$ as long as $t = o(k^2)$ (and in the interesting case $t = O(k)$), the only difference from the upper estimate (1.10.4) is a factor of 6.

1.11 On the non-commutative case

Our attention was focused on commutative groups, with special emphasis on integers. At several places, namely at Theorems 1.8.1 and 1.9.2, we mentioned the possibility of a non-commutative extension. We now explore the limits of this extension.

First we collect some examples that show how certain attempts of extension fail. These examples use a free group, which is "very non-commutative"; it is possible that for groups "nearer" to commutative ones in some sense, some results can be extended.

First recall some results that did not require commutativity. Thoerem 1.8.1 told us that

$$|A||Y-Z| \le |A-Y||A-Z|. \tag{1.11.1}$$

This had the following consequences (Corollary 1.8.2): If $|A| = m$, $|2A| \le \alpha m$, then $|-A+A| \le \alpha^2 m$ and $|A-A| \le \alpha^2 m$.

We first show that the two cases in the above corollary are not superfluous, in a non-commutative group, $|-A+A|$ and $|A-A|$ can be very different (of course, without the assumption on $2A$).

Indeed, take a free group generated by the elements a, b and put

$$A = \{ia+b : 1 \le i \le m\} \cup \{ia : 1 \le i \le m\}.$$

Then $|A| = 2m$ and

$$-A = \{-b-ja : 1 \le j \le m\} \cup \{-ja : 1 \le j \le m\}.$$

Here $A-A$ contains the $2m^2$ different elements $ia \pm b - ja$, while

$$-A+A = \{(i-j)a\} \cup \{(i-j)a+b\} \cup \{-b+(i-j)a\} \cup \{-b+(i-j)a+b\},$$

altogether $4m$ elements.

So if the sumset is small, both difference sets are small without commutativity. In the commutative case from Corollary 1.2.4 we obtain a similar implication: If $|A| = m$, $|A-A| \le \alpha m$, then $|2A| \le \alpha^2 m$. This also fails in non-commutative groups. As above, take a free group with generators a, b and put

$$A = \{ia+b : 1 \le i \le m\}.$$

Then both difference sets $A-A$ and $-A+A$ have $2m-1$ elements, while $|2A| = m^2$.

Between double and triple sums we had the following inequality without commutativity (Theorem 1.9.2):

$$|X+Y+Z|^2 \le |X+Y||Y+Z| \max_{y \in Y} |X+y+Z|. \tag{1.11.2}$$

We show by an example that the maximum cannot be omitted and cannot even be replaced by an average, even in the case of identical sets. As before, take a free group with generators a, b and put

$$X = Y = Z = \{a, 2a, \dots, na, b\}.$$

We have $|X| = n+1$, $|2X| = 4n$ and $|3X| > n^2$ since all the elements $ia+b+ja$, $1 \le i, j \le n$, are distinct. From the $n+1$ sets $X+y+X$, $y \in X$, only one is of size n^2, namely the one with $y = b$, all the others have $O(n)$ elements.

For a constrast, by applying Corollary 1.2.4, with similar values of $|X|$ and $|2X|$ in a commutative group we would have $|3X| \le 4^3 n$.

These examples suggest that commutativity is not only an assumption heavily used in Plünnecke's method, but a typical result will fail without it. The last example suggests a possible non-commutative replacement.

Problem 1.11.1 (A non-commutative Plünnecke?). Theorems 1.9.1 and 1.9.2 suggest a way to find non-commutative analogues of inequalities that, for commutative groups, were proved by Plünnecke's method. We formulate the simplest possible of them. Let A, B be finite sets in a non-commutative group and define α by

$$\max_{b \in B} |A + b + B| = \alpha |A|.$$

Must there exist a nonempty $X \subset A$ such that

$$|X + 2B| \le \alpha' |X|$$

with an α' depending only on α?

We rather expect a negative answer.

However, Plünnecke's method can be modified to handle some non-commutative situations. We give a simple example of this.

Theorem 1.11.2. *Let A, B_1, B_2 be sets in a (typically non-commutative group) G and write $|A| = m$, $|B_1 + A| = \alpha_1 m$, $|A + B_2| = \alpha_2 m$. There is an $X \subset A$, $X \ne \emptyset$, such that*

$$|B_1 + X + B_2| \le \alpha_1 \alpha_2 |X|. \tag{1.11.3}$$

Proof. We take four copies of G and build a 2-layered graph on them. V_0 contains the set A in one copy, V_1 contains the sets $B_1 + A$ and $A + B_2$ in different copies, and V_2 contains $B_1 + A + B_2$; edges are drawn in the natural way.

We claim that this graph is commutative. Indeed, take vertices such that $x \to y \to z_i$, $i = 1, \dots, k$. The edge $x \to y$ can go to either $B_1 + A$ or $A + B_2$; assume the first, the other is similar. Then $x \in A$, $y = b + x$ with some $b \in B_1$ and $z_i = b + x + c_i$ with $c_i \in B_2$. Then with $y_i = x + c_i$ we have $x \to y_i \to z_i$; observe that these replacing vertices are in the other half of V_1 than the original one.

The other side of commutativity goes similarly. If we have $x_i \to y \to z$, and $y \in B_1 + A$, then there are elements $b_i \in B_1$ such that $b_i + x_i = y$, and $c \in B_2$ such that $z = y + c$. The replacing edges are again $x_i \to x_i + c \to b_i + x_i + c = y + c = z$ in the other half of the graph.

Note that this was based on a special kind of commutativity: adding an element from the left and adding another from the right commute – this property bears the name "associativity".

Applying Plünnecke's graph Theorem 1.2.1 to this graph we obtain a set $X \subset A$ such that

$$|B_1 + X + B_2| \le (\alpha_1 + \alpha_2)^2 |X|. \tag{1.11.4}$$

If $\alpha_1 = \alpha_2 = \alpha$, this is $4\alpha^2$ rather than the α^2 claimed in (1.11.3), if they are rather different, it can be much worse. We can improve the situation by embedding G into a larger group $G' = G \times H_1 \times H_2$, where the H_i are cyclic groups, $|H_i| = n_i$, and we identify G with $G \times \{0\} \times \{0\}$. In G' we consider the sets $A' = A$, $B_i' = B_i \times H_i$. We have $\alpha_i' = \alpha_i n_i$, and an application of (1.11.4) yields an $X \subset A$ such that

$$|B_1' + X + B_2'| = n_1 n_2 |B_1 + X + B_2| \le (\alpha_1 n_1 + \alpha_2 n_2)^2 |X|.$$

If we select the n_i so that $\alpha_1 n_1 = \alpha_2 n_2$, this gives

$$|B_1 + X + B_2| \le 4\alpha_1\alpha_2 |X| \tag{1.11.5}$$

in the general case.

We can remove the factor 4 like in the previous proof. We take the 1-layered graph built on the layers A and $B_1 + A + B_2$. We are interested in the magnification ratio μ of this graph. We take the similar graph made from the sets A^k, B_i^k. This graph is the same as the k-th power of the previous graph, thus its magnification ratio is μ^k. An application of (1.11.5) gives $\mu^k \le 4(\alpha_1\alpha_2)^k$; taking k-th roots and making $k \to \infty$ we obtain (1.11.3). □

We mention, without proof, how this result can be generalized to several summands, with an extra condition.

Definition 1.11.3. A collection of sets $B_1, \dots, B_k$ in a (non-commutative) group is *exocommutative*, if for all $x \in B_i$, $y \in B_j$ with $i \ne j$ we have $x + y = y + x$.

Theorem 1.11.4. *Let $A, B_1, B_2, \dots, B_k, C_1, C_2, \dots, C_l$ be sets in a (typically non-commutative group) G and write $|A| = m$, $|B_i + A| = \alpha_i m$, $i = 1, \dots, h$, $|A+C_j| = \beta_j m$, $j = 1, \dots, l$. Assume that both $B_1, \dots, B_k$ and $C_1, \dots, C_l$ are exocommutative. Then there is an $X \subset A$, $X \ne \emptyset$, such that*

$$|B_1 + \ldots + B_k + X + C_i + \ldots + C_l| \le \alpha_1 \ldots \alpha_k \beta_1 \ldots \beta_l |X|. \tag{1.11.6}$$

The moral seems to be that it is hopeless to undestand sets such that $|2A| \le \alpha |A|$ in general groups. On the other hand, if we start with an assumption on a threefold sum, say $|A| = m$, $|3A| \le gam$, then an iterated application of Theorem 1.8.1 gives estimates for arbitary sum-difference combinations. For instance, putting $Y = Z = -2A$ we get $|-2A + 2A| \leq \alpha^2 m$, then with $Y = Z = A - 2A$ and putting $-A$ into the place of A we get an estimate for a sixfold sumset and so on. It is the step from 2 to 3 which fails in lack of commutativity.

Finally we add that a weaker conclusion can be drawn from the assumption $|2A| \leq \alpha m$, namely, that there is an $A' \subset A$, $|A'| > (1 - \varepsilon)m$, such that $3A'$ and hence each kA' is small.

To this end we apply an argument like in Section 1.7. Theorem 1.11.2 above assures the existence of a nonempty set X such that, if $|L + A| \leq \alpha m$ and $|A + R| \leq \beta m$, then $|L + X + R| \leq \alpha\beta|X|$.

Choose an $\epsilon > 0$, let $m = |A|$ and define $X_1 = X$. If $|X_1| > (1 - \epsilon)|A|$, we are done. If this is not the case, we apply Theorem 1.11.2 on $A_1 = A \setminus X_1$, with this procedure we obtain an $X_2 \in A_1$ with a similar property. If $X_1 \cup X_2$ is still not large enough, we continue with the procedure till we get $X' = X_1 \cup \ldots \cup X_h$ for some h and such that $|X'| > (1 - \epsilon)m$. To bound $|L + X' + R|$ we need to estimate

$$\frac{|L + A_i|}{|A_i|}$$

for all $1 \leq i \leq h$. Such a bound is $\alpha|A|/|A_h| \leq \alpha/\epsilon$, since $|A_h| \geq \epsilon|A|$ and $|L + A_i| \leq |L + A|$. Similarly,

$$\frac{|A_i + R|}{|A_i|} \leq \frac{\beta}{\epsilon}|A|$$

for all i. Hence, by adding all the pieces X_i, we obtain

$$|L + A' + R| \leq \frac{\alpha\beta}{\epsilon^2}|A'|.$$

We had pay an ϵ^2 price to get $|A'| > (1 - \varepsilon)\,|A|$; this can be somewhat improved with a more involved argument like in Section 1.7.

Chapter 2

Structure of sets with few sums

2.1 Introduction

We want to describe sets that have few sums. If $|A| = m$, then clearly $|A+A| \geq m$ in every group (with equality for cosets), which can be improved to $2m - 1$ for sets of integers (or torsion-free groups in general). What can we say if we know that $|A + A| \leq \alpha m$, where α is constant or grows slowly as $n \to \infty$? That is, we are looking for statements of the form

$$|A| = m, \ |A + A| \leq \alpha m \Longrightarrow (\dots).$$

Such a condition $(\dots)$ is *adequate*, if this implication can be reversed to some degree; that is, there is an implication in the other direction

$$(\dots) \Longrightarrow |A + A| \leq \alpha' m,$$

with $\alpha' = \alpha'(\alpha)$ depending only on α and not on m or other properties of the set.

Among such results we can distinguish on two grounds. First, the smaller the value of α', the better the description; next, subjectively, the more we learn on the structure of the set, the happier we are.

As an example, consider the following implications (see Chapter 1, Section 1.8):

$$|A| = m, \ |A + A| \leq \alpha m \Longrightarrow |A - A| \leq \alpha^2 m$$

and

$$|A| = m, \ |A - A| \leq \alpha m \Longrightarrow |A + A| \leq \alpha^2 m.$$

If we combine both expressions we get that

$$|A + A| \leq \alpha m \Longrightarrow |A - A| \leq \alpha^2 m \Longrightarrow |A + A| \leq \alpha^4 m, \ \alpha' = \alpha^4,$$

so this is an adequate description with a very good value of α', but it tells little about the structure of A and it is not surprising. Indeed,

$$a + b = c + d \Longleftrightarrow a - c = d - b,$$

so a coincidence between sums corresponds to a coincidence between differences. In particular, this shows that $|A+A|$ attains its maximal value $m(m+1)/2$ exactly when $|A-A|$ attains its maximal value $m(m-1)+1$. (Such sets, with no nontrivial coincidence between sums or differences, are often called *Sidon sets.*)

There is a similar connection between minimal values of these quantities. For sets of integers, the minimal value of both $|A + A|$ and $|A - A|$ is $2m - 1$, and equality occurs only for arithmetic progressions.

Still, the connection here is less obvious than it looks. We illustrate this by the case of near-maximal values. Suppose that $|A + A| \geq \kappa m^2$; does it follow that $|A - A| \geq \kappa' m^2$ with some κ' depending on κ? The answer is negative in a rather strong way: $|A + A| > m^2/2 - m^{2-\delta}$ and $|A - A| < m^{2-\delta}$ can happen with some constant $\delta > 0$. Similarly, $|A - A| > m^2/2 - m^{2-\delta}$ and $|A + A| < m^{2-\delta}$ is also possible [54].

A set of integers with a minimal sumset ($|A + A| = 2m - 1$) is necessarily an arithmetic progression. This easy result exhibits some stability. A set with a nearly minimal sumset is almost an arithmetic progression, as the following result shows.

Theorem 2.1.1 (G. Freiman [11]). *If $A \subset \mathbb{N}$, $|A| = m$, $|A + A| \leq 3m - 4$, then A is contained in an arithmetic progression of length $\leq |A + A| - m + 1 \leq 2m - 3$.*

The proof is given in the next chapter.

Beyond $3m$, however, a single arithmetic progression is insufficient, as the following example shows. Take

$$A = \{1, \ldots, m/2\} \cup \{t + 1, \ldots, t + m/2\}.$$

Graphically the set looks like:

. . . .

We have $|A + A| = 3m - 3$, and A cannot be covered by a progression shorter than $t + m/2$. The reason is that this set has a hidden two-dimensional structure as depicted below:

. . . .

. . . .

These sets are not isomorphic algebraically, but they behave analogously regarding the coincidence of sums. To describe such sets we need multi-dimensional or generalized arithmetic progressions.

Definition 2.1.2. Let $q_1, \ldots, q_d$ and a be elements of an arbitrary commutative group, $l_1, \ldots, l_d$ positive integers. A d-dimensional *generalized arithmetic progression* is a set of the form

$$P = P(q_1, \ldots, q_d; l_1, \ldots, l_d; a) = \{a + x_1 q_1 + \cdots + x_d q_d : 0 \le x_i \le l_i\} \quad (2.1.1)$$

(a projection of a cube). More exactly, we think of it as a set together with a fixed representation in the form (2.1.1); this representation is in general not unique. We call d the *dimension* of P, and by its *size* we mean the quantity

$$\|P\| = \prod_{i=1}^{d} (l_i + 1),$$

which is the same as the number of elements if all sums in (2.1.1) are distinct. In this case we say that P is *proper.*

Exercise 18. If P is a d-dimensional progression, then

$$|2P| < 2^d\,|P| \le 2^d \|P\|.$$

The principal result sounds as follows.

Theorem 2.1.3 (G. Freiman [11]). *If $A \subset \mathbb{Z}$, $|A| = n$, $|A + A| \le \alpha n$, then A is contained in a generalized arithmetic progression of dimension $\le d(\alpha)$ and size $\le s(\alpha) n$.*

This is an adequate description with the simplest possible structure: If $A \subset P$, then

$$|A + A| \le |P + P| < 2^d \|P\| \le 2^d sn,$$

$$\alpha' = 2^{d(\alpha)} s(\alpha).$$

More generally, we have

$$|kA| \le k^d \|P\|.$$

This shows that this dimension is closely connected with the rate of growth of $|kA|$ as a function of k.

For a comprehensive account of this theory up to 1996, see Nathanson's book [35].

Three basic questions arise here:

1) to find good bounds for $d(\alpha)$, $s(\alpha)$;

2) is this the "real" form?

3) how to extend this from $\mathbb{Z}$ to other groups.

Bounds: Due to works by the author [52, 55], Y. Bilu [2], M. C. Chang [5] we know that $d < \alpha$ (best possible) and $s < e^{\alpha^c}$. It is also known that a bound for s must be $\gg 2^\alpha$; probably the proper order is $e^{c\alpha}$.

The real form: Probably a flexible form (several covering sets, projections of lattice points in more general convex bodies) would give better bounds for α'.

Other groups. For sets situated in $\mathbb{Z}^m$ or in general commutative torsion-free groups verbatim the same result holds (and later we shall formulate and prove it in this setting).

In groups with torsion a new phenomenon arises, namely any coset has $|A + A| = |A|$. For groups with a strong torsion property this alone suffices to characterize sets with small sumsets.

Recall that the *exponent* of a group G is the smallest positive integer r such that $rg = 0$ for every $g \in G$.

Theorem 2.1.4. *Let G be a commutative group of exponent r, $A \subset G$, $|A| = m$, $|A + A| \le \alpha m$. A is contained in a coset of a subgroup of size $\le \alpha^2 r^{\alpha^4} m$.*

We shall start (in the next section) with the proof of this theorem, which is simple and highlights some aspects of the case of integers.

General commutative groups

In a general commutative group, a set with a small sumset can be covered by a combination of the two mentioned structures, cosets and generalized arithmetic progressions.

Theorem 2.1.5 (Green–Ruzsa [17]). *Let G be a commutative group, $A \subset G$, $|A| = m$, $|A + A| \le \alpha m$. A is contained in a set of the form $H + P$, where H is a subgroup, P is a generalized arithmetic progression, the dimension of P is $\le d(\alpha)$ and $|H||P| \le s(\alpha)m$.*

For the quantities d, s we have the following bounds: $d(\alpha) \ll \alpha^c$, $s(\alpha) \ll e^{\alpha^c}$.

Non-commutative groups

For general groups, we do not even have a decent conjecture. There is a structure theorem for $SL_2(\mathbb{R})$ (Elekes–Király[7]). Roughly speaking, it asserts that a set with a small sumset is contained in a few cosets of a commutative subgroup, and within a coset we have a generalized arithmetic progression structure.

2.2 Torsion groups

In this section we prove Theorem 2.1.4, in a superficially more general form.

Theorem 2.2.1. *Let $r \geq 2$ be an integer, and let G be a commutative group of exponent r. Let $A \subset G$ be a finite set, $|A| = m$. If there is another set $A' \subset G$ such that $|A'| = m$ and $|A+A'| \leq \alpha m$ (in particular, if $|A+A| \leq \alpha m$ or $|A-A| \leq \alpha m$), then A is contained in a subgroup H of G such that*

$$|H| \leq f(r, \alpha)m,$$

where

$$f(r, \alpha) = \alpha^2 r^{\alpha^4}.$$

Proof. Let $b_1, b_2, \ldots, b_k$ be a maximal collection of elements such that $b_i \in 2A - A$ and the sets $b_i - A$ are all disjoint. We have

$$b_i - A \subset 2A - 2A,$$

hence

$$\left| \bigcup (b_i - A) \right| = km \leq |2A - 2A| \leq \alpha^4 m$$

(the last inequality follows from Theorem 1.1.1). This implies $k \leq \alpha^4$.

Take an arbitrary $x \in 2A - A$. Since the collection $b_1, \ldots, b_k$ was maximal, there must be an i such that

$$(x - A) \cap (b_i - A) \neq \emptyset,$$

that is, $x - a_1 = b_i - a_2$ with some $a_1, a_2 \in A$, which means that

$$x = b_i + a_1 - a_2 \in b_i + (A - A).$$

Hence

$$2A - A \subset \bigcup (b_i + (A - A)) = B + A - A, \tag{2.2.1}$$

where $B = \{b_1, \ldots, b_k\}$.

Now we prove that

$$jA - A \subset (j-1)B + A - A \quad (j \geq 2) \tag{2.2.2}$$

by induction on j. By (2.2.1), this holds for $j = 2$. Now we have

$$\begin{aligned}
(j+1)A - A &= (2A - A) + (j-1)A \\
&\subset B + A - A + (j-1)A \text{ by (2.2.1)} \\
&= B + (jA - A) \\
&\subset B + (j-1)B + A - A \\
&= jB + A - A,
\end{aligned}$$

which provides the inductive step.

Let H and I be the subgroups generated by A and B, respectively. By (2.2.2) we have

$$jA - A \subset I + (A - A) \tag{2.2.3}$$

for every j. We have also

$$\bigcup (jA - A) = H, \tag{2.2.4}$$

which easily follows from the fact that the order of the elements of G is bounded. Relations (2.2.3) and (2.2.4) imply that

$$H \subset I + (A - A).$$

Since I is generated by k elements of order $\le r$ each, we have

$$|I| \le r^k \le r^{\alpha^4},$$

consequently

$$|H| \le |I||A - A| \le \alpha^2 r^{\alpha^4} m$$

(the estimate for $|A - A|$ follows again from Theorem 1.1.1). □

Remark. Take a group of the form $G = Z_r^n$, where Z_r is a cyclic group of order r, and a set $A \subset G$ of the form

$$A = (a_1 + G') \cup \cdots \cup (a_k + G')$$

with a subgroup G', where the cosets are all disjoint. Here $|A| = m = k|G'|$, and if all the sums $a_i + a_j$ lie in different cosets of G', then

$$|A + A| = \frac{k(k+1)}{2}|G'| = \alpha m, \quad \alpha = \frac{k+1}{2}.$$

The subgroup generated by A can have as many as $r^k|G'|$ elements, hence our function

$$f(r, \alpha) = \alpha^2 r^{\alpha^4}$$

cannot be replaced by anything smaller than

$$r^k = r^{2\alpha - 1}.$$

By recent improvements of the above argument by Green–Ruzsa and then Sanders, the above bound is now almost achieved.

The following conjecture of Katalin Marton would yield a more efficient covering in a slightly different form.

Conjecture 2.2.2. If $|A| = n$, $|A + A| \le \alpha n$, then there is a subgroup H of G such that $|H| \le n$ and A is contained in the union of α^c cosets of H, where the constant c may depend on r but not on n or α.

In the most optimistic form c would be $1+o(1)$.

This is equivalent to the following problem, which we think is interesting in its own right.

Conjecture 2.2.3 (Equivalent conjecture). Let G be as above, $f : G \to G$ a function such that $f(x+y)-f(x)-f(y)$ assumes at most α distinct values. Then f has a decomposition $f = g+h$, where g is a homomorphism and h assumes $\leq \alpha^c$ values.

The equivalence is meant in a loose sense, the values of c need not be the same. (The proof of this equivalence is unpublished.)

2.3 Freiman isomorphism and small models

Definition 2.3.1. Let G_1, G_2 be commutative groups, $A_1 \subset G_1$, $A_2 \subset G_2$. We say that a mapping $\varphi : A_1 \to A_2$ is a *homomorphism of order r in the sense of Freiman*, or an *F_r-homomorphism* for short, if for every $x_1, \dots, x_r, y_1, \dots, y_r \in A_1$ (not necessarily distinct), the equation

$$x_1 + x_2 + \cdots + x_r = y_1 + y_2 + \cdots + y_r \tag{2.3.1}$$

implies

$$\varphi(x_1) + \varphi(x_2) + \cdots + \varphi(x_r) = \varphi(y_1) + \varphi(y_2) + \cdots + \varphi(y_r). \tag{2.3.2}$$

We call φ an *F_r-isomorphism* if it is (1-1) and its inverse is a homomorphism as well, that is, (2.3.2) holds if and only if (2.3.1) does. If we say Freiman homomorphism or isomorphism without specifying r, then the first nontrivial case $r=2$ is meant.

Any affine linear function is an F_r-isomorphism for every r, and the non-degenerate ones are F_r-isomorphisms.

Prexercise. If one of two F-isomorphic sets contains an l-term arithmetic progression, then so does the other.

Prexercise. If A and B are F_r-isomorphic with $r = q(k+l)$, then $kA - lA$ and $kB - lB$ are F_q-isomorphic.

Prexercise. The F-homomorphic image of a d-dimensional arithmetic progression is also a d-dimensional arithmetic progression with the same "lengths" $l_1, \dots, l_d$.

A Freiman isomorphism preserves additive properties up to a point. We show that being a generalized arithmetic progression is such a property.

Lemma 2.3.2. *Let G, G' be commutative groups. If a set $P' \subset G'$ is the homomorphic image of a generalized arithmetical progression $P(q_1, \dots, q_d; l_1, \dots, l_d; a) \subset G$, then there are elements $q_1', \dots, q_d', a' \in G'$ such that*

$$P' = P(q_1', \dots, q_d'; l_1, \dots, l_d; a') \tag{2.3.3}$$

and the homomorphism is given by

$$\phi(a + x_1 q_1 + \cdots + x_d q_d) = a' + x_1 q_1' + \cdots + x_d q_d'. \tag{2.3.4}$$

Proof. Define a' and q_i' by

$$a' = \phi(a), \quad q_i' = \phi(a + q_i) - \phi(a).$$

We prove (2.3.4) by induction on $r = x_1 + \cdots + x_d$. For $r \le 1$ it is an immediate consequence of the definition. Assume that $r \ge 2$ and the statement holds for every smaller value. Consider an element

$$x = x_1 q_1 + \cdots + x_d q_d, \quad x_1 + \cdots + x_d = r.$$

Since $r \ge 2$, either there are subscripts $i \ne j$ such that $x_i \ge 1$ and $x_j \ge 1$, or there is a subscript for which $x_i \ge 2$. In the second case write $j = i$. In both cases the sums

$$y = x - x_i, \ z = x - x_j, \ u = x - x_i - x_j$$

are in P, their sums of coefficients are at most $r-1$ and they satisfy $x+u = y+z$. This implies $\phi(x) + \phi(u) = \phi(y) + \phi(z)$, that is, $\phi(x) = \phi(y) + \phi(z) - \phi(u)$. Substituting (2.3.4) for y, z and u into this equation we conclude that (2.3.4) holds for x as well, which completes the inductive step. □

Lemma 2.3.3. *Let G, G' be commutative groups, and let $A \subset G$, $A' \subset G'$ be F_r-isomorphic sets. Assume that $r = r'(k + l)$ with non-negative integers r', k, l. The sets $kA - lA$ and $kA' - lA'$ are $F_{r'}$-isomorphic.*

Proof. Let ϕ be the isomorphism between A and A'. For an

$$x \in kA - lA, \ x = a_1 + \cdots + a_k - b_1 - \cdots - b_l$$

we define naturally

$$\psi(x) = \phi(a_1) + \cdots + \phi(a_k) - \phi(b_1) - \cdots - \phi(b_l).$$

The facts that this depends only on x and not on the particular representation, and that ψ is an $F_{r'}$-isomorphism, follow immediately from the definition. □

With this concept we can formulate principle (iii) from the Introduction of Chapter 1 exactly.

Lemma 2.3.4. *Let A be a finite set in a torsion-free commutative group, and let r be any positive integer. There is a set $A' \subset \mathbb{Z}$ which is F_r-isomorphic to A.*

The proof, as also outlined there, consists of first applying the structure theorem of finitely generated torsion-free groups to reduce the general case to sets lying in $\mathbb{Z}^d$, and then a suitable projection to go to $\mathbb{Z}$.

We define the *Freiman dimension* of a set $A \subset \mathbb{R}^k$ as the largest d for which there is an isomorphic properly d-dimensional set.

Exercise 19. For a set $A \subset \mathbb{Z}^d$ the following are equivalent:

a) its Freiman dimension is d,

b) every Freiman homomorphism from A to any $\mathbb{R}^k$ is affine linear.

The first step towards finding the structure of a set will be to find a Freiman isomorphic image, or "model", which is comfortably sitting in a small group or interval.

Theorem 2.3.5. *Let A be a finite set in a torsion-free commutative group, $|A| = m$, $r \geq 2$ an integer and $|rA - rA| = n$.*

(a) *For every $q \geq n$ there exists a set $A' \subset A$, $|A'| \geq m/r$ which is F_r-isomorphic to a set T' of residues modulo q.*

(b) *There is a set $A^* \subset A$, $|A^*| \geq m/r^2$, which is F_r-isomorphic to a set T^* of integers,*

$$T^* \subset [0, n/r].$$

Proof. In view of the previous lemma we may assume that $A \subset \mathbb{Z}$.
The isomorphism in (a) will be given by a function

$$\varphi(a) = [\xi a] \pmod q$$

for a suitably chosen real number $\xi \in [0, q]$, and the set A' will be one of the r sets

$$A_j = \left\{ a \in A : \frac{j-1}{r} \leq \{\xi a\} < \frac{j}{r} \right\}, \quad j = 1, \ldots, r.$$

We claim that for a suitable choice of ξ the restriction of φ is an isomorphism on each set A_j; clearly at least one of them will have $\geq m/r$ elements.

This isomorphism means that for arbitrary $a_1, \ldots, a_r, b_1, \ldots, b_r \in A_j$ the congruence

$$[\xi a_1] + \cdots + [\xi a_r] \equiv [\xi b_1] + \cdots + [\xi b_r] \pmod q \tag{2.3.5}$$

should be equivalent to the equality

$$a_1 + \cdots + a_r = b_1 + \cdots + b_r.$$

First we show that this equality implies

$$[\xi a_1] + \cdots + [\xi a_r] = [\xi b_1] + \cdots + [\xi b_r],$$

and a fortiori the congruence (2.3.5) for every ξ. Indeed,

$$\sum \big([\xi a_i] - [\xi b_i]\big) = \xi \sum (a_i - b_i) - \sum \big(\{\xi a_i\} - \{\xi b_i\}\big). \tag{2.3.6}$$

If all the fractional parts are in an interval $[u, u + 1/r)$, then the absolute value of the last sum is < 1. The left side, as an integer with absolute value < 1, must be 0.

Assume now congruence (2.3.5). The left side of (2.3.6) is a multiple of q, and the right side is of the form $\xi t+\delta$, where $t \in rA-rA$ and $|\delta|<1$. We want to infer $t=0$, that is, we try to exclude all possible equalities of the type $kq=\xi t+\delta$, or

$$\xi = \frac{kq-\delta}{t}.$$

For a given value of t this is a collection of $t+1$ intervals of total length 2. If the union of these systems of intervals does not cover $[0,q]$, we can find a ξ which is not contained in any of them. The number of values of t that we have to take into account is $(n-1)/2$, since t and $-t$ induce the same collection of excluded intervals. Hence a sufficient condition is $2(n-1)/2 < q$, or $q \geq n$.

To prove part (b), we combine this map φ with $\psi : \mathbb{Z}_q \to \mathbb{Z}$, where ψ is the smallest non-negative representation of a residue class. We split the integers of the interval $[0, q-1]$ into r almost equal subintervals of type $[(i-1)q/r, iq/r)$, $i=1,\dots,r$. The r-fold sums from a fixed interval lie in an interval of length $<q$, thus they are incongruent modulo q unless they are equal. This division splits A' into r parts, and any can serve as A^*. In this way we can achieve

$$|A^*| \geq |A'|/r \geq m/r^2.$$

The isomorphic image of A^* lies in an interval of type $[(i-1)q/r, iq/r)$, and a shift takes it into $[0, q/r]$. For q we take the smallest guaranteed value $q=n$. □

Exercise 20. Let p be a prime, $A \subset \mathbb{Z}_p$, $|A|=n$, k a positive integer. Prove that if $p > k^n$, then there is a $t \in \mathbb{Z}_p$, $t \neq 0$, such that $||at/p|| \leq 1/k$ for all $a \in A$.

Exercise 21. Let $A \subset \mathbb{N}$, $|A|=n$. Prove that there is a Freiman isomorphic set contained in $[0, 4^n]$.

Exercise 22. Show that the bound in the previous exercise cannot be improved below 2^{n-2}.

2.4 Elements of Fourier analysis on groups

In this section we collect some basic facts about the Fourier transform which will be used in the next section. Detailed proofs are not given; instead the main statements are split into several exercises, which even the uninitiated reader may try to solve. It is not *necessary* to solve these exercises to understand the next section; the prerequisites are here in the form of definitions and statements, but it certainly helps.

A group will mean a commutative group; a *character* is a homomorphism $\gamma : G \to \mathbb{C}_1$, where $\mathbb{C}_1 = \{z : |z|=1\}$ (with multiplication). So if the operation in G is denoted additively, then $\gamma(x+y)=\gamma(x)\gamma(y)$. The characters of G form a group (under pointwise multiplication). We write mostly Γ to denote this group.

Its unity is $\gamma_0 \equiv 1$, the *principal character*. We write $\bar{\gamma}(g) = \overline{\gamma(g)}$; it is the inverse, and at the same time the pointwise complex conjugate of γ.

Characters of a cyclic group $\mathbb{Z}_q$ are simple. Indeed, if γ is a character and $\gamma(1) = \omega$, then $\gamma(n) = \omega^n$. Since $\gamma(q) = \gamma(0) = 1$, we see that ω must be a q-th root of unity, say $\omega = e^{2\pi ik/q}$ with some k, consequently

$$\gamma(n) = e^{2\pi ikn/q}.$$

If we restrict our attention to cyclic groups, which is the most important object for what follows, then we could just use the above functions and not even mention the word "character". We think, however, that this is the natural way of presentation (some reasons are given later).

The above formula shows that $\mathbb{Z}_q$ has exactly q characters, moreover they also form a cyclic group of order q. This is not so obvious for other groups.

We write $G_1 < G$ to denote that G_1 is a subgroup of G.

Exercise 23. Let $G_1 < G$, $g \in G \backslash G_1$, γ a character of G_1. γ can be extended to a character of the group G_2 generated by $G_1 \cup \{g\}$.

Exercise 24. This γ above can be extended to a character of G. Consequently, for any $g \in G$, $g \neq e$ (e =unity), there is a character γ with $\gamma(g) \neq 1$ (in other words, the charaters separate G).

Exercise 25. The only important property of $\mathbb{C}_1$ in the above exercises is that it is *divisible*. A group G is divisible if for every $g \in G$ and positive integer k, there is an $h \in G$ such that $h^k = g$. (We use multiplicative notation here for compatibility with $\mathbb{C}_1$.) Show that the previous exercises hold with the set of homomorphisms to any fixed divisible group in the place of $\mathbb{C}_1$.

Exercise 26. Extend the previous three exercises to infinite groups.

Exercise 27. $\sum_{g \in G} \gamma(g) = 0$ unless $\gamma = \gamma_0 \equiv 1$. Hint: Compare it to $\sum_{g \in G} \gamma(ag)$.

Exercise 28. $\sum_{\gamma \in \Gamma} \gamma(g) = 0$ unless $g = 0$.

Exercise 29. $|\Gamma| = |G|$. Hint: Consider $\sum_{g \in G} \sum_{\gamma \in \Gamma} \gamma(g)$.

Exercise 30. For a $g \in G$, we define a character g^* of Γ by $g^*(\gamma) = \gamma(g)$. The mapping $g \longrightarrow g^*$ embeds G into $\hat{\Gamma}$, the group of characters of Γ. Show that this is an isomorphism.

Exercise 31. The previous exercise fails for infinite groups. In fact, it is wrong for each infinite group.

Exercise 32. If $G = G_1 \times G_2$, then Γ is isomorphic to $\Gamma_1 \times \Gamma_2$.

Exercise 33. For finite groups, Γ is isomorphic to G.

Exercise 34. The previous exercise also provides a direct access Exercises 29 and 30. It is of limited value, since this isomorphy is not natural: we cannot find a way to define a 1-1 correspondence between G and Γ. Try to formulate this observation exactly and then prove it.

Exercise 35. For functions $\alpha, \beta : G \to \mathbb{C}$ we define a direct product by

$$(\alpha, \beta) = |G|^{-1} \sum_{g \in G} \alpha(g)\tilde{\beta(g)}.$$

This turns Γ into an orthonormal system: For $\gamma, \gamma' \in \Gamma$ we have $(\gamma, \gamma') = 0$ if $\gamma \neq \gamma'$, and $(\gamma, \gamma') = 1$ if $\gamma = \gamma'$.

Exercise 36. Every funtion α on G has a development into a character series $\alpha = \sum_{\gamma \in \Gamma} c_\gamma \gamma$. Express the coefficients c_γ. Find the development of the indicator function of an element.

Definition 2.4.1. Let $\varphi : G \to \mathbb{C}$ be a function on the group G. Its *Fourier transform* is the function $f : \Gamma \to \mathbb{C}$ defined by

$$f(\gamma) = \sum_{g \in G} \varphi(g)\gamma(g).$$

The Fourier transform is often denoted by $f = \hat{\varphi}$.

For a cyclic group $G = \mathbb{Z}_q$ the characters are the functions

$$\gamma_k(n) = e^{2\pi i k/n}, \quad k = 0, 1, \ldots, q-1.$$

Consequently, the Fourier transform of a function φ is given by

$$f(\gamma_k) = \sum_n e^{2\pi i k n/q} \varphi(n).$$

If we identify this character γ_k with its subscript $k \in \mathbb{Z}_q$, we can also say that the Fourier transform is

$$f(k) = \sum_n e^{2\pi i k n/q} \varphi(n),$$

which is frequently done when no other group is used. In this booklet we will distinguish G and Γ, elements and characters, for methodological reasons.

Given the Fourier transform of a function, we can reconstruct the function from it as follows.

Statement 2.4.2 (Fourier inversion formula). *Let φ be a function on G and $f = \hat{\varphi}$ its Fourier transform. We have*

$$\varphi(x) = \frac{1}{|G|} \sum_{\gamma \in \Gamma} f(\gamma)\overline{\gamma}(x).$$

Exercise 37. Prove the inversion formula. (Is this a new exercise or an old one?)

Exercise 38. How does the inversion formula look for the group $\mathbb{Z}_q$?

Another important fact is the analogue of the Parseval (or Plancherel) identity.

Statement 2.4.3 (Parseval formula). *Let φ be a function on G and $f = \hat{\varphi}$ its Fourier transform. We have*

$$\sum_{\gamma\in\Gamma} |f(\gamma)|^2 = |G| \sum_{x\in G} |\varphi(x)|^2 .$$

Exercise 39. Prove the Parseval formula.

Exercise 40. Let φ_1, φ_2 be functions on G, with Fourier transforms f_1, f_2. What is the connection between the direct products (φ_1, φ_2) and (f_1, f_2)?

The case of 0-1-valued functions is of special imporance for us. Let $A \subset G$ be any set, and consider its *indicator function*

$$\varphi(x) = \begin{cases} 1 & \text{if } x \in A, \\ 0 & \text{if } x \notin A. \end{cases}$$

Its Fourier transform is

$$f(\gamma) = \sum_{a\in A} \gamma(a). \tag{2.4.1}$$

With an abuse of terminology we shall call this the *Fourier transform of the set* A and denote it by $\hat{A}(\gamma)$.

Exercise 41. What does the Parseval formula tell for the Fourier transform of a set?

Exercise 42. What does the inversion formula tell for the Fourier transform of a set?

Exercise 43. If the Fourier transform of a set A is f, what is the transform of the set $-A$?

Let now A_1, A_2 be sets in G with Fourier transforms f_1, f_2. By using the definition (2.4.1) and multiplying we obtain

$$f_1(\gamma) f_2(\gamma) = \sum_{x\in G} r(x),$$

where

$$r(x) = |\{(a_1, a_2) : a_i \in A_i, a_1 + a_2 = x\}| ,$$

the number of representations of x as a sum with summands from our sets. The inversion formula now gives

$$r(x) = \frac{1}{|G|} \sum_{\gamma\in\Gamma} f_1(\gamma) f_2(\gamma) \overline{\gamma}(x),$$

and in principle this gives a complete description of the sumset. This is the basis of the usage of analytic methods in additive number theory (under various names, like generating functions, circle method, Hardy–Littlewood method, depending on particular appearances).

Exercise 44. If the Fourier transform of a set A is f, whose transform is $|f|^2$?

For the next exercises let A be a set of integers, $|A| = n$ and

$$\hat{A}(t) = \sum_{a \in A} e^{2\pi i a t}, \ t \in \mathbb{R}.$$

Exercise 45. What is the connection between this function $\hat{A}$ of a real variable for $A \subset \mathbb{Z}$ and the function $\hat{A}(\gamma)$ for $A \subset \mathbb{Z}_q$?

Exercise 46. What is the arithmetical meaning of the integral $\int_0^1 |\hat{A}(t)|^4 dt$? What is its minimal value?

Exercise 47. What is the maximal value of the integral in the previous exercise, and for which sets does it occur?

Exercise 48. How can one express the number of three-term arithmetical progressions in A (that is, the number of pairs a, d such that $a, a+d, a+2d \in A$) by the function $\hat{A}$?

Exercise 49. And what happens if we count only those where $d > 0$?

2.5 Bohr sets in sumsets

Definition 2.5.1. If G is a commutative group, $\gamma_1, \ldots, \gamma_k$ are characters of G and $\varepsilon_j > 0$, we write

$$B(\gamma_1, \ldots, \gamma_k; \varepsilon_1, \ldots, \varepsilon_k) = \{g \in G : |\arg \gamma_j(g)| \leq 2\pi\varepsilon_j \text{ for } j = 1, \ldots, k\}$$

and we call these sets *Bohr sets.* In particular, if $\varepsilon_1 = \cdots = \varepsilon_k = \varepsilon$, we shall speak of a *Bohr* (k, ε)*-set.* (We take the branch of arg that lies in $[-\pi, \pi)$.)

In locally compact groups these sets form a base for the Bohr topology; we shall work with finite groups, but we preserve the name that suggests certain ideas.

We shall work mainly with the simplest possible cyclic groups $\mathbb{Z}_q$. Here a typical character is of the form

$$\gamma(x) = e^{2\pi i u x/q}, \ u \in \mathbb{Z}_q,$$

so $\arg \gamma(x) = 2\pi \|ux/q\|$, where $\|t\| = \min(\{t\}, 1 - \{t\})$ denotes the *absolute fractional part* of t, its distance from the nearest integer. In these formulas we were tacitly cheating a bit; for $u, x \in \mathbb{Z}_q$ we replaced them by any integer in the corresponding residue class, and though ux/q can have many different values, it is unique modulo one, so the fractional part and the exponential are uniquely determined.

Hence a Bohr set in $\mathbb{Z}_q$ can be written as

$$B(u_1, \ldots, u_k; \varepsilon_1, \ldots, \varepsilon_k) = \{x \in \mathbb{Z}_q : \|u_j x/q\| \le \varepsilon_j \text{ for } j = 1, \ldots, k\}.$$

We shall see in the next section that Bohr sets are rather similar to multidimensional arithmetic progressions.

Lemma 2.5.2. *Let G be a finite commutative group, $|G| = q$. Let A be a nonempty subset of G and write $|A| = m = \beta q$. The set $D = 2A - 2A$ (the second difference set of A) contains a Bohr (k, ε)-set with some integer $k < \beta^{-2}$ and $\varepsilon = 1/4$.*

This is essentially a result of Bogolyubov [3] which he used to study the Bohr topology on the integers.

Proof. Let Γ denote the group of characters. For $\gamma \in \Gamma$ put

$$f(\gamma) = \sum_{a \in A} \gamma(a).$$

We have

$$\sum_{\gamma \in \Gamma} |f(\gamma)|^2 = mq = \beta q^2$$

(Parseval formula) and $f(\gamma_0) = m$ for the principal character $\gamma_0 (\equiv 1)$.

Recall that $\overline{f(\gamma)}$ is the series corresponding to the set $-A$. Multiplying two copies of f and two copies of $\overline{f}$ we find that

$$|f(\gamma)|^4 = \sum r(x)\gamma(x),$$

where $r(x)$ counts the quadruples $a_1, a_2, a_3, a_4 \in A$ such that $a_1 + a_2 - a_3 - a_4 = x$. A Fourier inversion now gives

$$r(x) = \frac{1}{q} \sum_{\gamma \in \Gamma} |f(\gamma)|^4 \gamma(x).$$

Therefore we have $x \in D$ for those elements x for which

$$\sum_{\gamma \in \Gamma} |f(\gamma)|^4 \gamma(x) \ne 0. \tag{2.5.1}$$

To estimate (2.5.1), we split the characters $\gamma \ne \gamma_0$ into two groups. We put those for which $|f(\gamma)| \ge \sqrt{\beta} q$ into Γ_1 and the rest into Γ_2. We claim that $x \in D$ whenever $\operatorname{Re} \gamma(x) \ge 0$ is satisfied for all $\gamma \in \Gamma_1$. Indeed, we have

$$\left| \sum_{\gamma \in \Gamma_2} |f(\gamma)|^4 \gamma(x) \right| < \beta q^2 \sum_{\gamma \in \Gamma_2} |f(\gamma)|^2 < \beta^2 m^2 q^2 = m^4,$$

consequently

$$\operatorname{Re}\sum_{\gamma\in\Gamma}|f(\gamma)|^4\gamma(x)\geq m^4+\operatorname{Re}\sum_{\gamma\in\Gamma_2}|f(\gamma)|^4\gamma(x)\geq m^4-\left|\sum_{\gamma\in\Gamma_2}|f(\gamma)|^4\gamma(x)\right|>0.$$

The condition $\operatorname{Re}\gamma(x)\geq 0$ is equivalent to $|\arg\gamma(g)|\leq\pi/2$, thus we have a Bohr $(k,1/4)$-set with $k=|\Gamma_1|$. We estimate k. We have

$$k\beta m^2\leq\sum_{\gamma\in\Gamma_1}|f(\gamma)|^2<\sum_{\gamma\in\Gamma}|f(\gamma)|^2=\beta q^2,$$

hence $k\leq(q/m)^2=\beta^{-2}$ as claimed. □

This theorem used four copies of the set A. A similar result as Theorem 2.5.2 holds for three sets, even for different ones.

Theorem 2.5.3. *If A_1,A_2,A_3 are subsets of G, a commutative group with $|G|=q$ and $|A_i|\geq\beta_i q$, then, for some t, $A_1+A_2+A_3\supset t+B(\gamma_1,\dots,\gamma_k,\eta)$, where k and η depend only on the densities β_i.*

The corresponding result for two copies does not hold, not even for the difference set $A-A$. The reason for this is the following. A Bohr set always contains a long arithmetic progression. This will be proved in a stronger form in the next section.

Prexercise. A Bohr (k,ε)-set in $\mathbb{Z}_q$ contains an arithmetical progression of length n^δ, where $\delta>0$ depends on k and ε.

However, the set $A-A$ may not contain an arithmetic progression of length q^δ, with $\delta=\delta(\beta)$, assuming $|A|\geq\beta n$. The maximal length of the arithmetic progression may be $<e^{\log q^{2/3+\epsilon}}$. On the other hand, it is known that it is $\gg e^{\log q^{1/2-\epsilon}}$ (Green and Bourgain).

2.6 Some facts from the geometry of numbers

We consider sets situated in a Euclidean space $\mathbb{R}^d$.

Definition 2.6.1. A set $L\subset\mathbb{R}^d$ is a *lattice* if it is a discrete subgroup and it is not contained in any smaller dimensional subspace.

Any such lattice is necessarily isomorphic to $\mathbb{Z}^d$; that is, there are linearly independent vectors $e_1,\dots,e_d\in\mathbb{R}^d$ such that

$$L=\{x_1e_1+\cdots+x_de_d : x_i\in\mathbb{Z}\}.$$

Definition 2.6.2. A set $F\subset\mathbb{R}^d$ is a *fundamental domain* of this lattice if the sets $F+x$, $x\in L$, cover R^d without overlap (one representant from each coset of L). (Sometimes overlaps of boundaries is permitted.)

An example is

$$F = \{x_1e_1 + \cdots + x_de_d : 0 \le x_i < 1\}.$$

Exercise 50. Prove that measurable fundamental domains all have the same volume.

On the example of the domain above one can see that this is the absolute value of the determinant formed by the vectors e_i, which is hence independent of the choice of the basis (e_i).

Definition 2.6.3. The common value of volumes of fundamental domains and absolute value of determinants of matrices formed by integral bases is called the *determinant of the lattice.*

Exercise 51. If $L \subset \mathbb{Z}^d$ is a lattice, its determinant is the same as its index in $\mathbb{Z}^d$ as a subgroup.

Definition 2.6.4. Let Q be a closed neighborhood of 0, and let L be a lattice in $\mathbb{R}^d$. The *successive minima* of Q with respect to the lattice are the smallest positive numbers $0 < \lambda_1 \le \cdots \le \lambda_d$ such that there are linearly independent vectors $a_1, \ldots, a_d \in L$, $a_i \in \lambda_i Q$.

Imagine this as follows. Take a small homothetic image εQ and blow it up slowly. First the only lattice point inside is the origin, then at λ_1 another appears. As we increase λ, it may happen that the next lattice points are multiples of a_1, like $2a_1$ at $2\lambda_1$, but at some point λ_2 we get another, which is not a multiple of the first and so on.

Exercise 52. Show that the first appearing vectors a_i may not form a basis of L.

We will need the following important theorem of Minkowski.

Lemma 2.6.5 (Minkowski's inequality for successive minima.). *Let Q be a closed neighbourhood of* 0, *and let L be a lattice in $\mathbb{R}^d$. Let $0 < \lambda_1 \le \cdots \le \lambda_d$ be the successive minima of Q with respect to L. We have*

$$\lambda_1 \cdots \lambda_d \le 2^d \frac{\det L}{\operatorname{vol} Q}. \tag{2.6.1}$$

2.7 A generalized arithmetical progression in a Bohr set

We show that Bohr sets contain large generalized arithmetical progressions. We will do this for cyclic groups only; for general groups see Green and Ruzsa [17].

Theorem 2.7.1. *Let q be a positive integer, $u_1, \ldots, u_d$ residues modulo q such that $(u_1, u_2, \ldots, u_d, q) = 1$, $\varepsilon_1, \ldots, \varepsilon_d$ real numbers satisfying $0 < \varepsilon_j < 1/2$. Write*

$$\delta = \frac{\varepsilon_1 \cdots \varepsilon_d}{d^d}. \tag{2.7.1}$$

There are residues $v_1, \dots, v_d$ *and non-negative integers* $l_1, \dots, l_d$ *such that the set*

$$P = \{v_1 x_1 + \dots + v_d x_d : |x_i| \le l_i\} \tag{2.7.2}$$

satisfies

$$P \subset B(u_1, \dots, u_d; \varepsilon_1, \dots, \varepsilon_d), \tag{2.7.3}$$

the sums in (2.7.2) *are all distinct and*

$$|P| = \|P\| = \prod (2l_j + 1) \ge \prod (l_j + 1) > \delta q. \tag{2.7.4}$$

Proof. Let L be the d-dimensional lattice of integer vectors $(x_1, \dots, x_d)$ satisfying

$$x_1 \equiv xu_1, \dots, x_d \equiv xu_d \pmod q$$

with some integer x. This lattice is the union of q translations of the lattice $(q\mathbb{Z})^d$ (here we need the coprimality condition, otherwise there may be coincidences), hence its determinant is q^{d-1}.

Let Q be the rectangle determined by $|x_j| \le \varepsilon_j$, $j = 1, \dots, d$, and let $\lambda_1, \dots, \lambda_d$ denote the successive minima of Q with respect to the lattice L. These are the smallest positive numbers such that there are linearly independent vectors $a_1, \dots, a_d \in L$, $a_i \in \lambda_i Q$. By Minkowski's inequality (2.6.1) we have

$$\lambda_1 \cdots \lambda_d \le 2^d \frac{\det L}{\operatorname{vol} Q} = \frac{q^{d-1}}{\varepsilon_1 \cdots \varepsilon_d}. \tag{2.7.5}$$

Write

$$a_i = (a_{i1}, \dots, a_{id}).$$

The condition $a_i \in \lambda_i Q$ means that $|a_{ij}| \le \lambda_i \varepsilon_j$. Since $a_i \in L$, there are residues v_i such that $a_{ij} \equiv v_i u_j \pmod q$. These are our v_j's and we put

$$l_i = \left[\frac{q}{d\lambda_i}\right].$$

First we show that $P \subset B$. Consider an $x \in P$, $x = x_1 v_1 + \dots + x_d v_d$. We have

$$xu_j = \sum x_i v_i u_j \equiv \sum x_i a_{ij} \pmod q,$$

consequently

$$\begin{aligned} \left\| \frac{xu_j}{q} \right\| &= \left\| \sum \frac{x_i a_{ij}}{q} \right\| \\ &\le \sum \left| \frac{x_i a_{ij}}{q} \right| \\ &\le \sum \frac{l_i \lambda_i \varepsilon_j}{q} \le \sum \frac{\varepsilon_j}{d} = \varepsilon_j. \end{aligned} \tag{2.7.6}$$

Next we show that these elements are all distinct. If $x_1, \dots, x_d$ and $y_1, \dots, y_d$ give the same sum, then with $z_j = x_j - y_j$ we have

$$\sum z_i v_i \equiv 0 \pmod q, \quad |z_i| \le 2l_i.$$

Multiplying this congruence by u_j we infer that

$$\sum z_i a_{ij} \equiv 0 \pmod q$$

for all j. Moreover, a calculation like above yields

$$\left|\sum z_i a_{ij}\right| \le \sum l_i \lambda_i \varepsilon_j \le 2\varepsilon_j q < q.$$

Consequently $\sum z_i a_{ij} = 0$ for every j, which means that $\sum z_i a_i = 0$; by view of the linear independence of the vectors a_i, $z_i = 0$ for all i.

Finally we prove (2.7.4). We have

$$l_i + 1 > \frac{q}{d\lambda_i},$$

hence

$$\prod (l_i + 1) > \frac{q^d}{d^d \lambda_1 \cdots \lambda_d} \ge \frac{q}{d^d} \varepsilon_1 \cdots \varepsilon_d = \delta q$$

by (2.7.5). □

It is easy to see that the result need not hold if $(u_1, \dots, u_d, q) > 1$; consider, for instance, the case $q = r^2$, $d = 1$, $u_1 = r$. It can be shown that a $(d+1)$-dimensional arithmetical progression can always be found in B.

Lemma 2.7.2. *Let q be a prime, and let A be a nonempty set of residues modulo q with $|A| = \beta q$. There are residues $v_1, \dots, v_d$ and non-negative integers $l_1, \dots, l_d$ such that the set*

$$P = \{v_1 x_1 + \cdots + v_d x_d : |x_i| \le l_i\} \tag{2.7.7}$$

satisfies $P \subset D = 2A - 2A$, the sums in (2.7.7) are all distinct and

$$\|P\| = \prod (2l_j + 1) \ge \prod (l_j + 1) > \delta q, \tag{2.7.8}$$

where $d \le \beta^{-2}$ and

$$\delta = (4d)^{-d} \le (\beta^2/4)^{1/\beta^2}. \tag{2.7.9}$$

Proof. This follows from a combination of Lemma 2.5.2 and Theorem 2.7.1. The assumption that q is a prime guarantees the coprimality assumption required in Theorem 3.1. (??2.7.1??) □

2.8 Freiman's theorem

We prove Freiman's Theorem 2.1.3 in the following form.

Theorem 2.8.1. *Let A, B be finite sets in a torsion-free commutative group satisfying $|A| = |B| = m$, $|A+B| \le \alpha m$. There are numbers d, s depending only on α such that A is contained in a generalized arithmetical progression of dimension at most d and size at most sm.*

Since a bound on $A+B$ immediately gives a bound on $2A$, the generalization to different sets is not important as long as we do not give bounds for d and s. Given a set in a torsion-free group we can find Freiman isomorphic sets in $\mathbb{Z}$, however, it is not completely obvious (though not very difficult) to deduce the existence of a covering progression from that of an isomorphic image. The form above is just the natural one in our treatment.

Proof. We apply Theorem 2.3.5 for $r = 8$ and a prime number $q > |rA - rA|$. By Chebyshev's theorem we can find such a prime with

$$q < 2|rA - rA| \le 2\alpha^{16}m;$$

the second inequality follows from Theorem 1.1.1. We obtain a set $A' \subset A$, which is F_8-isomorphic to a set T of residues modulo q, $|A'| \ge m/r = m/8$.

Applying Lemma 2.7.2 we find a d'-dimensional proper arithmetical progression $P \subset 2T - 2T$ of size $\ge \delta m$, where $d' = d'(\alpha)$ and $\delta = \delta(\alpha) > 0$ depend only on α.

By Lemma 2.3.3 the F_8-isomorphism between T and A' induces an F_2-isomorphism between $2T - 2T$ and $2A' - 2A'$. The image P' of P is a proper d'-dimensional arithmetical progression by Lemma 2.3.2 and we have $P' \subset 2A' - 2A' \subset 2A - 2A$.

Select a maximal collection of elements $a_1, \dots, a_t \in A$ such that the sets $P' + a_i$ are pairwise disjoint. We estimate t. Since these sets are all subsets of $A + P' \subset 3A - 2A$, we have

$$t \le \frac{|3A - 2A|}{\|P'\|} \le \frac{\alpha^5 m}{\delta m} = \alpha^5/\delta(\alpha).$$

For every $a \in A$ there is an a_i such that

$$(a + P') \cap (a_i + P') \ne \emptyset.$$

Thus there are $p, p' \in P'$ such that $a + p = a_i + p'$; that is, $a = a_i + p' - p$. This means that

$$A \subset \{a_1, \dots, a_t\} + P' - P'. \tag{2.8.1}$$

Since P' is a d-dimensional arithmetical progression, so is $P' - P'$, and obviously

$$\|P' - P'\| \le 2^d \|P'\| \le 2^d |2A - 2A| \le 2^d \alpha^4 m.$$

The set $\{a_1, \dots, a_t\}$ can be covered by the t-dimensional arithmetical progression

$$P(a_1, \dots, a_t; 1, \dots, 1; 0).$$

Hence the right side of (2.8.1) can be covered by an arithmetical progression of dimension $d = d' + t$ and size sm, $s = 2^d \alpha^4$. Since both t and d were bounded in terms of α, the proof is completed. □

2.9 Arithmetic progressions in sets with small sumset

We show that sets with small sumset contain long arithmetic progressions. This will be a (necessarily) conditional result, depending on our knowledge of arithmetic progressions in dense sets. The first such result is again due to Freiman [11, Theorem 2.30]. He considered three-term progressions only, since Szemerédi's theorem on long progressions was not yet available.

Let $r_k(n)$ denote the maximal number of integers that can be selected from the interval $[1, n]$ without including a k-term arithmetical progression and write

$$\omega_k(n) = n/r_k(n).$$

Szemerédi's celebrated theorem [61] tells us that $\omega_k(n) \to \infty$ for every fixed k. The best known estimates are due to Gowers [14, 15] for general k, and to Bourgain [4] for $k = 3$.

Theorem 2.9.1. *Assume that $|A| = n$ and A does not contain any k-term arithmetical progression. We have*

$$|A + A - A - A| \geq \frac{1}{4} \omega_k(n) n, \tag{2.9.1}$$

$$|A + B| \geq \frac{1}{\sqrt{2}} \omega_k(n)^{1/4} n^{1/4} |B|^{3/4} \tag{2.9.2}$$

for every set B,

$$|A + B| \geq \frac{1}{\sqrt{2}} \omega_k(n)^{1/4} n \tag{2.9.3}$$

for every set B such that $|B| = n$,

$$|A + A| \geq \frac{1}{\sqrt{2}} \omega_k(n)^{1/4} n, \tag{2.9.4}$$

$$|A - A| \geq \frac{1}{\sqrt{2}} \omega_k(n)^{1/4} n. \tag{2.9.5}$$

By Bourgain's result we have $\omega_3(n) \gg (\log n)^{1/2 - \varepsilon}$. Applying this estimate we obtain the following version of Freiman's theorem.

Corollary 2.9.2. *Assume that* $|A| = n$ *and* A *does not contain any three-term arithmetical progression. For every constant* $c < 1/8$ *and* $n > n_0(c)$ *we have*

$$|A + B| \geq \frac{1}{2} n(\log n)^c \tag{2.9.6}$$

for every set B *such that* $|B| = n$*, in particular*

$$|A + A| \geq \frac{1}{2} n(\log n)^c, \tag{2.9.7}$$

$$|A - A| \geq \frac{1}{2} n(\log n)^c. \tag{2.9.8}$$

Problem 2.9.3. Can the exponent $1/4$ in (2.9.4)–(2.9.5) be improved to 1 or at least to $1 - \varepsilon$?

Proof. Write $|A| = n$ and $|2A - 2A| = \beta n$. We apply the case $r = 2$ of Theorem 2.3.5, part (b). We get a set $A^* \subset A$, $|A^*| \geq n/4$ which is isomorphic to a set $T \subset [0, \beta n/2]$. By Lemma 2.3.2, T contains no k-term arithmetical progression.

Since in an interval of length n there can be at most $r_k(n)$ integers without k-term arithmetical progression and the interval $[0, \beta n/2]$ can be covered by $[1 + \beta/2]$ such intervals, we have

$$n/4 \leq |T| \leq [1 + \beta/2] r_k(n) \leq \beta r_k(n),$$

therefore

$$\beta \geq \frac{1}{4} \frac{n}{r_k(n)},$$

which is equivalent to (2.9.1).

To obtain (2.9.2) we apply Theorem 1.1.1 and (2.9.1):

$$|A + B| \geq |B|^{3/4} |2A - 2A|^{1/4} \geq \frac{1}{\sqrt{2}} |B|^{3/4} \omega_k(n)^{1/4} n^{1/4}.$$

Inequality (2.9.3) is the case $|B| = n$ of (2.9.2), while (2.9.4)–(2.9.5) are the cases $B = A$ and $B = -A$ of (2.9.3). □

Chapter 3

Location and sumsets

3.1 Introduction

This chapter is about questions of the following kind. Assume we have finite sets A, B in a group G. What can we say about $A + B$ if we know the structure of G, or we have some information about how these sets are situated within G? The "what" will be in most cases a lower estimate for the cardinality.

A familiar example is the classical Cauchy–Davenport inequality.

Theorem 3.1.1. *Let p be a prime, $A, B \subset \mathbb{Z}_p$ nonempty sets. We have*

$$|A + B| \geq \min(|A| + |B| - 1, p).$$

Prexercise. Prove the Cauchy–Davenport inequality by comparing $|A + B|$ and $|A' + B'|$ for suitably chosen sets of the form

$$A' = A \cup (B + t), \quad B' = B \cap (A - t).$$

For another example consider Freiman's Theorem 2.1.1: If $A \subset \mathbb{Z}$, $|A| = m$, $|A + A| \leq 3m - 4$, then A is contained in an arithmetic progression of length $\leq |A + A| - m + 1 \leq 2m - 3$.

Definition 3.1.2. The *reduced diameter* $\operatorname{diam} A$ of a set $A \subset \mathbb{Z}$ is the smallest u such that A is contained in an arithmetic progression $\{b, b+q, \dots, b+uq\}$. (Later we shall generalize and rename this concept.)

Now we can formulate Freiman's theorem equivalently as follows.

Theorem 3.1.3. *For any set $A \subset \mathbb{Z}$ with $|A| = m$ and* $\operatorname{diam} A = u$ *we have*

$$|2A| \geq \min(m + u, 3m - 3).$$

This illustrates that the distinction between "direct" and "inverse" or "structural" results is often only a case of style. This chapter will contain results that are more naturally expressed in the "direct" form.

First we consider finite groups, then lattices $\mathbb{Z}^d$, and then more general structures.

3.2 The Cauchy–Davenport inequality

Here we give a proof of Theorem 3.1.1. Several proofs are known, we present a well-known one as outlined in the prexercise above, mainly for the sake of presenting a method in the simplest form which will be used several times later.

This is based on two transformations:

(1) Translation. If we replace A, B by sets $A+x$, $B+y$, the cardinalities of $A, B, A+B$ remain unchanged.

(2) Transfusion (elements go from A to B). We replace A, B by $A' = A \cap B$ and $B' = A \cup B$. This operation does change the cardinalities but preserves their sum:

$$|A'| + |B'| = |A \cap B| + |A \cup B| = |A| + |B| . \tag{3.2.1}$$

It does not increase the sumset: we have

$$A' + B' \subset A + B. \tag{3.2.2}$$

This operation yields a new pair of sets if $A \not\subset B$ and $A \cap B \neq \emptyset$.

Proof of Theorem 3.1.1. Write $|A| = m$, $|B| = n$.

We use induction on m. The case $m = 1$ is obvious. Assume now we know the statement for every pair of sets where $1 \leq |A| \leq m-1$.

Given a pair of sets A, B with $|A| = m$, we try to make a transfusion. If we get a new pair A', B' with $1 \leq |A'| \leq m-1$, then (4.2.1) and (4.2.2) complete the inductive step. If this does not work, we have either $A \subset B$ or $A \cap B = \emptyset$.

Now combine this transfusion with a translation. If it never works, we know that for every x we have either $A + x \subset B$ or $(A+x) \cap B = \emptyset$.

Take now a $y \in A - A$, $y \neq 0$; such a y exists if A has at least two elements. Start with an x such that $(A+x) \cap B \neq \emptyset$ (any $x \in B - A$). Then $A + x \subset B$ by the above dichotomy, and then $(A+x+y) \cap B \neq \emptyset$ again: if $y = a' - a$, then

$$a + x + y = a' + x \in B.$$

By repeating this argument we see that all $x, x+y, x+2y, \ldots$ are in this category. This list contains all elements of $\mathbb{Z}_p$, that is, always $A + x \subset B$. Hence $B = \mathbb{Z}_p$ and the claim holds again evidently. □

A set is *sumfree* if it has no three elements such that $x+y=z$ (so we exclude $2x = z$ too; if we do not, the following exercises change only minimally).

Exercise 53. What is the size of the largest sumfree subset of $[1, n]$?

Exercise 54. What is the size of the largest sumfree subset of $\mathbb{Z}_p$, p prime?

Exercise 55. Every $A \subset \mathbb{N}$, $|A| = n$ has a sumfree subset of cardinality $\geq n/3$.

Exercise 56. Same problem with $n/3 + 1$ for n sufficiently large. (Bourgain's theorem, extremely difficult.)

Exercise 57. The set of positive integers has no partition into finitely many sumfree parts.

3.3 Kneser's theorem

We show how to extend the Cauchy–Davenport theorem to composite moduli and general commutative groups. A verbatim extension fails, since $A + A = A$ if A is a subgroup.

Definition 3.3.1. Let S be a nonempty set in a commutative group G. The *stabilizer* or *group of periods* of S is the set

$$\operatorname{stab} S = \{x \in G : x + S = S\}.$$

(This is clearly a subgroup of G.)

Theorem 3.3.2. *Let A, B be finite sets in a commutative group G, $S = A + B$ and $H = \operatorname{stab} S$. We have*

$$|A + B| \geq |A + H| + |B + H| - |H| . \tag{3.3.1}$$

If (3.3.1) *holds with strict inequality, then*

$$|A + B| \geq |A + H| + |B + H| \geq |A| + |B| . \tag{3.3.2}$$

This clearly implies the Cauchy–Davenport theorem, as in $\mathbb{Z}_p$ the only possibilities are $H = \{0\}$ or $H = \mathbb{Z}_p$.

Lemma 3.3.3. *Let S be a finite set in a group, $S = S_1 \cup S_2$. We have*

$$|S| + |\operatorname{stab} S| \geq \min \left(|S_i| + |\operatorname{stab} S_i|\right) . \tag{3.3.3}$$

Proof. The claim is obvious if $S_i = S$ for either i, so we assume they are proper (and consequently nonempty) subsets.

Write $\operatorname{stab} S_i = H_i$, $\operatorname{stab} S = H$. We may also assume that $H_0 = H_1 \cap H_2 = \{0\}$, since otherwise every set is a union of cosets of H_0 and the claim can be reduced to the corresponding claim in the factor group G/H_0.

Write $|H_i| = h_i$. Let $\overline{H} = H_1 + H_2$; clearly $\left|\overline{H}\right| = h_1 h_2$. Each coset of $\overline{H}$ is the union of h_2 cosets of H_1 as well as h_1 cosets of H_2.

We can rewrite (3.3.3) as

$$|S \setminus S_i| \geq h_i - |H| \text{ for some } i. \tag{3.3.4}$$

We shall see how to find lower estimates for $|S \setminus S_i|$.

Consider a typical nonempty intersection of S with a coset of $\overline{H}$, say $\overline{H} + x$. Some of the h_2 cosets of H_1 inside it, say k_1, are in S_1, and some k_2 of the h_1 cosets of H_2 are in S_2. From each coset of H_1 exactly k_2 elements are in S_2 and $h_1 - k_2$ in $S \setminus S_2$, hence

$$\left|(S \setminus S_2) \cap (\overline{H} + x)\right| = k_1(h_1 - k_2)$$

and similarly

$$\left|(S \setminus S_1) \cap (\overline{H} + x)\right| = k_2(h_2 - k_1).$$

If there is a coset of $\overline{H}$ such that $0 < k_1 < h_2$ and $0 < k_2 < h_1$, then we multiply the above equations to obtain

$$|S \setminus S_2|\,|S \setminus S_1| \geq k_1 k_2 (h_1 - k_2)(h_2 - k_1) \geq (h_1 - 1)(h_2 - 1),$$

hence at least one of the inequalities

$$|S \setminus S_i| \geq h_i - 1$$

is true and we are done.

If there is no such coset, but there is one in which $k_1 = 0 < k_2$ and a different one in which $k_2 = 0 < k_1$, then by using the first coset to estimate $S \setminus S_1$ and the second to estimate $S \setminus S_2$ we get

$$|S \setminus S_2|\,|S \setminus S_1| \geq h_1 h_2,$$

stronger than before.

Finally assume that one of the above possibilities is missing, say the first. In this case we claim that S is a union of cosets of H_1. We check this on each coset of $\overline{H} + x$. This happens obviously if $\overline{H} + x \subset S$. If this inclusion fails, then clearly $k_1 < h_2$ and $k_2 < h_1$, so one of them must vanish; we excluded $k_1 = 0 < k_2$, so $k_2 = 0$.

This means $H \supset H_1$ and then for $i = 1$ the right side of (3.3.4) is ≤ 0. □

Lemma 3.3.4. *Let S be a finite set in a group, $S = S_1 \cup S_2 \cup \cdots \cup S_k$. We have*

$$|S| + |\operatorname{stab} S| \geq \min\left(|S_i| + |\operatorname{stab} S_i|\right). \tag{3.3.5}$$

This follows from the previous one by an immediate induction.

Proof of Kneser's theorem. Fix a $b \in B$, and consider all possible finite sets $A_b, B_b \subset G$ with the properties

$$b \in B_b,\ A_b \supset A,\ A_b + B_b \subset A + B, |A_b| + |B_b| = |A| + |B|. \tag{3.3.6}$$

Such sets do exist, for instance, $A_b = A$, $B_b = B$. Fix from among them one for which $|B_b|$ is minimal. Put $S_b = A_b + B_b$. We have

$$\bigcup S_b = S.$$

Indeed, one inclusion follows from the second inclusion in (3.3.6), the other from $A_b + b \subset S_b$.

We try to find a pair with smaller $|B_b|$ by a transfusion:

$$B' = B_b \cap (A_b - t),\ A' = A_b \cup (B_b + t).$$

To preserve the first condition in (3.3.6) we need $b \in A_b - t$; that is,

$$t \in A_b - b.$$

The inclusions and the equality of cardinality sums hold automatically. The minimality assumption means that each such B' satisfies $B' = B_b$, that is, $B_b \subset A_b - t$, hence $B_b + t \subset A_b$. Forming the union of these inclusions we obtain

$$A_b \supset \bigcup_{t \in A_b - b} (B_b + t) = A_b + B_b - b.$$

This can be reformulated as

$$B_b - b \subset \operatorname{stab} A_b.$$

Clearly $\operatorname{stab} S_b \supset \operatorname{stab} A_b$, so for each b we have

$$|S_b| + |\operatorname{stab} S_b| \geq |S_b| + |B_b| \geq |A_b| + |B_b| = |A| + |B|$$

An application of the previous lemma to these sets S_b gives

$$|S| + |\operatorname{stab} S| \geq \min\left(|S_b| + |\operatorname{stab} S_b|\right) \geq |A| + |B|.$$

If we apply this inequality to the sets $A+H$ and $B+H$, since $A+H+B+H = A+B$ and $\operatorname{stab}(A+H+B+H) = H$, we obtain (3.3.1). To get inequality (3.3.2) observe that each quantity in (3.3.1) is a multiple of $|H|$, so if they are not equal, then the left exceeds the right at least by $|H|$. □

Exercise 58. What is the size of the largest sumfree subset of $\mathbb{Z}_n$, n composite?

3.4 Sumsets and diameter, part 1

In this section we prove Freiman's Theorem 3.1.3 in a generalized form.

Translate A so that its minimal element is 0, and divide each element by their greatest common divisor. After these operations we can write A as

$$A = \{a_1, \dots, a_m\},\ a_1 = 0,\ a_m = u$$

and we know that

$$\gcd(a_1, \dots, a_m) = 1.$$

Under these conditions the claim is $|2A| \ge \min(m+u, 3m-3)$.

This theorem can be extended to the addition of different sets in several ways; we mention two possibilities. In both let $A, B \subset \mathbb{Z}$, $A = \{a_1, \dots, a_m\}$, $B = \{b_1, \dots, b_n\}$ with $0 = a_1 < \dots < a_m = u$, $0 = b_1 < \dots < b_n = v$.

Theorem 3.4.1 (Freiman [10]). *If* $\gcd(a_1, \dots, a_m, b_1, \dots, b_n) = 1$ *and* $u \le v$, *then*

$$|A+B| \ge \min(m+v, m+n+\min(m,n)-3).$$

Theorem 3.4.2 (Lev and Smeliansky [31]). *If* $\gcd(b_1, \dots, b_n) = 1$ *and* $u \le v$, *then*

$$|A+B| \ge \min(m+v, n+2m-2-\delta), \tag{3.4.1}$$

where $\delta = 1$ *if* $u = v$ *and* $\delta = 0$ *if* $u < v$.

Proof. Let A', B' be the images of A, B in $\mathbb{Z}_v$; we have

$$|A'| = m' = m - \delta,\ |B'| = n' = n - 1.$$

Kneser's theorem tells us that

$$|A'+B'| \ge |A'+H| + |B'+H| - |H| \tag{3.4.2}$$

with $H = \operatorname{stab}(A'+B')$. Write $|H| = q$. We have $q|v$ (and then H consists exactly of the multiples of v/q). The choice of the two possibilities in (3.4.1) depends on whether $q = v$ or $q < v$.

In any case we have

$$|A+B| \ge |A'+B'| + m. \tag{3.4.3}$$

Indeed, $A+B$ has at least one element in each residue class of $A'+B'$. We can exhibit m classes when it has at least two, namely those of $a_1, \dots, a_m$ where a_i and $a_i + v$ are those elements if $u < v$. If $u = v$, this is only $m-1$ classes, but in the class of 0 there are three elements, 0, v and $2v$. If $A'+B' = \mathbb{Z}_v$, this gives us the required $v + m$.

If H is a proper subgroup, we will improve (3.4.3) as follows. Write

$$|A'+H| = kq,\ |B'+H| = lq.$$

We have $l \geq 2$; indeed, B cannot be in a proper subgroup by the assumption

$$\gcd(b_1, \ldots, b_n) = 1.$$

The set $A' + B'$ consists of $\geq k + l - 1 > k$ cosets, so there is one free of elements of A'. Fix such a coset. If $A + B$ has t elements with residues in this coset, we can improve (3.4.3) to

$$|A + B| \geq |A' + B'| + m + (t - q), \tag{3.4.4}$$

since in (3.4.3) only the excess in classes of A was counted.

This coset is the sum of a coset in $A' + H$ and one in $B' + H$. Assume that A and B have r and s elements with residues in these classes, respectively. Then $A + B$ has at least $r + s - 1$, so (3.4.4) implies

$$|A + B| \geq |A' + B'| + m + r + s - 1 - q. \tag{3.4.5}$$

In these classes A', B' have at most r and s elements, while $A' + H$, $B' + H$ have exactly q, so we have

$$|A' + H| \geq m' + q - r, \ \ |B' + H| \geq n' + q - s.$$

On substituting this into (3.4.2) and applying (3.4.5) we obtain

$$|A + B| \geq m' + n' + m - 1 = 2m + n - 2 - \delta. \qquad \square$$

3.5 The impact function

Let G be a semigroup (in most cases it will be a commutative group).

Definition 3.5.1. For a fixed finite set $B \subset G$ we define its *impact function* by

$$\xi_B(m) = \xi_B(m, G) = \min\{|A + B| : A \subset G, \ |A| = m\}.$$

This is defined for all positive integers if G is infinite, and for $m \leq |G|$ if G is finite.

This function embodies what can be told about cardinality of sumsets if one of the set is unrestricted up to cardinality. The name is a translation of Plünnecke's "Wirkungsfunktion", who first studied this concept systematically for density [40].

Some of the previous results, like the Cauchy–Davenport inequality, can be reformulated with this concept; some, like Lev and Smeliansky's Theorem 3.4.2 cannot, since about A other assumptions than its size are also used.

Exercise 59. Let G be a finite group. Prove the following "sort of concavity" of the impact function: For $2 \leq n < |G|$, $n \nmid |G|$ there is a number $1 \leq k \leq n - 1$ such that

$$\xi(n - k) + \xi(n + k) \leq 2\xi(n).$$

Exercise 60. Use the previous exercise to deduce the Cauchy–Davenport inequality.

Exercise 61. (= Exercise 6). Let A, B be finite sets in a (not necessarily commutative) torsionfree group. Show that

$$|A+B| \geq |A| + |B| - 1.$$

Exercise 62. In a finite group the graph of the impact function has a certain symmetry with respect to the line $x+y=|G|$. Formulate exactly and prove.

We show that the dependence on G can be omitted.

Theorem 3.5.2. *Let G' be a commutative group, G a subgroup of G', and let $B \subset G$ be a finite set. If G is infinite, we have*

$$\xi_B(m, G') = \xi_B(m, G) \tag{3.5.1}$$

for all m. If G is finite, say $|G| = q$, then for $m = kq + r$, $0 \leq r \leq q-1$, we have

$$\xi_B(m, G') = \xi_B(r, G) + kq. \tag{3.5.2}$$

Proof. Take an $A \subset G'$, $|A| = m$, with $|A+B| = \xi_B(m, G')$. Let $A = A_1 \cup \cdots \cup A_k$ be its decomposition according to cosets of G. For each $1 \leq i \leq k$ take an element x_i from the coset containing A_i so that the sets $A_i - x_i$ are pairwise disjoint; this is easily done as long as G is infinite. The set

$$A' = \bigcup (A_i - x_i)$$

satisfies $A' \subset G$, $|A'| = m$ and

$$|A' + B| \leq \sum |A_i - x_i + B| = \sum |A_i + B| = |A + B| = \xi_B(m, G'),$$

hence $\xi_B(m, G) \leq \xi_B(m, G')$. The inequality in the other direction is obvious.

In the finite case from all the sets A at which the minimum is attained select one for which k is minimal, and with k so fixed $\min |A_i|$ is minimal. We claim that all but one A_i are cosets of G; this clearly implies (3.5.2).

Assume $|A_1| \leq \cdots \leq |A_k|$ and $A_i \subset G + x_i$. We try to replace A_1, A_2 by sets

$$A_1' = A_1 \cap (A_2 - y), \; A_2' = (A_1 + y) \cup A_2$$

with suitable $y \in x_2 - x_1 + G$.

We claim that this operation does not change the cardinality of A and does not increase that of $A+B$. Indeed,

$$\begin{aligned} |A_1'| + |A_2'| &= |A_1' + y| + |A_2'| = |(A_1 + y) \cap A_2| + |(A_1 + y) \cup A_2| \\ &= |A_1 + y| + |A_2| = |A_1| + |A_2| \,. \end{aligned}$$

Write $A_i + B = C_i$, $A'_i + B = C'_i$. Then

$$C'_1 \subset C_1 \cap (C_2 - y), \; C'_2 = (C_1 + y) \cup C_2$$

and the comparison of cardinalities goes like for A_1, A_2.

If A_2 does not fill the complete coset, we can find y so that a prescribed element of A_1 be missing from A'_1 which would give an example with smaller $|A_1|$, or smaller k if $A'_1 = \emptyset$. □

This proof was adapted from arguments in Chapters 3 and 4 of Plünnecke's above-mentioned book [40].

In view of this result we will omit the ambient group G from the notation and write just $\xi_B(m)$ instead.

Let G be a torsion-free group. Take a finite $B \subset G$, and let G' be the subgroup generated by $B - B$, that is, the smallest subgroup such that B is contained in a single coset. Let $B' = B - a$ with some $a \in B$, so that $B' \subset G'$. The group G', as any finitely generated torsion-free group, is isomorphic to the additive group $\mathbb{Z}^d$ for some d. Let $\varphi : G' \to \mathbb{Z}^d$ be such an isomorphism and $B'' = \varphi(B')$. By Theorem 3.5.2 we have

$$\xi_B = \xi_{B'} = \xi_{B''},$$

so when studying the impact function we can restrict our attention to sets in $\mathbb{Z}^d$ that contain the origin and generate the whole lattice; we then study the set "in its natural habitat".

Definition 3.5.3. Let B be a finite set in a torsion-free group G. By the *dimension* of B we mean the number d defined above, and denote it by $\dim B$.

Observe that this dimension is not necessarily equal to the geometrical dimension. In the case when $B \subset \mathbb{R}^k$ with some k, this is its dimension over the field of rationals.

The reduced diameter makes sense exactly for one-dimensional sets.

3.6 Estimates for the impact function in one dimension

We give some estimates that use the diameter and cardinality.

It is possible to give an estimate using the diameter only.

Theorem 3.6.1. *Let B be a one-dimensional set in a torsion-free commutative group,* $\operatorname{diam} B = v \geq 3$.

(a) *For*

$$m > \frac{(v-1)(v-2)}{2}$$

we have $\xi_B(m) = m + v$.

(b) *If*

$$\frac{(k-1)(k-2)}{2} < m \le \frac{k(k-1)}{2}$$

with some integer $2 \le k < v$, *then* $\xi_B(m) \ge m + k$.

Equality holds for the set $B = \{0, 1, v\} \subset \mathbb{Z}$.

For $v \le 2$ we have obviously $\xi_B(m) = m + v$ for all m (such a set cannot be anything else than a $(v+1)$-term arithmetic progression).

This will be deduced from the following result, where the cardinality of B is also taken into account.

Theorem 3.6.2. *Let* B *be a one-dimensional set in a torsion-free commutative group,* $\operatorname{diam} B = v \ge 3$, $|B| = n$. *Define* w *by*

$$w = \min_{d|v,\, d\le n-2} d \left\lceil \frac{n-2}{d} \right\rceil. \tag{3.6.1}$$

For every m *we have*

$$\xi_B(m) \ge m + \min\left(v, \frac{w}{2} + \min_{t\in\mathbb{N}} \left(\frac{m}{t} + \frac{tw}{2}\right)\right). \tag{3.6.2}$$

The minimum is attained at one of the integers surrounding $\sqrt{2m/w}$. Unlike the previous theorem, typically we do not have examples of equality, and the extremal value and the structure of extremal sets are probably complicated. Also the value of w depends on divisibility properties of v and n. After the proof we give some less exact but simpler corollaries.

Proof. By Lemma 3.5.2 we may assume that $B \subset \mathbb{Z}$, its smallest element is 0 and it generates $\mathbb{Z}$; then its largest element is just v.

Lemma 3.6.3. *Let* B' *be the set of residues of elements of* B *modulo* v. *For every nonempty* $X \subset \mathbb{Z}_v$ *we have*

$$|X + B'| \ge \min(|X| + w, v). \tag{3.6.3}$$

Proof. By Kneser's theorem we have

$$|X + B'| \ge |X + H| + |B' + H| - |H|$$

with some subgroup H of the additive group $\mathbb{Z}_v$. Write $|H| = d$; clearly $d|v$. If $d = v$, we have $|X + H| = v$ and we are ready. Assume $d < v$. B' contains 0 and it generates $\mathbb{Z}_v$, hence it cannot be contained in H so we have $|B' + H| \ge 2|H| = 2d$. This gives the desired bound if $d > n - 2$. Assume $d \le n - 2$. Since $|B' + H|$ is a multiple of d and it is at least $|B'| = n - 1$, we obtain

$$|B' + H| \ge d\left\lceil \frac{n-1}{d} \right\rceil = d\left(1 + \left\lceil \frac{n-2}{d} \right\rceil\right) \ge d + w. \qquad \square$$

We resume the proof of Theorem 3.6.2. Take a set $A \subset \mathbb{Z}$, $|A| = m$. We are going to estimate $|A+B|$ from below.

For $j \in \mathbb{Z}_v$ let $u(j)$ be the number of integers $a \in A$, $a \equiv j \pmod{v}$, and let $U(j)$ be the corresponding number for the sumset $A+B$. We have

$$U(j) \geq u(j) + 1 \tag{3.6.4}$$

whenever $U(j) > 0$; this follows by adding the numbers $0, v$ to each element of A in this residue class if $u(j) > 0$, and holds obviously for $u(j) = 0$. We also have

$$U(j) \geq u(j-b) \tag{3.6.5}$$

for every $b \in B'$. Write

$$r(k) = \{j : u(j) \geq k\},$$

$$R(k) = \{j : U(j) \geq k\}.$$

Inequality (3.6.4) implies

$$R(k) \supset r(k-1) \ (k \geq 2), \tag{3.6.6}$$

and inequality (3.6.5) implies

$$R(k) \supset r(k) + B' \ (k \geq 1). \tag{3.6.7}$$

First case: $U(j) > 0$ for all j. In this case by summing (3.6.4) we get

$$|A+B| = \sum U(j) \geq v + \sum u(j) = |A| + v.$$

Second case: there is a j with $U(j) = 0$. Then we have $|R(k)| < v$ for every $k > 0$. An application of Lemma 3.6.3 to the sets $r(k)$ yields, by view of (3.6.7),

$$|R(k)| \geq |r(k)| + w \tag{3.6.8}$$

as long as $r(k) \neq \emptyset$. Let t be the largest integer with $r(t) > 0$. We have (3.6.8) for $1 \leq k \leq t$, and (3.6.6) yields

$$|R(k)| \geq |r(k-1)| \tag{3.6.9}$$

for all $k \geq 2$. Consequently, for $1 \leq k \leq t+1$ we have

$$|R(k)| \geq \frac{k-1}{t}|r(k-1)| + \left(1 - \frac{k-1}{t}\right)(|r(k)| + w). \tag{3.6.10}$$

Indeed, for $k = 1$ (3.6.10) is identical with (3.6.8), for $k = t+1$ it is identical with (3.6.9) and for $2 \leq k \leq t$ it is a linear combination of the two.

By summing (3.6.10) we obtain

$$|A+B| = \sum_{k\geq 1} |R(k)| \geq \sum_{k=1}^{t+1} |R(k)|$$

$$\geq \frac{t+1}{2}w + \left(1+\frac{1}{t}\right)\sum_{k=1}^{t} |r(k)| = \frac{t+1}{2}w + \left(1+\frac{1}{t}\right)|A|,$$

as claimed in (3.6.2). □

Corollary 3.6.4. *With the assumptions and notations of Theorem* 3.6.2 *we have*

$$\xi_B(m) \geq \min\left(m+v, \left(\sqrt{m}+\sqrt{w/2}\right)^2\right). \tag{3.6.11}$$

Proof. This follows from (3.6.2) and the inequality of arithmetic and geometric means. □

Proof of Theorem 3.6.1. Parts (a)–(b) of the theorem can be reformulated as follows: If $\xi_B(m) \leq m+k$ with some $k < v$, then $m \leq k(k-1)/2$. Theorem 3.6.2 yields (using only that $w \geq 1$) the existence of a positive integer t such that

$$\frac{m}{t} + \frac{t+1}{2} \leq k,$$

hence

$$m \leq kt - \frac{t(t+1)}{2}.$$

The right side, as a function of t, is increasing up to $k-1/2$ and decreasing afterwards; the maximal values at integers are assumed at $t = k-1$ and k, and both are equal to $k(k-1)/2$.

To show the case of equality in case (b), write $m = k(k-1)/2 - l$ with $0 \leq l \leq k-2$. The set A will contain the integers in the intervals $[iv, iv+k-3-i]$ for $0 \leq i \leq l-1$ and $[iv, iv+k-2-i]$ for $l \leq i \leq k-2$. □

In comparison to the results of Section 3.4 observe that they never give an increment exceeding $2n$, they are, however, better for small values of m.

Problem 3.6.5. Find a common generalization of Theorems 3.6.2 and 3.4.2.

3.7 Multi-dimensional sets

The first result that connects additive properties to geometrical dimension is perhaps the following theorem of Freiman.

Theorem 3.7.1 (Freiman [11, Lemma 1.14]). *Let $A \subset \mathbb{R}^d$ be a finite set, $|A| = m$. Assume that A is proper d-dimensional, that is, it is not contained in any affine hyperplane. Then*

$$|A + A| \geq (d+1)m - \frac{d(d+1)}{2}.$$

Proof. We use induction on m.

The starting case is $m = 2$. Then necessarily $d = 1$ and the claim is $|2A| \geq 3$, which is indeed true.

Assume now that the statement is true for m and all possible values of d (which are $1 \leq d \leq m-1$). We prove it for $m+1$. Let $A \subset \mathbb{R}^d$, $|A| = m+1$. Consider the convex hull of A, and let $a \in A$ be one of its vertices. Put $A' = A \setminus \{a\}$. We have $|A'| = m$, so the statement is true for A'.

The dimension of A' may be d or $d-1$.

Assume first that this dimension is d. Consider the supporting planes of the convex hull A'. Such a plane L intersects conv A' in one of its sides, whose vertices are elements of A', hence $|A' \cap L| \geq d$, and the rest of A' is on one side. At least one of these planes has the property that a is on the other side. Fix such a plane. Then none of the points of $a + (A' \cap L)$ is a point of $2A'$. Hence

$$\begin{aligned} |2A| &\geq |2A'| + |A' \cap L| + 1 \quad \text{(the $+1$ comes from the element $2a$)} \\ &\geq (d+1)m - \frac{d(d+1)}{2} + d + 1 = (d+1)(m+1) - \frac{d(d+1)}{2} + d + 1 \end{aligned}$$

as wanted.

Consider now the case when the dimension of A' is $d-1$. Then A' lies on a plane L and the point a is outside it, hence the sets $2A'$, $a + A'$ and $\{2a\}$ are all disjoint and again we obtain

$$\begin{aligned} |2A| &\geq |2A'| + |A'| + 1 \\ &\geq dm - \frac{(d-1)d}{2} + m + 1 = (d+1)(m+1) - \frac{d(d+1)}{2} + d + 1. \end{aligned}$$

□

This theorem is exact, equality can occur, namely it holds when A is a "long simplex", a set of the form

$$L_{dm} = \{0, e_1, 2e_1, \dots, (m-d)e_1, e_2, e_3, \dots, e_d\}. \tag{3.7.1}$$

In particular, if no assumption is made on the dimension, then the minimal possible cardinality of the sumset is $2m-1$, with equality for arithmetic progressions.

This result can be extended to sums of different sets. This extension is problematic from the beginning, namely the assumption "d-dimensional" can be interpreted in different ways. We can stipulate that both sets be d-dimensional, or only one, or, in the weakest form, make this assumption on the sumset only.

An immediate extension of Freiman's above result goes as follows.

Theorem 3.7.2 ([56, Corollary 1.1]). *If $A, B \subset \mathbb{R}^d$, $|A| \le |B|$ and $\dim(A+B) = d$, then we have*

$$|A+B| \ge |B| + d|A| - \frac{d(d+1)}{2}.$$

We can compare these results to the continuous case. Let A, B be Borel sets in $\mathbb{R}^d$; μ will denote the Lebesgue measure. The celebrated Brunn–Minkowski inequality asserts that

$$\mu(A+B)^{1/d} \ge \mu(A)^{1/d} + \mu(B)^{1/d}, \tag{3.7.2}$$

and here equality holds if A and B are homothetic convex sets, and under mild and natural assumptions this is the only case of equality. It can also be observed that the case $A = B$ is completely obvious here: we have

$$\mu(A+A) \ge \mu(2 \cdot A) = 2^d \mu(A).$$

Also the constant 2^d is much larger than the constant $d+1$ in Theorem 3.7.1. This is necessary, as there are examples of equality, however, one feels that this is an exceptional phenomenon and better estimations should hold for "typical" sets. A further difference is the asymmetrical nature of the discrete result and the symmetry of the continuous one. Finally, when $|A|$ is fixed, Theorem 3.7.2 gives a linear increment, while (3.7.2) yields

$$\mu(A+B) \ge \mu(B) + d\mu(A)^{1/d} \mu(B)^{1-1/d}.$$

In the next section we tell what can be said if we use cardinality as the discrete analogue of measure, and prescribe only the dimension of the sets. Later we try to find other spatial properties that may be used to study sumsets.

The main problems are perhaps the following. What are the best analogues of measure and dimension for discrete sets? How should a discrete analogue of the Brunn–Minkowski inequality look like? The partial answers explained below also suggest questions in the continuous case. Should we be satisfied with the usual concepts of measure and dimension for studying the addition of sets? We return to this in Chapter 5.

Mostly our sets will be in a Euclidean space $\mathbb{R}^d$, and $e_1, \dots, e_d$ will be the system of unit vectors. We can think of the *dimension* $\dim A$ of a set $A \subset \mathbb{R}^d$ as the dimension of the smallest affine hyperplane containing A, or as in Definition 3.5.3.

3.8 Results using cardinality and dimension

We consider finite sets in a Euclidean space $\mathbb{R}^d$.

Put

$$F_d(m,n) = \min\{|A+B| : |A| = m, |B| = n,\ \dim(A+B) = d\},$$
$$F'_d(m,n) = \min\{|A+B| : |A| = m, |B| = n,\ \dim B = d\},$$
$$F''_d(m,n) = \min\{|A+B| : |A| = m, |B| = n,\ \dim A = \dim B = d\}.$$

F_d is defined for $m+n \geq d+2$, F'_d for $n \geq d+1$ and F''_d for $m \geq d+1, n \geq d+1$. F_d and F''_d are obviously symmetric, while F'_d may not be (and, in fact, we will see that for certain values of m, n it is not), and they are connected by the obvious inequalities

$$F_d(m,n) \leq F'_d(m,n) \leq F''_d(m,n).$$

We determined the behavior of F_d and of F'_d for $m \leq n$. The more difficult problem of describing F''_d and F'_d for $m > n$ was solved by Gardner and Gronchi [13]; we shall quote their results later.

To describe F_d define another function G_d as follows:

$$G_d(m,n) = n + \sum_{j=1}^{m-1} \min(d, n-j), \quad n \geq m \geq 1,$$

and for $m > n$ extend it symmetrically, putting $G_d(m,n) = G_d(n,m)$. In other words, if $n - m \geq d$, then we have

$$G_d(m,n) = n + d(m-1).$$

If $0 \leq t = n - m < d$, then for $n > d$ we have

$$G_d(m,n) = n + d(m-1) - \frac{(d-t)(d-t-1)}{2} = n(d+1) - \frac{d(d+1)}{2} - \frac{t(t+1)}{2},$$

and for $n \leq d$

$$G_d(m,n) = n + \frac{(m-1)(2n-m)}{2}.$$

With this notation we have the following result.

Theorem 3.8.1 ([56, Theorem 1]). *For all positive integers m, n and d satisfying $m + n \geq d + 2$ we have*

$$F_d(m,n) \geq G_d(m,n).$$

Theorem 3.7.2 is an immediate consequence.

Theorem 3.8.1 is typically exact; the next theorem summarizes the cases when we have examples of equality.

Theorem 3.8.2 ([56, Theorem 2]). *Assume $1 \leq m \leq n$. We have*

$$F_d(m,n) = F'_d(m,n) = G_d(m,n)$$

unless either $n < d+1$ or $m \leq n - m \leq d$ (in this case $n \leq 2d$).

The construction goes as follows.

Assume $1 \leq m \leq n$, $n \geq d+1$. Let B be a long simplex, $B = L_{dn}$ as defined in (3.7.1).

If $n - m \geq d$, we put

$$A = \{0e_1, 1e_1, \ldots, (m-1)e_1\}.$$

This set satisfies $|A| = m$. The set $A + B$ consists of the vectors ie_1, $0 \le i \le n + m - d - 1$, and the vectors $ie_1 + e_j$, $0 \le i \le m - 1$, $2 \le j \le d$, consequently

$$|A + B| = n + d(m - 1) = G_d(m, n).$$

If $n - m = t < d$, write $t = d - k$ and assume $k \le m$. Now A is defined by

$$A = \{0e_1, 1e_1, \dots, (m - k)e_1\} \cup \{e_2, \dots, e_k\}.$$

This set satisfies $|A| = m$. The set $A+B$ consists of the vectors ie_1, $0 \le i \le 2(n-d)$, the vectors $ie_1 + e_j$, $0 \le i \le n - d$, $2 \le j \le d$, finally $e_i + e_j$, $2 \le i \le k$, $2 \le j \le d$, hence

$$\begin{aligned}|A + B| &= 2(n - d) + 1 + (d - 1)(n - d + 1) + d(k - 1) - \frac{k(k - 1)}{2} \\ &= n(d + 1) - \frac{d(d + 1)}{2} - \frac{t(t + 1)}{2} = G_d(m, n).\end{aligned}$$

These constructions cover all pairs m, n except those listed in Theorem 3.8.2. Observe that A is also a long simplex of lower dimension. For a few small values the exact bounds are yet to be determined.

We now describe Gardner and Gronchi's bound [13] for $F'_d(m, n)$. Informally, their main result (Theorem 5.1) asserts that the $|A + B|$ is minimalized when $B = L_{dn}$, a long simplex, and A is as near to the set of points inside a homothetic simplex as possible. More exactly, they define (for a fixed value of n) the weight of a point $x = (x_1, \dots, x_d)$ as

$$w(x) = \frac{x_1}{n - d} + x_2 + \dots + x_d.$$

This defines an ordering by writing $x < y$ if either $w(x) < w(y)$ or $w(x) = w(y)$ and for some j we have $x_j > y_j$ and $x_i = y_i$ for $i < j$.

Let D_{dmn} be the collection of the first m vectors with non-negative integer coordinates in this ordering. We have $D_{dnn} = L_{dn} = B$, and, more generally, $D_{dmn} = rB$ for any integer m such that

$$m = |rB| = (n - d)\binom{r + d - 1}{d} + \binom{r + d - 1}{d - 1}.$$

For such values of m we also have

$$|A + B| = |(r + 1)B| = (n - d)\binom{r + d}{d} + \binom{r + d}{d - 1}.$$

With this notation their result sounds as follows.

Theorem 3.8.3 (Gardner and Gronchi [13, Theorem 5.1]). *If $A, B \subset \mathbb{R}^d$, $|A| = m$, $|B| = n$ and $\dim B = d$, then we have*

$$|A + B| \ge |D_{dmn} + L_{dn}|.$$

For $m < n$ this reproves Theorem 3.8.2. For $m \geq n$ the extremal set D_{dmn} is also d-dimensional, thus this result also gives the value of F''_d.

Corollary 3.8.4. *For $m \geq n > d$ we have*

$$F''_d(m,n) = F'_d(m,n) = |D_{dmn} + L_{dn}|.$$

A formula for the value of this function is given in [13, Section 6]. We quote some interesting consequences.

Theorem 3.8.5 (Gardner and Gronchi [13, Theorem 6.5]). *If $A, B \subset \mathbb{R}^d$, $|A| = m \geq |B| = n$ and* $\dim B = d$, *then we have*

$$|A + B| \geq m + (d-1)n + (n-d)^{1-1/d}(m-d)^{1/d} - \frac{d(d-1)}{2}.$$

Theorem 3.8.6 (Gardner and Gronchi [13, Theorem 6.6]). *If $A, B \subset \mathbb{R}^d$, $|A| = m$, $|B| = n$ and* $\dim B = d$, *then we have*

$$|A + B|^{1/d} \geq m^{1/d} + \left(\frac{n-d}{d!}\right)^{1/d}.$$

This result is as close to the Brunn–Minkowski inequality as we can get by using only the cardinality of the summands.

3.9 The impact function and the hull volume

Let G be a torsion-free group. Take a finite $B \subset G$. At the end of Section 3.5 we defined a certain "natural image" of B as follows. Let G' be the subgroup generated by $B - B$ and $B' = B - a$ with some $a \in B$, so that $B' \subset G'$. The group G' is isomorphic to the additive group $\mathbb{Z}^d$ for some d. Let $\varphi : G' \to \mathbb{Z}^d$ be such an isomorphism and $B'' = \varphi(B')$. By Theorem 3.5.2 we know

$$\xi_B = \xi_{B'} = \xi_{B''},$$

so when studying the impact function we can restrict our attention to sets in $\mathbb{Z}^d$ that contain the origin and generate the whole lattice. We used this d as a definition for an "intrinsic dimension". This image has further usages.

Definition 3.9.1. Let B be a finite set in a torsion-free group G. By the *hull volume* of B we mean the volume of the convex hull of the set B'' described above and denote it by $\operatorname{hv} B$.

The set B'' is determined up to an automorphism of $\mathbb{Z}^d$. These automorphisms are exactly linear maps of determinant ± 1, hence the hull volume is uniquely defined.

Theorem 3.9.2. *Let B be a finite set in a torsion-free group G, $d = \dim B$, $v = \operatorname{hv} B$. We have*

$$\lim |kB|k^{-d} = v.$$

A proof can be found in [57, Section 11], though this form is not explicitly stated there. An outline is as follows. By using the arguments above we may assume that $B \subset \mathbb{Z}^d$, $0 \in B$ and B generates $\mathbb{Z}^d$. Let B^* be the convex hull of B. Then kB is contained in $k \cdot B^*$. The number of lattice points in $k \cdot B$ is asymptotically $\mu(k \cdot B^*) = k^d v$; this yields an upper estimate. To get a lower estimate one proves that with some constant p, kB contains all the lattice points inside translate of $(k-p) \cdot B^*$; this is Lemma 11.2 of [57].

This means that the hull volume can be defined without any reference to convexity and measure. Later we will show that this definition can even be extended to commutative semigroups.

It turns out that in $\mathbb{Z}^d$, hence in any torsion-free group, the dimension and hull volume determine the asymptotic behavior of the impact function.

Theorem 3.9.3. *Let B be a finite set in a torsion-free commutative group G, $d = \dim B$, $v = \operatorname{hv} B$. We have*

$$\lim \xi_B(m)^{1/d} - m^{1/d} = v^{1/d}.$$

This is the main result (Theorem 3.1) of [57]. In the same paper we announce the same result for non-necessarily torsion-free commutative groups without proof (Theorem 3.4). In a general semigroup $A + B$ may consist of a single element, so an attempt to an immediate generalization fails.

Problem 3.9.4. Does the limit $\lim \xi_B(m)^{1/d} - m^{1/d}$ exist in general commutative semigroups? Is there a condition weaker than cancellativity to guarantee its positivity?

Theorem 3.9.3 can be effectivized as follows (see Theorems 3.2 and 3.3 of [57]).

Theorem 3.9.5. *With the notations of the previous theorem, if $d \geq 2$ and $m \geq v$, we have*

$$\xi_B(m) \leq m + dv^{1/d}m^{1-1/d} + c_1 v^{2/d} m^{1-2/d},$$

$$\xi_B(m)^{1/d} - m^{1/d} \leq v^{1/d} + c_2 v^{2/d} m^{-1/d}$$

(c_1, c_2 depend on d). With $n = |B|$ for large m we have

$$\xi_B(m) \geq m + dv^{1/d}m^{1-1/d} - c_3 v^{\frac{d+3}{2d}} n^{-1/2} m^{1-\frac{3}{2d}},$$

$$\xi_B(m)^{1/d} - m^{1/d} \geq v^{1/d} - c_4 v^{\frac{d+3}{2d}} n^{-1/2} m^{-1/(2d)}.$$

Probably the real error terms are much smaller than these estimates. For $d = 1$ we have the obvious inequality $\xi_B(m) \le m + v$, with equality for large m because the integers $\xi_B(m) - m$ cannot converge to v otherwise. For $d = 2$ already $\sqrt{\xi_B(m)} - \sqrt{m}$ can converge to $\sqrt{v}$ from both directions.

Theorem 3.9.6. *The impact function of the set $B = \{0, e_1, e_2\} \subset \mathbb{Z}^2$ satisfies*

$$\sqrt{\xi_B(m)} - \sqrt{m} > \sqrt{v} \tag{3.9.1}$$

for all m.

The impact function of the set $B = \{0, e_1, e_2, -(e_1 + e_2)\} \subset \mathbb{Z}^2$ satisfies

$$\sqrt{\xi_B(m)} - \sqrt{m} < \sqrt{v} \tag{3.9.2}$$

for infinitely many m.

Inequality (3.9.1) was announced in [57] without proof as Theorem 4.1, and it is a special case of Gardner and Gronchi's Theorem 3.8.6. Inequality (3.9.2) is Theorem 4.3 of [57].

We cannot decide whether there is a set such that $\sqrt{\xi_B(m)} - \sqrt{m} < \sqrt{v}$ for all m.

3.10 The impact volume

Besides cardinality we saw the hull volume as a contender for the title "discrete volume". For both we had something resembling the Brunn–Minkowski inequality; for cardinality we had Gardner and Gronchi's Theorem 3.8.6, which has the (necessary) factor $d!$, and for the hull volume we have Theorem 3.9.3, which only holds asymptotically.

There is an easy way to find a quantity for which the analogue of the Brunn–Minkowski inequality holds exactly: we can make it a definition.

Definition 3.10.1. The d-dimensional *impact volume* of a set B (in an arbitrarily commutative group) is the quantity

$$\operatorname{iv}_d(B) = \inf_{m \in \mathbb{N}} \left(\xi_B(m)^{1/d} - m^{1/d} \right)^d .$$

Note that the d above may differ from the dimension of B, in fact, it need not be an integer. It seems, however, that the only really interesting case is $d = \dim B$.

The following statement list some immediate consequences of this definition.

Statement 3.10.2. *Let B be a finite set in a commutative torsion-free group.*

(a) $\operatorname{iv}_d(B)$ *is a decreasing function of d.*

(b) *If* $|B| = n$, *then*

$$\mathrm{iv}_1(B) = n - 1$$

and

$$\mathrm{iv}_d(B) \le \left(n^{1/d} - 1\right)^d \tag{3.10.1}$$

for every d.

(c) $\mathrm{iv}_d(B) = 0$ *for* $d > \dim B$.

(d) *For every pair* A, B *of finite sets in the same group and every* d *we have*

$$\mathrm{iv}_d(A+B)^{1/d} \ge \mathrm{iv}_d(A)^{1/d} + \mathrm{iv}_d(B)^{1/d}. \tag{3.10.2}$$

The price we have to pay for the discrete Brunn–Minkowski inequality (3.10.2) is that there is no easy way to compute the impact volume for a general set. We have the following estimates.

Theorem 3.10.3. *Let* B *be a finite set in a commutative torsion-free group,* $\dim B = d$, $|B| = n$. *We have*

$$\binom{n-d}{d!} \le \mathrm{iv}_d(B) \le \mathrm{hv}\, B, \tag{3.10.3}$$

with equality in both places if B *is a long simplex.*

The first inequality follows form Theorem 3.8.6 of Gardner and Gronchi, the second from Theorem 3.9.3.

Problem 3.10.4. What is the *maximal* possible value of $\mathrm{iv}_d(B)$ for n-element d-dimensional sets? Is perhaps the bound in (3.10.1) exact?

We now describe the impact volume for another important class of sets, namely cubes.

Theorem 3.10.5. *Let* $n_1, \ldots, n_d$ *be positive integers and let*

$$B = \{(x_1, \ldots, x_d) \in \mathbb{Z}^d : 0 \le x_i \le n_i\}. \tag{3.10.4}$$

We have

$$\mathrm{iv}_d(B) = \mathrm{hv}\, B = v = n_1 \cdots n_d.$$

Problem 3.10.6. Is it true that when B is the set of lattice points within a convex lattice polytope, then $\mathrm{hv}\, B$ and $\mathrm{iv}_d(B)$ are very near?

They may differ, as the second example in Theorem 3.9.6 shows.

We shall deduce Theorem 3.10.5 from the following one.

Theorem 3.10.7. *Let* $G = G_1 \times G_2$ *be a commutative group represented as the direct product of the groups* G_1 *and* G_2. *Let* $B = B_1 \times B_2 \subset G$ *be a finite set with* $B_1 \subset G_1$, $B_2 \subset G_2$. *We have*

$$\mathrm{iv}_d(B) \ge \mathrm{iv}_{d-1}(B_1)\mathrm{iv}_1(B_2). \tag{3.10.5}$$

Proof. Write $\mathrm{iv}_d(B) = v$, $\mathrm{iv}_{d-1}(B_1) = v_1$, $\mathrm{iv}_1(B_2) = v_2$ (which is $= |B_2| - 1$ if G_2 is torsion-free). We want to estimate $|A + B|$ from below for a general set $A \subset G$ with $|A| = m$.

First we transform them to some standard form; this will be the procedure what Gardner and Gronchi call compression. Let A_1 be the projection of A to G_1, and for an $x \in A_1$ write

$$A(x) = \{y \in G_2 : (x, y) \in A\}.$$

Let

$$A' = \{(x, i) : x \in A_1, i \in \mathbb{Z}, 0 \leq i \leq |A(x)| - 1\}$$

and

$$B' = \{(x, i) : x \in B_1, i \in \mathbb{Z}, 0 \leq i \leq v_2\}.$$

We have $A', B' \subset G' = G_1 \times \mathbb{Z}$.

Lemma 3.10.8. *We have*

$$|A'| = |A|, \ |A' + B'| \leq |A + B|. \tag{3.10.6}$$

Proof. The equality is clear. To prove the inequality, write $S = A+B$, $S' = A'+B'$. With the obvious notation, we will show that

$$|S'(x)| \leq |S(x)|$$

for each x. To this end observe that

$$S(x) = \bigcup_{x'+x''=x} (A(x') + B(x'')) = \bigcup_{x' \in x - B_1} A(x') + B_2,$$

hence

$$|S(x)| \geq \max_{x' \in x - B_1} |A(x') + B_2| \geq \max_{x' \in x - B_1} |A(x')| + v_2.$$

Similarly,

$$S'(x) = \bigcup_{x'+x''=x} (A'(x') + B'(x'')) = \bigcup_{x' \in x - B_1} [0, |A(x')| + v_2 - 1],$$

and so

$$|S'(x)| = \max_{x' \in x - B_1} |A(x')| + v_2. \qquad \square$$

Now we continue the proof of the theorem. Decompose A' into layers according to the value of the second component; write

$$A' = \bigcup_{i=0}^{k} L_i \times \{i\},$$

where $k = \max|A(x)|$, $L_i \subset G_1$. Write $|L_i| = m_i$. We have $L_0 \supset L_1 \supset \cdots \supset L_k$, consequently $m_0 \geq m_1 \geq \cdots \geq m_k$.

The set S' is the union of the sets $(L_i + B_1) \times \{i+j\}$, $0 \leq i \leq k$, $0 \leq j \leq v_2$. By the above inclusion it is sufficient to consider the L_i with the smallest possible i; that is,

$$S' = (L_0 + B_1) \times \{0, 1, \dots, v_2\} \cup \bigcup_{i=1}^{k} (L_i + B_1) \times \{i + v_2\}.$$

We obtain that

$$|S'| = v_2|L_0 + B_1| + \sum_{i=0}^{k} |L_i + B_1|. \tag{3.10.7}$$

To estimate the summands we use the $(d-1)$-dimensional impact of B_1. Recall that by definition this means that

$$|X + B_1| \geq \left(|X|^{\frac{1}{d-1}} + v_1^{\frac{1}{d-1}}\right)^{d-1}$$

for any set X. We apply this to the sets L_i to obtain

$$|L_i + B_1| \geq \left(m_i^{\frac{1}{d-1}} + v_1^{\frac{1}{d-1}}\right)^{d-1} \geq \frac{m_i}{m_0}\left(m_0^{\frac{1}{d-1}} + v_1^{\frac{1}{d-1}}\right)^{d-1};$$

the second inequality follows from $m_i \leq m_0$. By substituting this into (3.10.7) and recalling that $\sum m_i = m$ we obtain

$$|S| \geq \left(v_2 + \frac{m}{m_0}\right)\left(m_0^{\frac{1}{d-1}} + v_1^{\frac{1}{d-1}}\right)^{d-1}. \tag{3.10.8}$$

Consider the right side as a function of the real variable m_0. By differentiation we find that it assumes its minimum at

$$m_0 = v_1^{1/d}(m/v_2)^{1-1/d}.$$

(This minimum typically is not attained; this m_0 may be < 1 or $> m$, and it is generally not integer.) Substituting this value of m_0 into (3.10.8) we obtain the desired bound

$$|S| \geq \left(m^{1/d} + (v_1 v_2)^{1/d}\right)^d. \qquad \square$$

Problem 3.10.9. Does equality always hold in Theorem 3.10.7?

We expect a negative answer.

Problem 3.10.10. Can Theorem 3.10.7 be extended to an inequality of the form

$$\mathrm{iv}_{d_1+d_2}(B_1 \times B_2) \geq \mathrm{iv}_{d_1}(B_1)\mathrm{iv}_{d_2}(B_2)?$$

Proof of Theorem 3.10.5. To prove $\geq$ we use induction on d. The case $d = 1$ is obvious, and Theorem 3.10.7 provides the inductive step.

This means that with the cube B defined in (3.10.4) we have

$$|A + B| \geq \left(|A|^{1/d} + v^{1/d}\right)^d.$$

Equality can occur for infinitely many values of $|A|$, namely it holds whenever A is also a cube of the form

$$A = \{(x_1, \dots, x_d) \in \mathbb{Z}^d : 0 \leq x_i \leq kn_i - 1\}$$

with some integer k; we have $|A| = k^d v$, $|A + B| = (k+1)^d v$. It may be difficult to describe $\xi_B(m)$ for values of m which are not of the form $k^d v$. Possibly an argument like Gardner and Gronchi's for the simplex may work.

Observe that these special sets A are not homothetic to B; in particular, $A = B$ may not yield a case of equality. □

As Theorem 3.10.3 shows, the impact volume can be $d!$ times smaller than cardinality. The example we have of this phenomenon, the long simplex, is, however, "barely" d-dimensional, and we expect that a better estimates hold for a "substantially" d-dimensional set.

Definition 3.10.11. The *thickness* $\vartheta(B)$ of a set $B \subset \mathbb{R}^d$ is the smallest integer k with the property that there is a hyperplane P of $\mathbb{R}^d$ and $x_1, \dots, x_k \in \mathbb{R}^d$ such that $B \subset \bigcup_{i=1}^k P + x_j$.

Conjecture 3.10.12. For every $\varepsilon > 0$ and d there is a k such that for every $B \subset \mathbb{R}^d$ with $\vartheta(B) > k$ we have $\mathrm{iv}_d(B) > (1 - \varepsilon)|B|$.

This conjecture would yield a discrete Brunn–Minkowski inequality of the form

$$|A + B|^{1/d} \geq |A|^{1/d} + (1 - \varepsilon)|B|^{1/d}$$

assuming a bound on the thickness of B. Such an inequality is true at least in the special case $A = B$. This can be deduced from a result of Freiman ([11, Lemma 2.12]; see also Bilu [2]), which sounds as follows. If $A \subset \mathbb{R}^d$ and $|2A| < (2^d - \varepsilon)|A|$, then there is a hyperplane P such that $|P \cap A| > \delta|A|$, with $\delta = \delta(d, \varepsilon) > 0$.

3.11 Hovanskii's theorem

We saw examples where cardinalities of sumsets can behave wildly. We show that, in a rather general setting, if we keep on adding the same set persistently, then these irregularities fade.

Theorem 3.11.1. *Let A be a finite set in a commutative semigroup G. There is a polynomial f and an integer n_0 such that for $n > n_0$ we have*

$$|nA| = f(n).$$

This theorem is due to Hovanskii [25, 26]. A generalization to the effect that $|n_1A_1 + \cdots + n_kA_k|$ becomes a polynomial if all the n_i are large is given by Nathanson [36]. Another proof of this theorem was given by Nathanson and Ruzsa [37]. Below we give this proof for the case of one variable only.

Unfortunately there is no way to tell this polynomial (except the leading term, see [37]) and the threshold n_0.

Proof. We assume that G has a zero element. If it does not, extend it by a new element (this is only a notational convenience).

Let $A = \{a_1, \dots, a_m\}$. The elements of nA are all sums of the form

$$b = \sum x_i a_i, \ \sum x_i = n,$$

where the coefficients x_i are non-negative integers and $0a_i$ is the zero of G. We shall consider these coefficients together in the form of a vector $\mathbf{x} = (x_1, \dots, x_m)$ with non-negative integer coordinates.

Several vectors may induce the same b. From the possible representations we shall select the *lexicographically first.* We write $\mathbf{x} \prec \mathbf{y}$ and say that $\mathbf{x}$ *precedes* $\mathbf{y}$ if there is an i, $1 \le i \le m$, such that $x_1 = y_1, \dots, x_{i-1} = y_{i-1}$, $x_i < y_i$. By the *rank* $r(\mathbf{x})$ of a vector we mean the sum of its coordinates.

We say that a vector $\mathbf{x}$ is *useless*, if there is a $\mathbf{y} \prec \mathbf{x}$ of the same rank such that $\sum x_i a_i = \sum y_i a_i$, and we call it *useful*, if no such $\mathbf{y}$ exists. With this terminology, $|nA|$ is the number of useful vectors of rank n.

Write $\mathbf{x} \le \mathbf{y}$ if $x_i \le y_i$ for each coordinate, and $\mathbf{x} < \mathbf{y}$ if $\mathbf{x} \le \mathbf{y}$ and $\mathbf{x} \ne \mathbf{y}$. If $\mathbf{x}$ is useless and $\mathbf{x} \le \mathbf{x}'$, then $\mathbf{x}'$ is also useless. Indeed, take a $\mathbf{y}$ of the same rank as $\mathbf{x}$, such that $\sum x_i a_i = \sum y_i a_i$ and $\mathbf{y} \prec \mathbf{x}$. Then by adding $\sum (x_i' - x_i) a_i$ to both sides of this equation we find a vector, namely $\mathbf{y}' = \mathbf{y} + \mathbf{x}' - \mathbf{x}$, that precedes $\mathbf{x}'$, has the same rank and induces the same product.

We say that $\mathbf{z}$ is *primitive useless*, if it is useless and there is no useless $\mathbf{x}$ satisfying $\mathbf{x} < \mathbf{z}$. Clearly a vector $\mathbf{x}$ is useless if and only if there is a primitive useless $\mathbf{z}$ such that $\mathbf{z} \le \mathbf{x}$.

By definition, the primitive useless vectors are all incomparable with respect to the relation $<$. It is a well-known (and easy) fact that any collection of incomparable vectors (with non-negative integer coordinates) must be finite.

Exercise 63. Prove this finiteness claim.

Hence there are only finitely many primitive useless vectors, say $\mathbf{z}_1, \dots, \mathbf{z}_k$. By the sieve formula we have

$$|nA| = \sum_{j=0}^{k} (-1)^j \sum_{i_1, \dots, i_j} B(n; i_1, \dots, i_j),$$

where

$$B(n; i_1, \dots, i_j) = \#\{\mathbf{x} : r(\mathbf{x}) = n, \mathbf{x} \ge \mathbf{z}_{i_1}, \dots, \mathbf{x} \ge \mathbf{z}_{i_j}\}.$$

The system of inequalities $\mathbf{x} \geq \mathbf{z}_{i_t}$ is equivalent to a single inequality $\mathbf{x} \geq \mathbf{z}$, where each coordinate of $\mathbf{z}$ is the maximum of the corresponding coordinates of the vectors $\mathbf{z}_{i_t}$. The number of vectors $\mathbf{x}$ satisfying $\mathbf{x} \geq \mathbf{z}$ and $r(\mathbf{x}) = n$ is 0 if $r(\mathbf{z}) > n$, and it is equal to the number of vectors of rank $n - r(\mathbf{z})$ otherwise. This latter is equal to

$$\binom{n - r(\mathbf{z}) + k - 1}{k - 1},$$

a polynomial in n. Hence all the summands $B(n; \dots)$ are polynomials for large n, thus so is $|nA|$. □

Remark. A corresponding result will not hold without the assumption of commutativity; indeed, if the elements of A generate a free semigroup, then we have $|nA| = k^n$. In a non-commutative semigroup, $|nA|$ need not be monotonically increasing. However, we cannot decide the following.

Problem 3.11.2. Let S be a non-commutative group. Suppose that there are positive constants c, C such that $|nA| \leq Cn^c$. Does it follow that $|nA|$ is a polynomial for large n?

This theorem enables us to define dimension and volume in semigroups in a way that is compatible with our notions in $\mathbb{Z}^d$.

Definition 3.11.3. Let B be a finite set in a commutative semigroup, and let vk^d be the leading term of the polynomial which coincides with $|kB|$ for large k. By the *dimension* of B we mean the degree d of this polynomial, and by the *hull volume* we mean the leading coefficient v.

Chapter 4

Density

4.1 Asymptotic and Schnirelmann density

A finite set is naturally measured by its cardinality. A set of reals is naturally measured by its Lebesgue measure (non-measurable sets do exist, just we never meet them). There is no similarly universal way to measure and compare infinite sets of integers. The most naturally defined one is the asymptotic density.

For a set A of integers we shall use the same letter to denote its counting function

$$A(x) = |A \cap [1, x]| .$$

We allow A to contain 0 or negative numbers, but they are not taken into account in the counting function.

Definition 4.1.1. The *asymptotic density* of a set A of integers is defined by

$$d(A) = \lim_{x \to \infty} A(x)/x,$$

if this limit exists. The *lower* and *upper* (asymptotic) densities are the corresponding lower and upper limits, respectively:

$$\underline{d}(A) = \liminf_{x \to \infty} A(x)/x, \quad \overline{d}(A) = \limsup_{x \to \infty} A(x)/x.$$

Exercise 64. If $\underline{d}(A) > 0$, is there always an $A' \subset A$ with $d(A') > 0$?

Exercise 65. If $\underline{d}(A) + \underline{d}(B) > 1$, then $A + B$ contains all but finitely many positive integers.

Exercise 66. Let α, β, γ be positive real numbers such that $\alpha + \beta \leq \gamma \leq 1$. Construct sets of positive integers such that $d(A) = \alpha$, $d(B) = \beta$, $d(A + B) = \gamma$.

As we mentioned in the introduction, combinatorial additive theory grew out of the classical one, by Schnirelmann's approach to the Goldbach problem. Goldbach's conjecture asserts that any integer > 3 can be expressed as a sum of two or three primes, depending on parity. Schnirelmann proved the weaker result that there is a bound k so that every large enough integer is a sum of at most k primes.

The best universe to work with will be the set $\mathbb{N}_0$ of non-negative integers.

Definition 4.1.2. A set $A \subset \mathbb{N}_0$ is an *additive basis of order* h, if $hA = \mathbb{N}_0$, that is, every positive integer can be expressed as a sum of h integers from A.

A set $A \subset \mathbb{N}_0$ is an *asymptotic basis of order* h, if every sufficiently large integer can be expressed as a sum of h integers from A, that is, $\mathbb{N}_0 \setminus hA$ is finite.

The smallest such integer h is called the *exact order* or *exact asymptotic order* of A, respectively.

So the proper wording is that the set P of primes forms an asymptotic basis. To be a basis, a set must contain 0 and 1.

To this end Schnirelmann established that integers that can be written as a sum of two primes have positive density; and every set having positive density is a basis. An exact form of the first claim is simply

$$\underline{d}(2P) > 0, \tag{4.1.1}$$

a result which is (in hindsight) not too difficult to prove by sieve methods. Today we know that almost all even integers can be written as a sum of two primes, hence $d(2P) = 1/2$.

To formulate the second claim exactly Schnirelmann introduced a different notion of density.

Definition 4.1.3. The *Schnirelmann density* of a set A of integers is the number

$$\sigma(A) = \inf_{n \in \mathbb{N}} A(n)/n.$$

This is a less natural concept than asymptotic density. Asymptotic density is translation invariant and it is invariant under the exclusion or inclusion of finitely many elements; Schnirelmann density does not have either property, in fact, $\sigma(A) = 0$ if $1 \notin A$.

Exercise 67. $\sigma(A) > 0$ if and only if $1 \in A$ and $\underline{d}(A) > 0$.

Exercise 68. Let $\sigma(A) = \alpha$. Show the existence of an $A' \subset A$ such that $\sigma(A') = \alpha$, but by omitting any single element the density of the remaining set will be $< \alpha$.

Exercise 69. Let $B \subset \mathbb{N}_0 = \mathbb{N} \cup \{0\}$ and a number $\alpha \in (0, 1)$ be given, and define β as

$$\beta = \inf\{\sigma(A + B) : \sigma(A) \geq \alpha\}.$$

Show the existence of a set A satisfying $\sigma(A) = \alpha$, $\sigma(A + B) = \beta$.

Exercise 70. Show that we have always

$$\underline{d}(A) = \sup \sigma(A - n).$$

Show that we cannot replace the supremum by maximum.

Prexercise. If $\sigma(A) + \sigma(B) > 1$ and $0 \in A$, then $A + B \supset \mathbb{N}$. (See Theorem 4.2.2 below.)

In these terms Schnirelmann's result sounds as follows.

Theorem 4.1.4. *If $0 \in A$ and $\sigma(A) > 0$, then A is a basis.*

This theorem will be proved and an estimate for the order of this basis in terms of $\sigma(A)$ will be given in the next sections.

Exercise 71. How does it follow from (4.1.1) and the theorem above that P is an asymptotic basis?

4.2 Schirelmann's inequality

Schnirelmann deduced his Theorem 4.1.4 from the following inequality.

Theorem 4.2.1. *Let A and B be sets of non-negative integers with positive Schnirelmann densities $\sigma(A) = \alpha$ and $\sigma(B) = \beta$, respectively. If $0 \in A \cup B$, then*

$$\sigma(A + B) \geq \alpha + \beta - \alpha\beta. \tag{4.2.1}$$

Proof. Without restricting generality we can assume $0 \in A$. Put $C = A + B$; we are going to estimate $C(n)$ for an arbitrary positive integer n. Let

$$1 = b_1 < \cdots < b_k \leq n$$

be the elements of B in $[1, n]$. We have $k = B(n) \geq \beta n$. Since $0 \in A$, these numbers are also in C. Further elements of C are given by

$$b_1 + (A \cap [1, b_2 - b_1 - 1]),\ b_2 + (A \cap [1, b_3 - b_2 - 1]), \ldots,$$
$$b_{k-1} + (A \cap [1, b_k - b_{k-1} - 1]),\ b_k + (A \cap [1, n - b_k]).$$

(The last block may be empty if $b_k = n$.) We estimate the number of elements in a typical block by

$$|A \cap [1, m]| = A(m) \geq \alpha m.$$

(This is also true for $m = 0$, which may be the case for the last block.) Adding these estimates for the blocks above we obtain

$$\alpha\big((b_2 - b_1 - 1) + (b_3 - b_2 - 1) + \cdots + (b_k - b_{k-1} - 1) + (n - b_k)\big) = \alpha(n - k).$$

Consequently,

$$C(n) \geq k + \alpha(n - k) = \alpha n + (1 - \alpha)k \geq \alpha n + (1 - \alpha)\beta n = (\alpha + \beta - \alpha\beta)n. \qquad \square$$

Clearly this inequality also holds in the degenerate case when $\alpha = 0$, provided $0 \in A$.

Prexercise. Construct sets A, B satisfying $0 \in A$, $0 < \sigma(A), \sigma(B) < 1$ and

$$\sigma(A + B) = \sigma(A) + \sigma(B) - \sigma(A)\sigma(B).$$

Prexercise. Show that for every pair of sets satisfying the previous exercise the values of $\sigma(A)$ and $\sigma(B)$ are always rational.

We can write (4.2.1) in the symmetric form

$$1 - \sigma(A + B) \leq \bigl(1 - \sigma(A)\bigr)\bigl(1 - \sigma(B)\bigr).$$

An iterated application then gives

$$1 - \sigma(hA) \leq \bigl(1 - \sigma(A)\bigr)^h, \tag{4.2.2}$$

which will become small but not quite 0. We complement this inequality with the following result.

Theorem 4.2.2. *Let A and B be sets of non-negative integers with positive Schnirelmann densities $\sigma(A) = \alpha$ and $\sigma(B) = \beta$, respectively. If $\alpha + \beta \geq 1$ and $0 \in A \cup B$, then $A + B \supset \mathbb{N}$.*

Proof. Assume $0 \in A$ and take a positive integer n. We want to prove that $n \in A + B$. If $n \in B$, we are done, so assume now that $n \notin B$. This implies

$$B(n - 1) = B(n) \geq \beta n.$$

Consider the pairs $(i, n - i)$ with $1 \leq i \leq n - 1$. In $A(n - 1)$ cases we have $i \in A$, and in $B(n - 1)$ cases we have $n - i \in B$. Since

$$A(n - 1) + B(n - 1) \geq \alpha(n - 1) + \beta n > n - 1,$$

at least once both happen. □

We can now prove Schnirelmann's theorem on bases.

Proof of Theorem 4.1.4. Assume that $\sigma(A) = \alpha > 0$. Take an integer h such that $(1 - \alpha)^h < 1/2$. Then $\sigma(hA) > 1/2$ according to (4.2.2), and so $2hA \supset \mathbb{N}$ by the previous theorem. □

The above argument estimates the order of this basis by $(\log 4)/\alpha$. We will see that the optimal estimate is $1/\alpha$.

Exercise 72. Given an integer h, construct a set A such that $0 \in A$, $\sigma(A) = 1/h$ and the exact order of A is h.

4.3 Mann's theorem

In Schnirelmann's theorem the role of the sets A, B is asymmetric: one of them contains 0 and the other need not. We now show how this inequality can be improved under the symmetric condition $0 \in A \cap B$, which is also better suited for the repeated addition of the same set.

Theorem 4.3.1 (Mann). *If* $0 \in A \cap B$, *then*

$$\sigma(A+B) \geq \min\bigl(1, \sigma(A)+\sigma(B)\bigr).$$

By iteration, if $0 \in A$, then $\sigma(kA) \geq \min\bigl(1, k\sigma(A)\bigr)$ and thus, next corollary follows.

Corollary 4.3.2. *If* $0 \in A$ *and* $\sigma A = \alpha > 0$, *then* A *is a basis of order* $\leq 1/\alpha$.

Exercise 73. Let α, β, γ be positive real numbers such that $\alpha + \beta \leq \gamma \leq 1$. Construct sets of positive integers such that $\sigma(A) = \alpha$, $\sigma(B) = \alpha$, $\sigma(A+B) = \gamma$. (See Lepson [30].)

This theorem is similar to the Cauchy–Davenport inequality: superadditivity save an obstruction, which in our case consists in densities being bounded by 1. The proof will also be based on a transfusion method. However, while a transfusion preserves the sum of cardinalities, it does typically change the sum of densities. It does not change the value of $A(n)+B(n)$ for any n, and this suggests the following approach.

Definition 4.3.3. The *joint (Schnirelmann) density* of the sets $A_1, \dots, A_k$ is defined by

$$\sigma(A_1, \dots, A_k) = \inf \frac{A_1(n) + \cdots + A_k(n)}{n}.$$

Now Theorem 4.3.1 will follow from the version below.

Theorem 4.3.4. *If* $0 \in A \cap B$, *then*

$$\sigma(A+B) \geq \min\bigl(1, \sigma(A, B)\bigr).$$

Formally Schnirelmann density is a limit, a thing related to infinity, but it gives information for every $A(n)$; it is perhaps not surprising that the above theorem will be proved in a finite setting.

Theorem 4.3.5. *Let* $0 \leq \gamma \leq 1$, *let* n *be a positive integer and let* A, B *be sets such that* $0 \in A \cap B$. *Put* $C = A + B$. *If*

$$A(k) + B(k) \geq \gamma k \quad \textit{for } 1 \leq k \leq n, \tag{4.3.1}$$

then

$$C(k) \geq \gamma k \quad \textit{for } 1 \leq k \leq n. \tag{4.3.2}$$

Exercise 74. Deduce Theorem 4.3.5 from Theorem 4.3.4.

We present another slight (but useful) generalization.

Theorem 4.3.6. *Let* $0 \le \gamma \le 1$, $0 \le \delta \le 1-\gamma$, *let* n *be a positive integer and let* A, B *be sets such that* $0 \in A \cap B$. *Put* $C = A + B$. *If*

$$A(k) + B(k) \ge \gamma k - \delta \text{ for } 1 \le k \le n, \tag{4.3.3}$$

then

$$C(k) \ge \gamma k - \delta \text{ for } 1 \le k \le n. \tag{4.3.4}$$

The other interesting case is $\delta = 1-\gamma$, which can be reformulated as follows.

Theorem 4.3.7 (Van der Corput [6]). *Let* $0 \le \gamma \le 1$, *let* n *be a positive integer and let* A, B *be sets such that* $0 \in A \cap B$. *Put* $C = A + B$. *If*

$$1 + A(k) + B(k) \ge \gamma(k+1) \text{ for } 1 \le k \le n, \tag{4.3.5}$$

then

$$1 + C(k) \ge \gamma(k+1) \text{ for } 1 \le k \le n. \tag{4.3.6}$$

Proof. Suppose the above statement is false; then among the counterexamples there is one with the smallest value of n, and with n fixed, with the minimal value of $B(n)$. We consider this example now. We may assume that

$$A, B \subset [0, n],$$

since omitting the element outside this range does not change the assumptions or the conclusion.

If $n = 1$, then either $1 \in A \cup B$ and then $C(1) = 1 \ge \gamma - \delta$, or $1 \notin A \cup B$ and $\gamma = 0$, so $C(1) = 0 \ge -\delta$. So assume $n \ge 2$.

If $B = \{0\}$, then the statement is obviously true. Assume B has also positive elements.

We try to make a translation-transfusion in the following form: we try to replace A, B by

$$A' = A \cup (B + t), \; B' = B \cap (A - t)$$

with suitable t. Any such pair of sets satisfies $A' + B' \subset A + B$. Hence this will also be a counterexample, provided it satisfies conditions $0 \in A \cap B$ and (4.3.3). The first is equivalent to $t \in A$; we will return to the second.

This pair of sets will contradict the minimality assumption if $B' \ne B$; that is,

$$B' \not\subset A - t.$$

Such values of t do exist, for instance, the maximal element of A has this property. From such values of t we choose the minimal one. The minimality of t means that

$$a \in A, \quad a < t \Longrightarrow B \subset A - a \Longrightarrow B + a \subset A. \tag{4.3.7}$$

A consequence of (4.3.7) (with $a = 0$) is

$$B \subset A.$$

Another one is

$$A(x) = C(x) \geq \gamma x - \delta \text{ for } x < t.$$

Indeed, (4.3.7) means that any sum of the form $b + a$, $b \in B, a \in A$, with $a < t$ is in A, and this includes all cases when $a + b < t$. Furthermore, restricting the inclusion (4.3.7) for a fixed $b \in B$ and $a \leq x < t$ we see that

$$b + \big(A \cap [0, x]\big) \subset A \cap [b, b + x].$$

Comparing the cardinalities we obtain

$$A(b + x) - A(b - 1) \geq A(x) + 1 = C(x) + 1 \geq \gamma x - \delta + 1 \tag{4.3.8}$$

for $x < t$.

Our aim is to show that

$$A'(k) + B'(k) \geq \gamma k - \delta \text{ for } 1 \leq k \leq n. \tag{4.3.9}$$

From the definition of A', B' we immediately see that

$$A'(k) + B'(k) \geq A(k) + B(k - t),$$

hence this holds if $B(k - t) = B(k)$, in particular, if $t = 0$. So we may assume that $t > 0$ and $B(k - t) < B(k)$. This means that there are elements of B in the interval $(k - t, k]$; let b' be the smallest of them. Write $k = b' + x$, $0 \leq x < t$.

We have

$$\begin{aligned} A'(k) + B'(k) &\geq A(k) + B(k - t) = A(k) + B(b' - 1) \\ &= \big(A(b' - 1) + B(b' - 1)\big) + \big(A(k) - A(b' - 1)\big). \end{aligned}$$

We estimate the first term by the induction hypothesis, the second one by (4.3.8):

$$A'(k) + B'(k) \geq \big(\gamma(b' - 1) + \delta\big) + \big(\gamma(k - b') + 1 - \delta.\big) = \gamma k + 1 - \gamma - 2\delta \geq \gamma k - \delta;$$

in the last step we need the assumption $\gamma + \delta \leq 1$. □

4.4 Schnirelmann's theorem revisited

In Schnirelmann's Theorem 4.1.4 equality can hold for *certain* values of α and β (Exercise 4.2). Lepson [30] showed that in Mann's Theorem 4.3.1 equality can hold for any α and β (Exercise 73 above).

By writing

$$S(\alpha,\beta) = \inf\{\sigma(A+B) : \sigma(A)=\alpha,\ \sigma(B)=\beta,\ 0\in A\}$$

and

$$M(\alpha,\beta) = \inf\{\sigma(A+B) : \sigma(A)=\alpha,\ \sigma(B)=\beta,\ 0\in A\cap B\},$$

we can restate Schnirelmann's, Mann's and Lepson's results as

$$\alpha+\beta-\alpha\beta \le S(\alpha,\beta) \le M(\alpha,\beta) = \min(\alpha+\beta,1). \tag{4.4.1}$$

If $\alpha+\beta>1$, then we have $S(\alpha,\beta)=M(\alpha,\beta)=1$, (Schnirelmann's Theorem 4.2.2), thus in both inequalities of (4.4.1) equality can actually occur.

In this section we give a formula for $S(\alpha,\beta)$ and describe the cases of equality in inequalities (4.4.1). These results are from Hegedűs–Piroska–Ruzsa [22].

Theorem 4.4.1. *For all α,β we have*

$$S(\alpha,\beta) = \inf_{n\ge 0} \frac{\lceil \alpha n\rceil + \lceil \beta(n+1)\rceil}{n+1}. \tag{4.4.2}$$

Definition 4.4.2. Let α,β be positive real numbers satisfying $\alpha+\beta\le 1$. We call (α,β) a *Schnirelmann pair* if $S(\alpha,\beta)=\alpha+\beta-\alpha\beta$, and a *Mann pair* if $S(\alpha,\beta)=\alpha+\beta$.

Theorem 4.4.3. *The numbers (α,β) form a Schnirelmann pair if and only if they can be expressed as*

$$\alpha=\frac{k}{n},\quad \beta=\frac{1}{n+1}$$

with certain integers $n\ge 2$ and $1\le k\le n-1$.

Proof of Theorem 4.4.1. Denote the right side of (4.4.2) by γ. First we show that $S(\alpha,\beta)\ge\gamma$. Since $\sigma(B)>0$, we have $1\in B$. Write $B'=B-1$; thus $0\in B'$. We will apply Theorem 4.3.7 to the sets A,B'; the requirement that both contain 0 is hence fulfilled.

Next we show that the sets A,B' satisfy (4.3.5). Indeed, by the definition of the Schnirelmann density we have $A(k)\ge\alpha k$, and since it must be an integer, we have

$$A(k)\ge\lceil\alpha k\rceil.$$

We have

$$B'(k)=|B'\cap[1,k]|=|B\cap[2,k+1]|=B(k+1)-1\ge\beta(k+1)-1,$$

and again this is an integer, thus

$$B'(k)\ge\lceil\beta(k+1)\rceil-1.$$

On adding these inequalities we find that

$$1 + A(k) + B'(k) \geq \lceil \alpha k \rceil + \lceil \beta(k+1) \rceil \geq \gamma(k+1)$$

by the definition of γ.

An application of Theorem 4.3.7 to the sets A, B' yields that their sum $C' = A + B'$ satisfies

$$1 + C'(n) \geq \gamma(n+1)$$

for all n.

Since $C = A + B$ is connected to C' via $C = C' + 1$, we conclude that

$$C(n) = |C \cap [1,n]| = |C' \cap [0, n-1]| = 1 + C'(n-1) \geq \gamma n$$

for all n, which is equivalent to saying $\sigma(C) \geq \gamma$.

To show that $S(\alpha, \beta) \leq \gamma$, suppose first that the infimum in the definition (1.4) is a minimum, and let n be any integer satisfying

$$\gamma = \frac{\lceil \alpha n \rceil + \lceil \beta(n+1) \rceil}{n+1}.$$

Consider the sets

$$A_0 = \{0, 1, \dots, \lceil \alpha n \rceil\} \cup \{n+1, n+2, \dots\}$$

and

$$B_0 = \{1, \dots, \lceil \beta(n+1) \rceil\} \cup \{n+2, n+3, \dots\}.$$

These sets satisfy

$$\sigma(A_0) = \frac{\lceil \alpha n \rceil}{n} \geq \alpha$$

and

$$\sigma(B_0) = \frac{\lceil \beta(n+1) \rceil}{n+1} \geq \beta.$$

We can select subsets $A \subset A_0$ and $B \subset B_0$ such that $\sigma(A) = \alpha$, $\sigma(B) = \beta$ and $0 \in A$. These sets satisfy

$$A + B \subset A_0 + B_0 = \{1, 2, \dots, \lceil \alpha n \rceil + \lceil \beta(n+1) \rceil\} \cup \{n+2, \dots\},$$

consequently (by evaluating the counting function at $n+1$) we find that

$$\sigma(A+B) \leq \sigma(A_0 + B_0) \leq \frac{\lceil \alpha n \rceil + \lceil \beta(n+1) \rceil}{n+1} = \gamma$$

as wanted.

Suppose next that the infimum is not attained. In this case we have

$$s = \inf_{n \geq 0} \frac{\lceil \alpha n \rceil + \lceil \beta(n+1) \rceil}{n+1} = \lim_{n \to \infty} \frac{\lceil \alpha n \rceil + \lceil \beta(n+1) \rceil}{n+1} = \alpha + \beta,$$

hence the example of equality in Mann's theorem serves also as an example for $S(\alpha, \beta) \leq \gamma$. □

Prof of Theorem 4.4.3. By Theorem 4.4.1, α and β form a Schnirelmann pair if and only if

$$\inf_{n\geq 0} \frac{\lceil \alpha n\rceil + \lceil \beta(n+1)\rceil}{n+1} = \alpha+\beta-\alpha\beta. \tag{4.4.3}$$

Since the limit of the left side of (4.4.3) is $\alpha+\beta$, in this case there must be an n such that

$$\frac{\lceil \alpha n\rceil + \lceil \beta(n+1)\rceil}{n+1} = \alpha+\beta-\alpha\beta.$$

Observe that the value of the left side for $n=0$ is 1, so we must have $n\geq 1$. Write

$$\lceil \alpha n\rceil = k, \quad \lceil \beta(n+1)\rceil = l.$$

We have $\alpha n \leq k$ and $\beta(n+1)\leq l$, hence $k\neq 0$, $l\neq 0$ and

$$\alpha \leq k/n, \quad \beta \leq l/(n+1). \tag{4.4.4}$$

By the monotonicity of the function $\alpha+\beta-\alpha\beta$ in both variables (in our domain), we have

$$\alpha+\beta-\alpha\beta \leq \frac{k}{n} + \frac{l}{n+1} - \frac{k}{n}\frac{l}{n+1} = \frac{k+l}{n+1} - \frac{k(l-1)}{n(n+1)}.$$

Since $l\geq 1$, the last expression is always $\leq (k+l)/(n+1)$, and equality can hold only if $l=1$ and both inequalities in (4.4.4) hold with equality. This means that $\alpha = k/n$ and $\beta = l/(n+1) = 1/(n+1)$ as claimed. □

We mention, without proof, some results on Mann pairs.

Theorem 4.4.4. *If α and β form a Mann pair, then they are either both rational or both irrational. A pair of rational numbers, say $\alpha = p/q$, $\beta = r/s$, is a Mann pair if and only if they satisfy*

$$\{\alpha(1-n)\} + \{-\beta n\} \geq \alpha \tag{4.4.5}$$

for every integer $1\leq n\leq \operatorname{lcm}[q,s]$. A pair of irrational numbers is a Mann pair if and only if there are integers k, l, m such that

$$\alpha k + \beta l = m, \quad 0<k<1/\alpha,\ 0\leq k-l<1/\alpha. \tag{4.4.6}$$

The description of rational Mann pairs is less satisfactory than that of irrational ones, though it provides a finite algorithm for each pair of rational numbers. The following can be observed.

Statement 4.4.5. *Let α, β be rational numbers, and write $\alpha/\beta = a/b$ with $(a,b)=1$. If there are integers satisfying (4.4.6), then (α,β) is a Mann pair. In particular, if $\alpha \leq 1/(a+b)$, then it is a Mann pair.*

The difficulty is that the set P will be a lattice in the rational case, and there seems no easy way to decide when a lattice intersects a triangle.

We note that condition (4.4.6) is not necessary in the rational case. This is seen by the examples $\alpha = 4/11$, $\beta = 5/11$ or $\alpha = 8/65$, $\beta = 2/13$.

4.5 Kneser's theorem, density form

In the previous sections we discussed addition theorems based on Schnirelmann density. An analogous result was found for the more natural concept of asymptotic density by Kneser [27]. This is more complicated than the privous ones and its proof is difficult. We state it without proof; a proof can be found in Halberstam and Roth's monograph [21].

Theorem 4.5.1 (Kneser). *Let A and B be sets of positive integers. Either*

$$\underline{d}(A+B) \geq \underline{d}(A) + \underline{d}(B),$$

or there exists positive integers q, k, l such that $q \geq k+l-1$ and

(a) *A is contained in k residue classes modulo q,*

(b) *B is contained in l residue classes modulo q,*

(c) *$A+B$ is equal to $k+l-1$ residue classes modulo q except a finite set.*

A density cannot exceed 1; in the above formulation the case when $\underline{d}(A) + \underline{d}(B) > 1$ is included as the extremal case $q = k = l = 1$.

A typical example of the second case is $A = \{1, \dots, k \mod q\}$, the first k residue classes modulo q and $B = \{1, \dots, l \mod q\}$, the first l classes modulo q.

Exercise 75. Suppose $\underline{d}(A)+\underline{d}(B) = 1$. Show that $A+B$ has an asymptotic density, and find its possible values.

4.6 Adding a basis: Erdős' theorem

The previous results gave estimates for the density of a sumset using the density of summands. Sometimes a density increment occurs also when we add a set of density 0. The first example of this phenomenon was given by Hinchin in 1933. He proved that for the set Q of non-negative squares we have $\sigma(A+Q) > \sigma(A)$ whenever $0 < \sigma(A) < 1$. A few years later Erdős proved that every basis has this property. In this section we give an account of this result.

Theorem 4.6.1. *Let $B \subset \mathbb{Z}$ be a basis of order k and let $A \subset \mathbb{Z}$. Then*

$$\sigma(A+B) \geq \sigma(A) + \frac{\sigma(A)\,(1-\sigma(A))}{2k}.$$

Proof. Write $\alpha = \sigma(A)$ and let $C = A+B$. We are going to estimate $C(n)$.

We will try to find a $b \in B$ with a large proportion of $(A+b)$ outside A; this will make $A \cup A+b$ large. For that purpose define

$$f(t) = |((A+t) \setminus A) \cap [1, n]|.$$

This function has a subadditivity property. Indeed, we have

$$\begin{aligned}(A+x+y)\setminus A &\subset ((A+x+y)\setminus(A+x))\cup((A+x)\setminus A)\\ &= ((A+y)\setminus A)+x\cup((A+x)\setminus A),\end{aligned}$$

and by comparing the cardinalities we easily find

$$f(x+y)\le f(x)+f(y).$$

Observe that we have $f(0)=0$. We are going to calculate the average of f. We have

$$\begin{aligned}\sum_{t=1}^{n-1} f(t) &= |\{(a,t):1\le a<a+t\le n, a\in A, a+t\notin A\}|\\ &= |\{(a,x):1\le a<x\le n, a\in A, x\notin A\}|\end{aligned}$$

by introducing $x=a+t$.

Since

$$\begin{aligned}\{(a,x),\ 1\le a<x\le n,\ a\in A,\ x\notin A\}\cup\{(a,x),\ 1\le a<x\le n,\ a\in A,\ x\in A\}\\ =\{(a,x),\ 1\le a<x\le n,\ a\in A\},\end{aligned}$$

and the cardinality of this last set can be expressed by counting $\{(a,x),$ $1\le a<x\le n,\ a\in A\}$ over x as $\sum A(x-1)$, we find

$$\sum_{t=1}^{n-1} f(t)=\sum_{x=1}^{n} A(x-1)-|\{(a,x),a\in A,x\in A,1\le a<x\le n\}|.$$

Using the definition of the Schnirelmann density we can conclude that $A(x-1)\ge\alpha(x-1)$. As the second part is equal to $A(n)(A(n)-1)/2$, we can bound $\sum f(t)$ by

$$\sum_{t=1}^{n-1} f(t)\ge\alpha\frac{n(n-1)}{2}-\frac{A(n)(A(n)-1)}{2}.$$

This inequality implies that there exists a t_0 for which this $f(t_0)$ is large:

$$f(t_0)\ge\frac{1}{n-1}\left(\frac{\alpha n(n-1)}{2}-\frac{A(n)(A(n)-1)}{2}\right).$$

Since B is a basis of order k, we can write $t_0=b_1+\cdots+b_k$, for some $b_j\in B$. From the subadditivity property we conclude that $f(t_0)\le\sum f(b_i)$, consequently there is a $b=b_i$ for which

$$f(b)\ge\frac{1}{k}\frac{1}{n-1}\left(\frac{\alpha n(n-1)}{2}-\frac{A(n)(A(n)-1)}{2}\right).$$

In particular, as $C(n) \geq A(n) + f(b)$ for any single b, we get

$$C(n) \geq A(n) + \frac{1}{k}\frac{1}{n-1}\left(\frac{\alpha n(n-1)}{2} - \frac{A(n)\,(A(n)-1)}{2}\right).$$

Since the right-hand side, as a function of $A(n)$, is increasing up to $A(n) \leq k(n-1)+1/2$, we get a lower estimate by replacing each occurrence of $A(n)$ by its lower bound αn, and we obtain

$$C(n) \geq \alpha n + \frac{1}{k(n-1)}\left(\frac{\alpha n(n-1)}{2} - \frac{\alpha n(\alpha n-1)}{2}\right) \geq \alpha n + \frac{\alpha\,(1-\alpha)\,n}{2k}.$$

As this fact is true for all n, the estimate for the Schnirelmann density follows. □

With a minimal modification of the proof a similar result can be obtained for asymptotic lower density.

Theorem 4.6.2. *Let $B \subset \mathbb{Z}$ be an asymptotic basis of order k and let $A \subset \mathbb{Z}$. Then*

$$\underline{d}(A+B) \geq \underline{d}(A) + \frac{\underline{d}(A)\,(1-\underline{d}(A))}{2k}.$$

4.7 Adding a basis: Plünnecke's theorem, density form

If we add a basis of order k to a set of density $\alpha > 0$, Erdős' theorem in the previous section estimates the density of the sumset essentially by $\alpha(1+1/(2k))$ for small values of α. Plünnecke [41] gave a much stronger estimate, one which goes to infinity after division by α as $\alpha \to \infty$.

Theorem 4.7.1. *If $A, B \subset \mathbb{N}_0$, $0 \in B$, then*

$$\sigma(A+B) \geq \sigma(A)^{1-\frac{1}{k}}\sigma(kB)^{\frac{1}{k}}. \tag{4.7.1}$$

In particular, if $kB = \mathbb{N}_0$, then $\sigma(A+B) \geq \sigma(A)^{1-\frac{1}{k}}$.

Exercise 76. Prove that

$$\alpha^{1-1/k} > \alpha + \frac{\alpha(1-\alpha)}{k}$$

for all $\alpha \in (0,1)$, hence Plünnecke's inequality is stronger than that of Erdős for all possible values of $\sigma(A)$.

Plünnecke's theorem gives us the correct order of magnitude.

Exercise 77. Construct a basis B of order k, such that, for all α, there exists a set A, with $\sigma(A) \geq \alpha$, such that $\sigma(A+B) < C\alpha^{1-\frac{1}{k}}$, for some $C \geq 0$.

Proof. Before starting the proof we remark that a result which is weaker by a constant factor can be easily deduced from the finite Plünnecke inequality given in Chapter 1. Indeed, the case $j = 1$ of Corollary 1.2.4, or the case $l = 0$ of Theorem 1.1.1 gives us the following. If A, B are finite sets and $|A| = m$, $|A + B| = \alpha m$, then $|kB| \leq \alpha^k m$. By substituting $\alpha = |A + B| / |A|$ and rearranging, this can be written as

$$|A + B| \geq |A|^{1-1/k} |B|^{1/k}. \tag{4.7.2}$$

This is analogous to (4.7.1), just we have cardinality here and density there.

Let now A, B be infinite sets such that A and kB have positive Schnirelmann density. Let $C = A + B$; we want to estimate $C(n)$. To this end we apply (4.7.2) for the sets

$$A' = A \cap [1, \lceil n/2 \rceil], \;\; B' = B \cap [0, [n/2]].$$

We have $|A'| \geq \sigma(A)n/2$. Since the set kB' contains every element of kB up to $n/2$ (and contains 0), we have $|kB'| \geq \sigma(kB)n/2$. Now an application of (4.7.2) gives

$$C(n) \geq |A' + B'| \geq \sigma(A)^{1-\frac{1}{k}} \sigma(kB)^{\frac{1}{k}} n/2.$$

Since this holds for every n, we have a lower estimate for $\sigma(A + B)$, which is half of the one claimed in (4.7.1).

To remove this factor we apply an induction argument like for Mann's theorem, and we use Plünnecke's method for a trimmed additive graph.

We write $\alpha = \sigma(A)$, $\beta = \sigma(kB)$, $C = A + B$ and $\gamma = \alpha^{1-\frac{1}{k}} \beta^{\frac{1}{k}}$. We want to show that $C(n) \geq \gamma n$ for all n.

We reformulate this in the following finite form.

Let $n \in \mathbb{N}$. If $A(m) \geq \alpha m$ for all $m \leq n$, $0 \in B$ and $B_k(m) = |kB \cap [1, m]| \geq \beta m$ for all $m \in \mathbb{N}$, then $C(m) \geq \gamma m$ for all $m \leq n$.

We will proceed by induction on n. Since the case $n = 1$ is clear, suppose that $n > 1$ and that the claim above has been proved for all $n' < n$. Let

$$A_1 = A \cap [1, n'],$$
$$A_2 = A \cap [n' + 1, n] - n' \subset [1, n - n'].$$

Observe that $A_1(m) \geq \alpha m$ is satisfied for all $m \leq n'$. If, for some $n' < n$, it also happens that $A_2(m) \geq \alpha m$ for all $m \leq n - n'$, then we apply the induction hypothesis for the set pairs A_1, B and A_2, B. Since

$$C \supset (A_1 + B) \cup (A_2 + n' + B),$$

we get the desired conclusion for C.

Now consider the case when this does not happen. This means that for all $n' < n$ there is some $m \leq n - n'$ such that

$$|A \cap [n' + 1, n' + m]| < \alpha m.$$

We claim that the above inequality is satisfied for $m = n - n'$; i.e.,

$$|A \cap [n'+1, n]| < \alpha(n - n'). \tag{4.7.3}$$

Put $n_1 = n'$. We find m_1 such that

$$|A \cap [n_1+1, n_1+m_1]| < \alpha m_1.$$

If $m_1 = n - n'$, we are done. If not, we take $n_2 = n' + m_1$ and obtain m_2 such that

$$|A \cap [n_2+1, n_2+m_2]| < \alpha m_2.$$

We iterate this process and when it stops, we add all the inequalities to get (4.7.3).

Now we build a restricted addition graph (like described in Section 1.2) as follows. It consists of the layers $V_0 = A \cap [1, n], V_1 = (A + B) \cap [1, n], \dots, V_h = (A + hB) \cap [1, n]$, with edges going from each $x \in V_{i-1}$ to $x + b \in V_i$ as usual. One can easily check that it is commutative, which allows us to apply Plünnecke's inequality (Theorem 1.2.1). We use μ_j to denote the magnification ratios of this graph. This inequality tells us that $\mu_k \leq \mu_1^k$. For μ_1 we use the obvious estimate $\mu_1 \leq C(n)/A(n)$ to deduce

$$\mu_k \leq \left(\frac{C(n)}{A(n)}\right)^k. \tag{4.7.4}$$

Now we find a lower bound for μ_k. Let $X \subset A$ be such that $|\text{im}(X, V_k)| = \mu_k |X|$, and let $n' + 1$ be its first element, so that $X \subset [n'+1, n]$. From (4.7.3) we infer that

$$|X| \leq \alpha(n - n')$$

if $n' > 0$, while for $n' = 0$ we have $|X| \leq A(n)$.

We also know that

$$|(X + kB) \cap [1, n]| \geq |(n' + 1 + kB) \cap [n'+1, n]| \geq \beta(n - n').$$

Combining the two inequalities we get

$$\mu_k \geq \min\left\{\frac{\beta}{\alpha}, \frac{\beta n}{A(n)}\right\}.$$

Now (4.7.4) completes the proof. □

We can easily deduce a similar asymmetric combining lower asymptotic density and Schnirelmann density.

Theorem 4.7.2. *Let $A, B \subset \mathbb{Z}$ and let k be a positive integer. We have*

$$\underline{d}(A + B) \geq \underline{d}(A)^{1-\frac{1}{k}} \sigma(kB)^{\frac{1}{k}}.$$

This can be deduced from the previous theorem using the connection between lower density and Schnirelmann density as expressed in Exercise 70.

Let $\epsilon > 0$. To prove this result we find a t such that $\sigma(A-t) \geq \underline{d}(A-\epsilon)$. This implies that

$$\underline{d}(A+B) \geq \sigma(A+B-t) \geq (\underline{d}(A)-\epsilon)^{1-\frac{1}{k}} \sigma(kB)^{\frac{1}{k}}.$$

The result follows by letting ϵ go to 0.

The theorem is also true with the lower density everywhere but the proof is more involved.

4.8 Adding the set of squares or primes

Let Q be the set of squares: $Q = \{n^2, n \in \mathbb{N}_0\}$. They form a basis of order 4, that is, $4Q = \mathbb{N}_0$. We also know that $3Q$ contains all numbers except those of the form $4^a(8b-1)$, for some a and b. This easily implies that the Schnirelmann density of the set of threefold sums of squares is positive: $\sigma(3Q) > 0$.

Exercise 78. Calculate $d(3Q)$ and $\sigma(3Q)$.

If we have $A \subset \mathbb{Z}$ with $\sigma(A) = \alpha$, then using Plünnecke's density theorems from the previous section we see that $\sigma(A+Q) \geq \alpha^{\frac{3}{4}}$, and for small α we can improve this to $\sigma(A+Q) > c\alpha^{\frac{2}{3}}$. This is still not the best possible; the real exponent is $\frac{1}{2}$.

Theorem 4.8.1 (Plünnecke [39]). *Let A be a subset of the integers with $\sigma(A) = \alpha$, and Q the set of squares. We have*

$$\sigma(A+Q) \geq c\alpha^{\frac{1}{2}}$$

for some absolute constant $c > 0$.

To see that this exponent is sharp, set $A = \{1, q+1, q+2, \dots\}$, with a large q. This set has Schnirelmann density $\sigma(A) = \frac{1}{q}$. Since up to q the only elements of $A+Q$ are the integers of the form k^2+1, we easily find that

$$\sigma(A+Q) = \frac{1+[\sqrt{q}]}{q} \sim \frac{1}{\sqrt{q}} = \sqrt{\sigma(A)}.$$

For asymptotic density the increase is larger, a similar result holds with arbitrarily small exponent.

Theorem 4.8.2 ([47]). *For every $\epsilon > 0$ there exists a constant c_ϵ depending on ϵ such that if $\underline{d}(A) = \alpha$, then*

$$\underline{d}(A+Q) \geq c_\epsilon \alpha^\epsilon.$$

A good configuration that gives approximately the correct order of magnitude is a residue class modulo some q. Put $A = q\mathbb{Z}$, so that $\underline{d}(A) = \frac{1}{q}$. The sumset $A+Q$ contains the quadratic residues modulo q. If the prime factorization of q is $q = p_1 \dots p_k$, then the number of quadratic residues is

$$\left(\frac{p_1+1}{2}\right) \dots \left(\frac{p_k+1}{2}\right) \approx \frac{q}{2^k}.$$

To minimize it we take the product of the first primes. By the prime number theorem this will be essentially the primes $p_i < \log q$, and their number is approximately $(\log q)/(\log\log q)$. So the density of $A+Q$ is approximately

$$\underline{d}(A+Q) \approx \frac{1}{q}\frac{q}{2^{\frac{\log q}{\log\log q}}} = q^{-\frac{\log_q 2 \log q}{\log\log q}}.$$

The exponent goes to zero as q grows.

Let P be the prime numbers, and let $P' = P - 2$. It is known that there exists a k such that $kP' = \mathbb{N}_0$ for some k ($k = 7$?). It is also known that $d(2P) = 1/2$, $d(3P) = 1$ and that every large number can be written as a sum of four primes. These give estimates for the density of $A+P$ by the Plünnecke inequalities; similarly to the case of squares, this is far from the reality.

Theorem 4.8.3 ([47, 48]). *Let A be a subset of integers. There is a positive constant c with the following properties (valid for q sufficiently large).*

(a) *If $\sigma(A) = \frac{1}{q}$, then $\sigma(A+P') \geq \frac{c}{\log q}$.*

(b) *If $\underline{d}(A) = \frac{1}{q}$, then $\underline{d}(A+P) \geq \frac{c}{\log\log q}$.*

The examples to show that this is the correct order of magnitude are similar to the case of squares. For Schnirelmann density use the same set $A = \{1, q+1, q+2, \dots\}$. This set has Schnirelmann density $\sigma(A) = \frac{1}{q}$. Since up to q the only elements of $A+P'$ are neighbors of primes, we find that

$$\sigma(A+Q) = \frac{\pi(q+1)}{q} \sim \frac{1}{\log q}.$$

For asymptotic density again we use the multiples of q. The numbers in $A+P$ are sets coprime to q with finitely many exceptions. Then

$$d(A+P) = \frac{\varphi(q)}{q} = \left(1-\frac{1}{p_1}\right) \cdots \left(1-\frac{1}{p_k}\right) \sim \frac{c}{\log\log q}$$

if again we take the product of the first primes.

4.9 Essential components

We say that B is an *essential component* if B is such that $\sigma(A+B) > \sigma(A)$ for every $0 < \sigma(A) < 1$.

In the previous sections we met examples of essential components: first sets of positive density (theorems of Schnirelmann and Mann), then bases (theorems of Erdős and Plünnecke). The first example of an essential component that is not a basis was given by Linnik [32]; this set was too thin to be a basis. Clearly if B is a basis of order k, it must satisfy $B(x) > x^{1/k}$, hence a set such that $B(x) = O(x^{\varepsilon})$ for every positive ε cannot be a basis.

The following theorem tells exactly how thin an essential component can be.

Theorem 4.9.1 ([46]). (a) *For every $\epsilon > 0$ there exists an essential component with $B(n) < c(\log n)^{1+\epsilon}$.*

(b) *There is no essential component B with $B(n) < c(\log n)^{1+o(1)}$.*

Chapter 5

Measure and topology

5.1 Introduction

In this chapter we mention some loosely connected things. The common feature is that we now leave the safe familiar world of finite sets. Our excursions are in two different directions.

The first is to measures. Measure is a close analogue of cardinality; the same questions we asked for finite sets can be formulated for measures of sets of reals, or in $\mathbb{R}^d$, or in a more general setting. We already mentioned a classical example, the Brunn–Minkowski inequality: For measurable sets in $\mathbb{R}^d$ we have

$$\mu(A+B)^{1/d} \geq \mu(A)^{1/d} + \mu(B)^{1/d}. \tag{5.1.1}$$

This illustrates some basic differences. While measure is a more sophisticated concept than cardinality of a finite sets, the results are often simpler and sometimes also easier to prove.

The second excursion is to topology. We think of the integers as a discrete set; however, other topologies on them do exist, and some have a relevance to our subject. We will mainly discuss the connection of the Bohr topology to additive properties.

5.2 Raikov's theorem and generalizations

A natural analog of adding integers modulo q is the addition of reals modulo 1. The analogue of the Cauchy–Davenport inequality is the following theorem, due to Raikov [42] from 1939.

In what follows we consider subsets of $[0, 1)$, addition is meant modulo 1. Problems of mesurability are not in the focus of our interest, so assume that every set mentioned is compact or open. Lebesgue measure is denoted by μ.

Theorem 5.2.1. *For $A, B \subset [0, 1)$ we have*

$$\mu(A+B) \geq \min\bigl(1, \mu(A) + \mu(B)\bigr). \tag{5.2.1}$$

Prexercise. Deduce Raikov's theorem from the Cauchy–Davenport inequality.

a) Approximate a general set by a union of intervals.

b) Take a prime p and compare $\mu(A)$ to the cardinalites of the sets

$$A(x, p) = \{j \in \mathbb{Z}_p : x + j/p \in A\}. \tag{5.2.2}$$

(Naturally j/p and $x + j/p$ are interpreted modulo 1.)

Of the two approaches suggested above, a) is more natural but also has more cumbersome details. We describe method b).

Proof. Write $A + B = S$. Take a prime p and construct sets of residues from the sets A, B, S as described in (5.2.2). For every x, y we have

$$S(x, p) \supset A(x - y, p) + B(y, p).$$

An application of the Cauchy–Davenport inequality to these sets now gives

$$|S(x, p)| \geq \min\bigl(|A(x - y, p)| + |B(y, p)| - 1, p\bigr).$$

We can reformulate this as follows. Either $S(x, p) = \mathbb{Z}_p$, or

$$|S(x, p)| \geq |A(x - y, p)| + |B(y, p)| - 1 \tag{5.2.3}$$

for each y.

The average of such a cardinality is connected to the measure of the set in an immediate way:

$$\int_0^1 |A(x, p)dx = p\mu(A).$$

(Why?) By integrating both sides of (5.2.3) with respect to y we obtain

$$|S(x, p)| \geq p(\mu(A) + \mu(B)) - 1.$$

This holds unless $S(x, p) = \mathbb{Z}_p$; hence the following inequality always holds:

$$|S(x, p)| \geq \min\bigl(p(\mu(A) + \mu(B)) - 1, p\bigr).$$

Now integrating this inequality with respect to x and dividing by p we get

$$\mu(A + B) \geq \min\bigl(1, \mu(A) + \mu(B)\bigr) - 1/p.$$

As this holds for every prime p, (5.2.1) follows. □

Raikov's theorem was generalized by Macbeath [33] for the n-dimensional torus, by Shields [60] for connected commutative compact second countable groups, by Kneser [28] for commutative locally compact groups and by Kemperman [24] for non-compact groups. We state below his result in less than complete generality to avoid discussing certain aspects of non-commutative groups and measurability.

Let G be a locally compact topological group. If G is compact, or commutative, and in some other cases too, it has an invariant measure μ, called *Haar measure*. Invariance means that we have

$$\mu(A + x) = \mu(x + A) = \mu(A)$$

for every measurable set and every $x \in G$. Without any condition, we can only claim that there is a right-invariant μ_r satisfying $\mu_r(A + x) = \mu_r(A)$ and a left-invariant μ_l satisfying $\mu_l(x + A) = \mu_l(A)$. If these coincide, that is, an invariant Haar measure exists, the group is called *unimodular*. We shall state the unimodular case. Also we restrict our attention to measurable sets.

Theorem 5.2.2. *Let G be a compact, connected group, $A, B \subset G$ measurable sets such that $A + B$ is also measurable. We have*

$$\mu(A + B) \geq \min\bigl(\mu(A) + \mu(B), \mu(G)\bigr).$$

Theorem 5.2.3. *Let G be a locally compact, non-compact group which does not have any proper compact-open subgroup. Let $A, B \subset G$ be measurable sets such that $A + B$ is also measurable. We have*

$$\mu(A + B) \geq \mu(A) + \mu(B).$$

5.3 The impact function

Let G be a group with a Haar measure μ, and $B \subset G$ a measurable set. We define the *impact function* of B analogously to the finite situation

$$\xi_A(x) = \inf\{\mu(A + B) : B \subset G, \mu(B) = x\}.$$

Exercise 79. Let G be the interval $[0, 1)$ with addition modulo 1. Prove the concavity of the impact function: for $0 < y < x < 1$ and $x + y \leq 1$ we have

$$\xi(x - y) + \xi(x + y) \leq 2\xi(x).$$

Exercise 80. Use the previous exercise to give another proof of Raikov's theorem.

This method also can be extended to every locally compact group.

Theorem 5.3.1. *If G does not have any proper compact-open subgroup, then ξ is a continuous concave function on its whole domain.*

See [53]; there this result is also applied to deduce Kemperman's theorems from the previous section.

Another curious property of the impact function is its symmetry. To avoid an exception we redefine the impact function at 0 by continuity:

$$\xi(0) = \lim_{x \to 0+} \xi(x).$$

Theorem 5.3.2. *Assume that G is compact, commutative and connected. The smallest value of x for which $\xi(x) = \mu(G)$ is $x = \mu(G) - \xi(0)$. The graph of $\xi(x)$ on the interval $[0, \mu(G) - \xi(0)]$ is symmetric to the line $x + y = \mu(G)$.*

5.4 Meditation on convexity and dimension

Let A, B be Borel sets in $\mathbb{R}^d$. The Brunn–Minkowski inequality (5.1.1) estimates $\mu(A+B)$ in a natural way, with equality if A and B are homothetic convex sets.

This can be exptessed in terms of the impact function as

$$\xi_B(a) \geq \left(a^{1/d} + \mu(B)^{1/d}\right)^d,$$

and this is the best possible estimate in terms of $\mu(B)$ only.

To measure the degree of non-convexity, one can try to use the measure of the convex hull beside the measure of the set. This is analogous to the hull volume, and it is sufficient to describe the asymptotic behavior of ξ.

Theorem 5.4.1 ([59, Theorem 1]). *For every bounded Borel set $B \subset \mathbb{R}^d$ of positive measure we have*

$$\lim_{a \to \infty} \xi_B(a)^{1/d} - a^{1/d} = \mu(\operatorname{conv} B)^{1/d}.$$

This is the continuous analogue of Theorem 3.9.3, and there is an analogue to the effective version Theorem, 3.9.5, as well.

Note that by considering sets homothetic to $\operatorname{conv} B$ we immediately obtain

$$\xi_B(a)^{1/d} \leq a^{1/d} + \mu(\operatorname{conv} B)^{1/d},$$

thus we need only to give a lower estimate. This is as follows.

Theorem 5.4.2 ([59, Theorem 2]). *Let $\mu(B) = b$, $\mu(\operatorname{conv} B) = v$. We have*

$$\xi_B(a)^{1/d} \geq a^{1/d} + v^{1/d}\left(1 - c(v/b)^{1/2}(v/a)^{1/(2d)}\right),$$

$$\xi_B(a) \geq a + dv^{1/d}a^{1-1/d}\left(1 - c(v/b)^{1/2}(v/a)^{1/(2d)}\right)$$

with a suitable positive constant c depending on d.

If $v > b$, we get a nontrivial improvement over the Brunn–Minkowski inequality for $a > a_0(b, v)$. It would be desirable to find an improvement also for small values of a, or, even more, to find the best estimate in terms of $\mu(B)$ and $\mu(\operatorname{conv} B)$.

The exact bound and the structure of the extremal set may be complicated. This is already so in the case $d = 1$, which was solved in [51]. Observe that in one dimension $\mu(\operatorname{conv} B)$ is the diameter of B.

Theorem 5.4.3 ([51, Theorem 2]). *Let $B \subset \mathbb{R}$, and write $\mu(B) = b$, $\mu(\operatorname{conv} B) = v$. If*

$$a \geq \frac{v(v-b)}{2b} + \frac{b\{v/b\}(1 - \{v/b\})}{2}, \tag{5.4.1}$$

then $\xi_B(a) = a + v$. If (5.4.1) does not hold, then let k be the unique positive integer satisfying

$$\frac{k(k-1)}{2} \leq \frac{a}{b} < \frac{k(k+1)}{2},$$

and define δ by

$$\frac{a}{b} = \frac{k(k-1)}{2} + \delta k.$$

We have

$$\xi_B(a) \geq a + (k + \delta)b,$$

and equality holds if $B = [0, b] \cup \{v\}$.

A set A such that $\xi_B(a) = \mu(A + B)$ for the above set B is given by

$$A = [0, (k - 1 + \delta)b] \cup [v, v + (k - 2 + \delta)b] \cup \cdots \cup [(k-1)v, (k-1)v + \delta b].$$

A less exact, but simple and still quite good, lower bound sounds as follows.

Corollary 5.4.4 ([51, Theorem 1]). *Let $B \subset \mathbb{R}$, and write $\mu(B) = b$, $\mu(\operatorname{conv} B) = v$. We have*

$$\xi_B(a) \geq \min\left(a + v, (\sqrt{a} + \sqrt{b/2})^2\right).$$

A comparison with the two-dimensional Brunn–Minkowski inequality gives the following interpretation: Initially a long one-dimensional set B tries to behave as if it were a two-dimensional set of area $b/2$.

It can be observed that Corollary 5.4.4 is weaker than the obvious inequality

$$\mu(A + B) \geq \mu(A) + \mu(B) \tag{5.4.2}$$

for small a. For small values of a, Theorem 5.4.3 yields the following improvement of (5.4.2).

Corollary 5.4.5 ([51, Corollary 3.1]). *If $a \leq b$, then we have*

$$\mu(A + B) \geq \min(2a + b, a + v).$$

If $b < a \le 3b$, then we have

$$\mu(A+B) \ge \min\Big(\frac{3}{2}(a+b), a+v\Big).$$

Problem 5.4.6. How large must $\mu(A+B)$ be if $\mu(A)$, $\mu(B)$, $\mu(\operatorname{conv} A)$ and $\mu(\operatorname{conv} B)$ are given?

What are the minima of $\mu(A+A)$ and $\mu(A-A)$ for fixed $\mu(A)$ and $\mu(\operatorname{conv} A)$?

The results above show that for $d = 1$ (like in the discrete case, but for less obvious reasons) the limit relation becomes an equality for $a > a_0$. Again, this is no longer the case for $d = 2$.

An example of a set $B \subset \mathbb{R}^2$, such that

$$\xi_B(a)^{1/2} < a^{1/2} + v^{1/2}$$

will hold for certain arbitrarily large values of a, is as follows.

Let $0 < c < 1$ and let B consist of the square $[0, c] \times [0, c]$ and the points $(0, 1), (1, 0)$ and $(1, 1)$. Hence $b = c^2$ and $v = 1$.

For an integer $n \ge 1$ put

$$A_n = [0, n] \times [0, n] \cup \bigcup_{j=0}^{n} [j, j+c] \times [n, n+c] \cup \bigcup_{j=0}^{n-1} [n, n+c] \times [j, j+c].$$

Thus A_n consists of a square of side n and $2n+1$ small squares of side c, hence

$$\mu(A_n) = n^2 + (2n+1)b.$$

We can easily see that $A_n + B = A_{n+1}$. Hence by considering the set $A = A_n$ we see that for a number a of the form $a = n^2 + (2n+1)b$ we have

$$\xi_B(a) \le \mu(A_{n+1}) = (n+1)^2 + (2n+3)b < \big(\sqrt{a} + 1\big)^2.$$

A more detailed calculation leads to

$$\xi_B(a)^{1/2} \le a^{1/2} + 1 - ca^{-1}$$

(for these special values of a).

If we tried to define an impact volume in the continuous case, we would recover the volume, at least for compact sets. Still, the above results and questions suggest that ordinary volume is not the best tool to understand additive properties. Perhaps one could try to modify the definition of impact volume by requiring $\mu(A) \ge \mu(B)$. So put

$$\operatorname{iv}_*(B) = \inf_{a \ge \mu(B)} \Big(\xi_B(a)^{1/d} - a^{1/d}\Big)^d.$$

Problem 5.4.7. Find a lower estimate for $\operatorname{iv}_*(B)$ in terms of $\mu(B)$ and $\mu(\operatorname{conv} B)$.

5.5 Topologies on integers

For most of the time we are happy with the integers as a discrete set. However, other important topologies do exist on them. An example is the p-adic topology, which can be compactified to give us p-adic integers. The Čech–Stone compactification also has applications in combinatorial number theory.

In what follows we will always use commutative groups ($\mathbb{Z}$ and extensions); for non-commutative groups the following definitions and claims have to be modified a litle.

Among all topologies we will be interested in those where addition behaves nice. We consider two possible interpretations, a stronger and a weaker one.

Definition 5.5.1. Let G be a group and $\mathcal{T}$ a topology on it. We say that $(G, \mathcal{T})$ is a *topological group*, if addition and substraction are continuous in $\mathcal{T}$; that is, $f(x, y) = x - y$ is jointly continuous in both variables. It is a *semitopological group*, if $x - y$ is continuous in each variable separately.

The weaker condition means two things: First, if U is a neighborhood of 0, then so is $-U$; next, if U is a neighborhood of an element x, then $U + a$ is a neighborhood of $x + a$. The stronger condition, in addition, requires that for any neighborhood U of 0 there is another neighborhood U' such that $U' - U' \subset U$.

Exercise 81. Construct a topology on $\mathbb{Z}$ which makes it a semitopological group but not a topological group.

When defining a topology on $\mathbb{Z}$, we shall typically proceed as follows. We define a basis of neighborhood of 0. Then, to make $\mathbb{Z}$ at least a semitopological group, we define neighborhoods of other integers by translation; that is, U will be a neighborhood of x if $U - x$ is a neighborhood of 0. In most cases it will be trivial that this indeed defines a topology, and we shall not give the easy details. If there is a hidden difficulty, we shall point it out.

A topological group may be *complete*, which means that every sequence (assuming a countability condition) or every generalized sequence (indexed by a general ordered set, not necessarily by positive integers) which satisfies the Cauchy condition must be convergent.

A topological group can always be *completed* by assigning a new element (a limit) to every (generalized) Cauchy sequence which does not already have one. This procedure is used to build p-adic integers.

If we are lucky, this completion is compact and then we can embed our group into a compact group; that is, we can *compactify* it.

Exercise 82. Construct a group topology on $\mathbb{Z}$, other than the discrete one, which cannot be compactified.

We can sometimes compare two topologies as follows.

Definition 5.5.2. Let $\mathcal{T}$, $\mathcal{T}'$ be two topologies on the same set. We say that $\mathcal{T}'$ is *finer* than $\mathcal{T}$, or $\mathcal{T}$ is *coarser* than $\mathcal{T}'$, if every set which is open in $\mathcal{T}$ is also open in $\mathcal{T}'$.

The finest topology of all is the discrete one. The coarsest is the one in which only the empty set and the whole space are open, a pretty uninteresting one.

In what follows we will find the answer to the following questions.

Question 1. What is the coarsest topology on $\mathbb{Z}$ in which all characters are continuous?

Recall that a character is a homomorphism into the circle $\{z \in \mathbb{C} : |z| = 1\}$. Now if γ is a character and $\gamma(1) = \omega = e^{2\pi it}$, then necessarily

$$\gamma(n) = \omega^n = e^{2\pi itn}$$

for all integers.

Question 2. What is the finest topology on $\mathbb{Z}$ which can be compactified to make it a compact topological group?

We shall see that the answer to these questions will be the same, and we shall call it the *Bohr topology.*

We first answer the first question.

By the multiplicative property of characters, continuity everywhere is equivalent to continuity at 0. This means that the following sets must be neighborhoods of 0 for each character γ and $\varepsilon > 0$:

$$\{n \in \mathbb{Z} : |\gamma(n) - \gamma(0)| < \varepsilon\}.$$

If we express γ as $\gamma(n) = e^{2\pi itn}$, we see that this set is the same as

$$\{n \in \mathbb{Z} : \|tn\| < \delta\},$$

where ε and δ are connected by the equation

$$\varepsilon = \left|e^{2\pi i\delta} - 1\right| = 2\sin \pi\delta.$$

If such a set has to be a neighborhood of 0, then so has any finite intersection of such sets. We saw similar objects in Section 2.5, which we now repeat.

Definition 5.5.3. If G is a commutative group, $\gamma_1, \ldots, \gamma_k$ are characters of G and $\varepsilon_j > 0$, we write

$$B(\gamma_1, \ldots, \gamma_k; \varepsilon_1, \ldots, \varepsilon_k) = \{g \in G : |\arg \gamma_j(g)| < 2\pi\varepsilon_j \text{ for } j = 1, \ldots, k\}$$

and call these sets *Bohr sets.* In particular, if $\varepsilon_1 = \cdots = \varepsilon_k = \varepsilon$, we shall speak of a *Bohr* (k, ε)*-set.* (We take the branch of arg that lies in $[-\pi, \pi)$.)

By view of the above, a Bohr set in $\mathbb{Z}$ can also be written as

$$B(u_1, \ldots, u_k; \varepsilon_1, \ldots, \varepsilon_k) = \{x \in \mathbb{Z} : \|u_j x\| < \varepsilon_j \text{ for } j = 1, \ldots, k\},$$

where now the parameters u_j are real numbers taken modulo 1.

Definition 5.5.4. The *Bohr topology* on a commutative group is the topology in which a set is a neighborhood of a point x if and only if it contains a set of the form $x + N$, where B is a Bohr set.

Exercise 83. The Bohr sets are open in the Bohr topology.

Exercise 84. The Bohr topology turns $\mathbb{Z}$ into a topological group.

Exercise 85. This group can be compactified.

Exercise 86. A sequence is convergent in the Bohr topology only if it is constant from a point on.

5.6 The finest compactification

Now we answer Question 2 of the previous section.

Let G be a compact group, and let U be a neighborhood of 0 in G. The collection of open sets

$$\{U + x : x \in G\}$$

covers G, hence so does a finite subcollection. Hence U has the property that there are finitely many elements $x_1, \ldots, x_k \in G$ such that

$$\bigcup (U + x_i) = G.$$

Definition 5.6.1. A set A in a group G is *syndetic*, if there are finitely many elements $x_1, \ldots, x_k \in G$ such that

$$\bigcup (A + x_i) = G.$$

So a neighborhood of 0 must be syndetic.

Exercise 87. A set $A \subset \mathbb{Z}$ is syndetic if and only if it is unbounded both from above and from below and has bounded gaps; that is,

$$A = \{\ldots, a_{-1}, a_0, a_1, a_2, \ldots\}, \; a_{k+1} - a_k < c$$

with some c.

Exercise 88. Bohr sets are syndetic.

Let U be a neighborhood of 0 in any conditionally compact topology (this expression means that it has a compactification). We can find an open set U_1 such that $U_1 - U_1 \subset U$. Then, there is an open set U_2 such that $U_2 - U_2 \subset U_1$, and so on. So, U can be a neighborhood of 0 only if there is a chain of syndetic sets $U_1, U_2, \ldots$ such that $U_{i+1} - U_{i+1} \subset U_i$ for all i.

Note that the Bohr sets have this property, because

$$B(\alpha_1, \ldots, \alpha_k, \varepsilon/2) - B(\alpha_1, \ldots, \alpha_k, \varepsilon/2) \subset B(\alpha_1, \ldots, \alpha_k, \varepsilon).$$

It is possible to show directly (without any reference to characters) that this requirement defines a class of sets which can serve as the basis of a topology, and it is indeed conditionally compact. Instead we shall prove that any set with this property contains a Bohr set. Moreover, we do not need an infinite chain for this, two steps are sufficient.

Note that for a sequence $U, U_1, U_2, \ldots$ of sets $U_1, U_2, \ldots$ such that $U_{i+1} - U_{i+1} \subset U_i$ we have

$$(U_2 - U_2) - (U_2 - U_2) = 2U_2 - 2U_2 \subset U.$$

So the exact formulation of the above claim sounds as follows.

Theorem 5.6.2 (Bogolyubov [3]). *If $A \subset \mathbb{Z}$ is a syndetic set, then $2A - 2A$ contains a Bohr set (in other words, it is a neighborhood of 0 in the Bohr topology).*

This will be proved in a stronger form in the next section.

5.7 Banach density

To put Bogolyubov's theorem into proper perspective we define some new concepts of density.

Definition 5.7.1. The *lower* and *upper Banach densities* of a set A of integers are defined by

$$d_*(A) = \lim_{n\to\infty} \min_x \frac{|A \cup [x+1, x+n]|}{n},$$
$$d^*(A) = \lim_{n\to\infty} \max_x \frac{|A \cup [x+1, x+n]|}{n}.$$

Exercise 89. These limits do exist.

Exercise 90. Show that for any set $A \subset \mathbb{Z}$, $d_*(A) \le \underline{d}(A) \le \overline{d}(A) \le d^*(A)$.

Exercise 91. $d_*(A) > 0$ if and only if A is syndetic.

Exercise 92. Let P be the set of primes. Show that $d^*(P) = 0$.

The stronger form of Bogolyubov's theorem requires only positive upper Banach density.

Theorem 5.7.2 (Bogolyubov). *If $d^*(A) > 0$, then there exist $\alpha_1, \ldots, \alpha_k$, and $\varepsilon > 0$, with k and ε depending only on $d^*(A) > 0$, such that $B(\alpha_1, \ldots, \alpha_k, \varepsilon) \subset 2A - 2A$*

Proof. We split the proof into three steps.

Step 1 (modular case). It corresponds to Lemma 2.5.2, which we repeat below.

> **Lemma 5.7.3.** *Let G be a finite commutative group, $|G| = q$. Let A be a nonempty subset of G and write $|A| = m = \beta q$. The set $D = 2A - 2A$ (the second difference set of A) contains a Bohr (k, ε)-set with some integer $k < \beta^{-2}$ and $\varepsilon = 1/4$.*

Step 2 (finite case).

> **Lemma 5.7.4.** *If $A \subset [t, t+l]$, $|A| \geq \beta l$, then there are $\alpha_1, \ldots, \alpha_k, \varepsilon$, with k, ε depending only on β, such that $2A - 2A \supseteq B(\alpha_1, \ldots, \alpha_k, \varepsilon) \cap [-l, l]$.*

To see this let $q > 4l$ and let $A \subset [t, t+l]$. We denote by $A' \subset \mathbb{Z}_q$ the set of residue classes of A modulo q, which satisfies $|A'| \geq \beta' l$ with $\beta' = \beta/5$. Then, by the modular case, there exist $k(\beta), u_1, \ldots, u_k$ and $\varepsilon(\beta)$ such that

$$2A' - 2A' \supseteq \left\{ x : \left\| \frac{xu_i}{q} \right\| < \varepsilon \right\}.$$

This means that for any n such that $\|n\alpha_i\| < \varepsilon$, where $\alpha_i = u_i/q$, we can find $a_1, a_2, a_3, a_4 \in A$ such that $n \equiv a_1 + a_2 - a_3 - a_4 \bmod q$ with $|a_1 + a_2 - a_3 - a_4| < 2l$. If $|n| < 2l$, they can be congruent only if they are equal. So, for all $n \in B(\alpha_1, \ldots, \alpha_k, \varepsilon) \cap [-l, l]$ we have $n = a_1 + a_2 - a_3 - a_4$, which completes the second step of the proof.

Last step (density case). Let β satisfy $d^*(A) > \beta > 0$. Given $l \in \mathbb{N}$, we can find t such that $|A \cup [t, t+l]| \geq \beta l$. The finite case provides a finite collection $\alpha_1, \ldots, \alpha_k, \varepsilon$ such that $2A - 2A \supseteq B(\alpha_1, \ldots, \alpha_k, \varepsilon) \cap [-l, l]$. Note that k and ε are fixed, while $\alpha_1, \ldots, \alpha_k$, which we can assume to belong to $[0, 1]$, depend on l.

For each l, we define

$$C_l = \{(\alpha_1, \ldots, \alpha_k) \in [0,1]^k \ : \ B(\alpha_1, \ldots, \alpha_k, \varepsilon) \cap [-l, l] \subseteq 2A - 2A\}.$$

For all l, C_l is a compact set, nonempty by the previous argument, and $C_{l+1} \subset C_l$. Then,

$$\bigcap_{l \in \mathbb{N}} C_l \neq \emptyset.$$

So there is at least one element $(\alpha_1, \ldots, \alpha_k)$ in the intersection. This defines the Bohr neighborhood in $2A - 2A$ we were looking for. □

5.8 The difference set topology

By Bogolyubov's theorem, with four copies of A we get a Bohr neighborhood of 0, namely by forming $A + A - A - A$. With some modification three copies also suffice. Denote

$$k \cdot A = \{ka, a \in A\}.$$

Theorem 5.8.1 (Bergelson–Ruzsa [1]). *Let A be a set of integers with $d^*(A) > 0$, and let r, s, t be integers such that $r + s + t = 0$. The set $r \cdot A + r \cdot A + t \cdot A$ is a Bohr neighborhood of 0. In particular, the set $2A - 2 \cdot A = A + A - 2 \cdot A$ is a Bohr neighborhood of 0.*

The question whether two copies of A is enough to have a Bohr neighborhood inside is difficult, and the answer may depend on the interpretation of density.

Unsolved problem. If A has positive lower Banach density ($d_*(A) > 0$), is $A - A$ a Bohr neighborhood of 0?

Here we used the strongest assumption with lower Banach density, and in Bogolyubov's theorem the weakest one with upper Banach density was sufficient. The medium ones with asymptotic density do not give anything new, for problems about the diffence set they behave as lower Banach density. An exact formulation is as follows.

Theorem 5.8.2 ([44]). *Assume $d^*(A) > 0$. Then there is an $A' \subset \mathbb{N}$ such that $d(A') > 0$ and $A' - A' \subset A - A$.*

However, upper Banach density is different.

Theorem 5.8.3 (Kříž [29]). *There exists an A with $d(A) > 0$ such that there is no A', with $d_*(A) > 0$ and $A' - A' \subset A - A$. Consequently, $A - A$ is not a Bohr neighborhood of 0.*

The following result guarantees a somewhat weaker property.

Theorem 5.8.4 (Følner [8, 9]). *If $d^*(A) > 0$, then there exists a $B = B(\alpha_1, \dots, \alpha_k, \epsilon)$ such that $d((A - A) \setminus B) = 0$.*

Unsolved problem. If A has positive upper Banach density, is $A - A$ a neighborhood of something?

So we know that difference sets of sets of positive density are not necessarily neighborhoods of 0 in the Bohr topology. They are, however, neighborhoods in some topology: we can use them to define a new topology.

Definition 5.8.5. We say that $V \subset \mathbb{Z}$ is a neighborhood of 0 in the *difference set topology* if there exists a set A with $d^*(A) > 0$ such that $A - A \subset V$. V' is said to be a neighborhood of $n \in \mathbb{Z}$ if $V = V' - n$ is a neighborhood of 0.

Exercise 93. The difference set topology is indeed a topology (not easy!).

Easier exercise: Formulate what this means using only sets and density, no topological concept.

Exercise 94. Is the difference set topology a group topology?

Definition 5.8.6. The *syndetic difference topology* is defined similarly to the difference set topology, but now we say that $V \subset \mathbb{Z}$ is a neighborhood of 0 if there exists an A with $d_*(A) > 0$ such that $A - A \subset V$.

With these concepts we can reformulate Kříž' theorem as the syndetic difference topology is different from the difference set topology, and we cannot decide whether it is the same as the Bohr topology.

Definition 5.8.7. The *combinatorial difference topology* is defined as follows. Let $A_1, \dots A_k$ be subsets of the integers such that $\mathbb{Z} = \bigcup_{i=1}^{n} A_i$, then $\bigcup_i (A_i - A_i)$ is a neighborhood of 0.

Exercise 95. The syndetic difference topology and the combinatorial difference topology are, indeed, topological spaces.

Exercise 96. If V is an open set in the syndetic difference topology so is in the combinatorial difference topology. Hence the syndetic difference topology and the combinatorial difference topology are identical.

Exercises

The next collection of exercises is deliberately vague. The number of points is my estimate for the difficulty. This naturally depends on expertise; some of them were told in the course; then naturally the difficulty turns to 0.

1) (3 points) Prove the Cauchy–Davenport inequality by comparing $|A+B|$ and $|A'+B'|$ for suitably chosen sets of the form

$$A' = A \cup (B+t), \quad B' = B \cap (A-t).$$

In a group G define the *impact function* of a set $A \subset G$ by the formula

$$\xi_A(n) = \min\{|A+B| : B \subset G, |B| = n\}.$$

The subscript will be generally omitted.

2) (2 points) Prove the following "sort of concavity" of the impact function: For $2 \le n < |G|$ there is a number $1 \le k \le n-1$ such that

$$\xi(n-k) + \xi(n+k) \le 2\xi(n).$$

3) (2 points) Use the previous exercise to deduce the Cauchy–Davenport inequality.

4) (2 points) Let A, B be finite sets in a (not necessarily commutative) torsion-free group. Show that

$$|A+B| \ge |A| + |B| - 1.$$

5) (2 points) In a finite group the graph of the impact function has a certain symmetry with respect to the line $x + y = |G|$. Formulate exactly and prove.

6) (3 points) Let $A, B \subset \mathbb{Z}_p$ be nonempty sets, p prime, and let

$$T = \{a+b : a \in A, b \in B, ab \ne 1\}.$$

Prove that

$$|T| \geq \min(p, |A| + |B| - 3)$$

and this is best possible.

Next come some problems about measure. We take subsets of the interval $[0, 1)$, addition is meant modulo 1. μ denotes Lebesgue measure, and we assume that every set is measurable (but meditations are welcome about how to use inner and outer measure to make things work for every set). We define the impact function analogously to the finite situation

$$\xi_A(x) = \inf\{\mu(A + B) : B \subset G, \mu(B) = x\}.$$

7) (3 points) Prove the concavity of the impact function: For $0 < y < x < 1$ and $x + y \leq 1$ we have

$$\xi(x - y) + \xi(x + y) \leq 2\xi(x).$$

8) Prove Raikov's theorem: For $A, B \subset [0, 1)$ we have

$$\mu(A + B) \geq \min(1, \mu(A) + \mu(B)).$$

a) (1 point) Deduce it from the previous exercise.

b) (2 points) Use Cauchy–Davenport and approximate a general set by a union of intervals.

c) (2 points) Use Cauchy–Davenport by taking a prime p and comparing $\mu(A)$ to the cardinalites of the sets

$$A(x, p) = \{j : x + j/p \in A\}.$$

A set is *sumfree* if it has no three elements such that $x+y = z$ (so we exclude $2x = z$ too; if we do not, the following exercises change only minimally).

9) (warm-up) What is the size of the largest sumfree subset of $[1, n]$?

10) (2 points) What is the size of the largest sumfree subset of $\mathbb{Z}_p$, p prime?

11) (2 points) What is the size of the largest sumfree subset of $\mathbb{Z}_n$, n composite?

12) (3 points) Every $A \subset \mathbb{N}$, $|A| = n$, has a sumfree subset of cardinality $\geq n/3$.

13) (10 points) Same problem with $n/3 + 1$ for n sufficiently large.

14) (3 points) The set of positive integers has no partition into finitely many sumfree parts.

15) (2 points) Let $A, B \subset \mathbb{Z}$ such that $|A| \geq 2$, $|B| \geq 2$ and $|A+B| = |A|+|B|-1$. Show that A, B are arithmetic progressions with a common difference.

The next collection of exercises is deliberately vague. There are obvious answers (for instance, $|A-A| < n^2$ in the next: 0 point), not obvious but doable (1–3 points) and finding the exact bound would be an important new result.

16) (0–∞ points) Let $|A| = n$ and assume that $|2A| \leq \alpha n$. Find a bound for $|A-A|$.

17) (0–∞ points) Let $|A| = n$ and assume that $|2A| \leq \alpha n$. Find a bound for $|3A|$.

18) (0–∞ points) Let $|A| = n$ and assume that $|A-A| \leq \alpha n$. Find a bound for $|2A|$.

Next come exercises about density. $\underline{d}$, $\overline{d}$ denote lower and upper densities, σ is the Schnirelmann density.

19) (1 point) If $\underline{d}(A) + \underline{d}(B) > 1$, then $A+B$ contains all but finitely many positive integers.

20) (2 points) Suppose $\underline{d}(A) + \underline{d}(B) = 1$. Show that $A+B$ has an asymptotic density, and find its possible values.

21) (2 points) Show that we have always

$$\underline{d}(A) = \sup \sigma(A-n).$$

Show that we cannot replace the supremum by maximum.

22) (1 point) Construct sets A, B satisfying $0 \in A$, $0 < \sigma(A), \sigma(B) < 1$ and

$$\sigma(A+B) = \sigma(A) + \sigma(B) - \sigma(A)\sigma(B).$$

23) (3 points) Show that for every pair of sets satisfying the previous exercise the values of $\sigma(A)$ and $\sigma(B)$ are always rational.

24) (2 points) If $\underline{d}(A) > 0$, does there necessarily exist a set $B \subset A$ with $d(B) > 0$?

25) (2 points) Let α, β, γ be positive real numbers such that $\alpha + \beta \leq \gamma \leq 1$. Construct sets of positive integers such that $d(A) = \alpha$, $d(B) = \alpha$, $d(A+B) = \gamma$.

26) (2 points) For $A \subset \mathbb{N}$ define

$$A_*(n) = \min |A \cap [m+1, m+n]|, \quad A^*(n) = \max |A \cap [m+1, m+n]|.$$

Show that the limits

$$d_*(A) = \lim A_*(n)/n, \quad d^*(A) = \lim A^*(n)/n$$

always exist. They are called the *lower and upper Banach density*, respectively, of the set A.

27) (2 points) If $d_*(A) > 0$, is there always an $A' \subset A$, such that $d(A') > 0$?

28) (2 points) Is there a quadratic polynomial f with integer coefficients such that every integer has a representation of the form $f(x) - f(y)$ with integer x, y?

29) (2 points) Let $A \subset \mathbb{Z}$, $|A| = n$, and for $1 \leq k \leq n$ let S_k be the set of all sums $a_1 + \cdots + a_k$, where $a_i \in A$ and all the k summands a_i are distinct. Find the minimum of $|S_k|$ (for fixed k, n over all sets of n integers) and describe the cases of equality.

30) (2 points) With the notations of the previous exercise show that for $1 \leq k < n$ we have

$$\frac{|S_{k+1}|}{|S_k|} \leq \frac{n}{k+1}.$$

Research problem: Does the stronger inequality

$$\frac{|S_{k+1}|}{|S_k|} \leq \frac{n-k}{k+1}$$

hold if n is sufficiently large for a given k (perhaps for $n > 2k$)?

31) (2 points) With the notations of the previous exercises show that for all k we have

$$\frac{|(k+1)A|}{|kA|} \leq \frac{n+k}{k+1}.$$

Research problem: Can equality hold for $k = 2$ and fail for $k = 1$? (I thought I have an example, but cannot find or reconstruct it.)

32) (2 points) Let X, Y, Z be finite sets in any commutative group. Show that

$$|X||Y - Z| \leq |X - Y||X - Z|.$$

33) (3 points) Let X, Y, Z be finite sets in any commutative group. Show that

$$|X + Y + Z|^2 \leq |X + Y||X + Z||Y + Z|.$$

34) Let

$$A = \{(x_1, \ldots, x_d) \in \mathbb{Z}^d : x_i \geq 0, \sum x_i \leq n\}.$$

a) (1 point) Calculate $|A|$.

b) (1 point) What are the elements of $A - A$?

c) (2 points) What are the limits of $|A + A|/|A|$ and $|A - A|/|A|$ as $n \to \infty$ for fixed d?

The *covering index*, $\operatorname{cov} A$, of a set $A \subset \mathbb{Z}$ is the smallest integer k (if it exists) such that k suitable translations cover $\mathbb{N}$:

$$\mathbb{N} \subset \bigcup (A + b_i)$$

for suitable $b_1, \ldots, b_k \in \mathbb{Z}$. If there is no such k, we say that the covering index is infinite.

35) (1 point) Let $A \subset \mathbb{N}$, $A = \{a_1, a_2, \ldots\}$, $a_1 < a_2 < \cdots$. The following are equivalent: $d_*(A) > 0$; $\operatorname{cov} A < \infty$; $a_{i+1} - a_i$ is bounded.

36) (2 points) Find a set with $\operatorname{cov} A = 2$ and such that $a_{n+1} - a_n = 1234$ infinitely often.

37) (2 points) For a set A of integers let $D(A)$ be the set of positive differences. Prove that if $\overline{d}(A) > 1/k$ with a positive integer k, then $\operatorname{cov} D(A) \le k - 1$.

38) (2 points) Show that $\operatorname{cov} D(A) \le \operatorname{cov} A$. Describe those sets where $\operatorname{cov} D(D(A)) = \operatorname{cov} A$. Show that if we repeat the operation D starting from a set of positive upper density, then in finitely many steps we arrive at the set of multiples of some integer.

Enhanced version (3 points): Describe the cases of equality in the inequality $\operatorname{cov} D(A) \le \operatorname{cov} A$.

39) (2 points) Given an $\alpha \in (0, 1)$, what are the possible values of $\operatorname{cov} A$ for sets such that $d_*(A) = \alpha$?

40) (2 points) Let $\sigma(A) = \alpha$. Show the existence of an $A' \subset A$ such that $\sigma(A') = \alpha$, but by omitting any single element the density of the remaining set will be $< \alpha$.

41) (2 points) Let $B \subset \mathbb{N}_0 = \mathbb{N} \cup \{0\}$ and let a number $\alpha \in (0, 1)$ be given, and define β as

$$\beta = \inf\{\sigma(A + B) : \sigma(A) \ge \alpha\}.$$

Show the existence of a set A satisfying $\sigma(A) = \alpha$, $\sigma(A + B) = \beta$.

42) (2 points) Let $A \subset \mathbb{R}^d$ be a finite set, $|A| = n$. Assume that A is *proper d-dimensional*, that is, it is not contained in any affine hyperplane of dimension $< d$. Prove that

$$|2A| \ge (d + 1)n - \frac{d(d + 1)}{2}.$$

43) (2 points) Let A, B be finite sets (in any commutative group), $|A| = n$, $|A + B| = \lambda n$. Show that there is a set T such that $|T| \le \lambda$ and $B \subset T + (A - A)$.

44) (2 points) Let A be a finite set (in any commutative group), $|A| = n$, $|2A| = \lambda n$. From Plünnecke's theorem we know that $|kA| \leq \lambda^k n$. For fixed λ this is an exponential function of k. Find a bound of the form $f(k, \lambda)n$, where $f(k, \lambda)$ is, for fixed λ, a polynomial of k.

45) (3 points) Let A be a finite set (in any commutative group). Prove that $|kA|$ is, for $k > k_0$, actually equal to some polynomial of k (Hovanskii's theorem). (The polynomial and the value of k_0 depend on the set A.)

46) (2 points) Let $A \subset \mathbb{Z}$. Show that

$$k|(k+1)A| \geq (k+1)|kA| - 1.$$

47) (3 points) Can the case $k = 2$ of the previous exercise be generalized to different sets as follows:

$$2|X + Y + Z| \geq |X + Y| + |Y + Z| + |Z + X| - 1?$$

This would be an additive counterpart to exercise 33.

48) (3 points) Can the case $k = 2$ of exercise 46 be generalized to $\mathbb{Z}_p$ like the Cauchy–Davenport inequality:

$$2|3A| \geq \min(3|2A| - 1, 2p)?$$

49) (3 points) Prove *elementarily* (without using Szemerédi's or van der Waerden's theorem) that whenever $d(A) > 0$, the set $A - A$ contains arbitrarily long arithmetical progressions.

50) (2 points) Prove that the set P of primes has $d^*(P) = 0$.

For the next exercises let A be a set of integers, $|A| = n$ and

$$\hat{A}(t) = \sum_{a \in A} e^{2\pi i a t}.$$

51) (2 points) What is the arithmetical meaning of the integral $\int_0^1 |\hat{A}(t)|^4 t$? What is its minimal value?

52) (3 points) What is the maximal value of the integral in the previous exercise, and for which sets does it occur?

53) (2 points) How can one express the number of three-term arithmetical progressions in A (that is, the number of pairs a, d such that $a, a + d, a + 2d \in A$) by the function $\hat{A}$?

54) (1 point) And what happens if we count only those where $d > 0$?

Let A, B be sets in commutative groups. A mapping $\varphi : A \to B$ is a *Freiman homomorphism* of order k, (F_k-homomorphism) if for every $a_1, \dots, a_k, a'_1, \dots, a'_k \in A$, such that $a_1 + \dots + a_k = a'_1 + \dots + a'_k$, we have $\varphi(a_1) + \dots + \varphi(a_k) = \varphi(a'_1) + \dots + \varphi(a'_k)$. A *Freiman isomorphism* is a 1-1 Freiman homomorphism whose inverse is also a Freiman homomorphism. If the order is not given, it is understood to be 2.

55) (1 point) If one of two F-isomorphic sets contains an l-term arithmetic progression, then so does the other.

56) (1 point) If A and B are F_r-isomorphic with $r = q(k + l)$, then $kA - lA$ and $kB - lB$ are F_q-isomorphic.

A d-dimensional arithmetic progression is a set of the form

$$P(q_1, \dots, q_d; l_1, \dots, l_d; a) = \{a + x_1 q_1 + \dots + x_d q_d : 0 \le x_i \le l_i\}.$$

57) (2 points) The F-homomorphic image of a d-dimensional arithmetic progression is also a d-dimensional arithmetic progression with the same "lengths" $l_1, \dots, l_d$.

58) (2 points) Let P be a d-dimensional arithmetic progression. Show that $|2P| < 2^d |P|$.

59) (1 point) Let p be a prime, $A \subset \mathbb{Z}_p$, $|A| = n$, k a positive integer. If $p > n^k$, then there is a $t \in \mathbb{Z}_p$, $t \neq 0$, such that $||at/p|| \le 1/k$ for all $a \in A$.

60) (3 points) Let $A \subset \mathbb{N}$, $|A| = n$. Prove that there is a Freiman isomorphic set contained in $[0, 4^n]$.

61) (2 points) Show that the bound in the previous exercise cannot be improved below 2^{n-2}.

We define the *Freiman dimension* of a set $A \subset \mathbb{R}^k$ as the largest d for which there is an isomorphic properly d-dimensional set.

62) (2 points) For a set $A \subset \mathbb{Z}^d$ the following are equivalent:
a) its Freiman dimension is d,
b) every Freiman homomorphism from A to any $\mathbb{R}^k$ is affine linear.

Bibliography

[1] V. Bergelson and I. Z. Ruzsa, *Sumsets in difference sets*, Israel J. Math., to appear.

[2] Y. Bilu, *Structure of sets with small sumset*, Structure theory of set addition, Astérisque, vol. 258, Soc. Mat. France, 1999, pp. 77–108.

[3] N. N. Bogolyubov, *Some algebraical properties of almost periods*, Zap. Kafedry Mat. Fiziki Kiev **4** (1939), 185–194.

[4] J. Bourgain, *On triples in arithmetic progression*, Geom. Funct. Anal. **9** (1999), 968–984.

[5] Mei-Chu Chang, *A polynomial bound in Freiman's theorem*, preprint.

[6] J. van der Corput, *On sets of integers I, II, III*, Proc. K. ned. Akad. Wet. **50** (1947), 252–262, 340–350, 429–435.

[7] György Elekes and Zoltán Király, *On the combinatorics of projective mappings*, Algebraic Combin. **14** (2001), 183–197.

[8] E. Følner, *Generalization of a theorem of Bogoliuboff to topological Abelian groups. With an appendix on Banach mean values in non-Abelian groups*, Math. Scandinavica **2** (1954), 5–18.

[9] ______, *Note on a generalization of a theorem of Bogoliuboff*, Math. Scandinavica **2** (1954), 224–226.

[10] G. Freiman, *Inverse problems in additive number theory VI. On the addition of finite sets III*, Izv. Vyss. Ucebn. Zaved. Matematika **3 (28)** (1962), 151–157 (in Russian).

[11] ______, *Foundations of a structural theory of set addition*, Amer. Math. Soc., 1973.

[12] G. Freiman and V. P. Pigaev, *The relation between the invariants r and t (russian)*, Kalinin. Gos. Univ. Moscow (1973), pp. 172–174.

[13] R. J. Gardner and P. Gronchi, *A Brunn-Minkowski inequality for the integer lattice*, Trans. Amer. Math. Soc. **353** (2001), 3995–4024.

[14] W. T. Gowers, *A new proof of Szemerédi's theorem for arithmetic progressions of length four*, Geom. Funct. Anal. **8** (1998), 529–551.

[15] ———, *A new proof of Szemerédi's theorem*, Geom. Funct. Anal. **11** (2001), 465–88.

[16] A. Granville, *An introduction to additive combinatorics*, Additive Combinatorics (Providence, RI, USA), CRM Proceedings and Lecture Notes, vol. 43, Amer. Math. Soc., 2007, pp. 1–27.

[17] B. J. Green and I. Z. Ruzsa, *Freiman's theorem in an arbitrary Abelian group*, J. London Math. Soc. **75** (2007), 163–175.

[18] K. Gyarmati, S. Konyagin, and I. Z. Ruzsa, *Double and triple sums modulo a prime*, Additive Combinatorics (Providence, RI, USA) (A. Granville, M.B. Nathanson, and J. Solymosi, eds.), CRM Proceedings and Lecture Notes, vol. 43, Amer. Math. Soc., 2007, pp. 271–277.

[19] K. Gyarmati, M. Matolcsi, and I. Z. Ruzsa, *Plünnecke's inequality for different summands*, in preparation.

[20] ———, *A superadditivity and submultiplicativity property for cardinalities of sumsets*, Combinatorica, to appear.

[21] H. Halberstam and K. F. Roth, *Sequences*, 2nd ed., Springer, 1983.

[22] P. Hegedűs, G. Piroska, and I. Z. Ruzsa, *On the Schnirelmann density of sumsets*, Publ. Math. Debrecen **53** (1998), 333–345.

[23] F. Hennecart, G. Robert, and A. Yudin, *On the number of sums and differences*, Structure theory of set addition, Astérisque, vol. 258, Soc. Mat. France, 1999, pp. 173–178.

[24] J. H. B. Kemperman, *On products of sets in a locally compact group*, Fund. Math. **56** (1964), 51–68.

[25] A. G. Khovanskii, *Newton polyhedron, Hilbert polynomial, and sums of finite sets*, Funct. Anal. Appl. **26** (1992), 276–281.

[26] ———, *Sums of finite sets, orbits of commutative semigroups, and hilbert functions*, Funct. Anal. Appl. **29** (1995), 102–112.

[27] M. Kneser, *Abschätzungen der asymptotischen Dichte von Summenmengen*, Math. Z. **58** (1953), 459–484.

[28] ———, *Summenmengen in lokalkompakten Abelschen Gruppen*, Math. Z. **66** (1956), 88–110.

[29] I. Kříž, *Large independent sets in shift-invariant graphs*, Graphs Combin. **3** (1987), 145–158.

[30] B. Lepson, *Certain best possible results in the theory of Schnirelmann density*, Proc. Amer. Math. Soc. **1** (1950), 592–594.

[31] V. F. Lev and P. Smeliansky, *On addition of two distinct sets of integers*, Acta Arith. **70** (1995), 85–91.

[32] Yu. V. Linnik, *On Erdős's theorem on the addition of numerical sequences*, Mat. Sbornik **10 (52)** (1942), 67–78.

[33] A. M. Macbeath, *On measure of sum sets II.*, Proc. Cambridge Philos. Soc. **49** (1953), 40–43.

[34] J. L. Malouf, *On a theorem of Plünnecke concerning the sum of a basis and a set of positive density*, J. Number Theory **54** (1995), 12–22.

[35] M. B. Nathanson, *Additive number theory: Inverse problems and the geometry of sumsets*, Springer, 1996.

[36] ______, *Growth of sumsets in abelian semigroups*, Semigroup Forum **61** (2000), 149–153.

[37] M. B. Nathanson and I. Z. Ruzsa, *Polynomial growth of sumsets in abelian semigroups*, J. Theor. Nombres Bordeaux **14** (2002), 553–560.

[38] Oystein Ore, *Theory of graphs*, Amer. Math. Soc., Providence, RI, USA, 1962 (reprints 1967, 1974).

[39] H. Plünnecke, *Über die Dichte der Summe zweier Mengen, deren eine die Dichte Null hat*, J. Reine Angew. Math. **205** (1960), 1–20.

[40] ______, *Eigenschaften und Abschätzungen von Wirkungsfunktionen*, Gesellschaft für Mathematik und Darenverarbeitung, Bonn, 1969.

[41] ______, *Eine zahlentheoretische Anwendung der Graphtheorie*, J. Reine Angew. Math. **243** (1970), 171–183.

[42] D. A. Raikov, *On the addition of point sets in the sense of Schnirelmann*, Mat. Sb. **5** (1939), 425–440.

[43] C. A. Rogers and G. C. Shephard, *The difference body of a convex body*, Arch. Math. **8** (1957), 220–233.

[44] I. Z. Ruzsa, *On difference sets*, Studia Sci. Math. Hungar. **13** (1978), 319–326.

[45] ______, *On the cardinality of $A + A$ and $A - A$*, Combinatorics Proc. Fifth Hungarian Colloq. (Keszthely 1976), Colloq. Math. Soc. J. Bolyai, vol. 18, North-Holland, Amsterdam, 1978, pp. 933–938.

[46] ______, *Essential components*, Proc. London Math. Soc. **54** (1987), 38–56.

[47] ______, *An additive property of squares and primes*, Acta Arith. **49** (1988), 281–289.

[48] ______, *An additive problem for powers of primes*, J. Number Theory **33** (1989), 71–82.

[49] ———, *An application of graph theory to additive number theory*, Sci., Ser. A Math. Sci. (N.S.) **3** (1989), 97–109.

[50] ———, *Addendum to: An application of graph theory to additive number theory*, Sci., Ser. A Math. Sci. (N.S.) **4** (1990/91), 93–94.

[51] ———, *Diameter of sets and measure of sumsets*, Monatsh. Math. **112** (1991), 323–328.

[52] ———, *Arithmetical progressions and the number of sums*, Period. Math. Hungar. **25** (1992), 105–111.

[53] ———, *A concavity property for the measure of product sets in groups*, Fund. Math. **140** (1992), 247–254.

[54] ———, *On the number of sums and differences*, Acta Math. Hungar. **59** (1992), 439–447.

[55] ———, *Generalized arithmetical progressions and sumsets*, Acta Math. Hungar. **65** (1994), 379–388.

[56] ———, *Sum of sets in several dimensions*, Combinatorica **14** (1994), 485–490.

[57] ———, *Sets of sums and commutative graphs*, Studia Sci. Math. Hungar. **30** (1995), 127–148.

[58] ———, *Sums of finite sets*, Number Theory, New York Seminar 1991–1995 (D. V. Chudnovsky, G. V. Chudnovsky, and M. B. Nathanson, eds.), Springer, New York, 1996, pp. 281–293.

[59] ———, *The Brunn-Minkowski inequality and nonconvex sets*, Geom. Dedicata **67** (1997), 337–348.

[60] A. Schields, *Sur la mesure d'une somme vectorielle*, Fund. Math. **42** (1955), 57–60.

[61] E. Szemerédi, *On sets of integers containing no k elements in arithmetic progression*, Acta Arith. **27** (1975), 299–345.

Part III

Thematic seminars

Chapter 1

A survey on additive and multiplicative decompositions of sumsets and of shifted sets

Christian Elsholtz

1.1 Introduction

Let $\mathcal{A}$ and $\mathcal{B}$ denote sets of integers and let $\mathcal{A}+\mathcal{B} = \{a+b : a \in \mathcal{A}, b \in \mathcal{B}\}$ be the sumset and $\mathcal{AB} = \{ab : a \in \mathcal{A}, b \in \mathcal{B}\}$ the product set.

We are interested in the additive and multiplicative structure of particular sets of integers $\mathcal{S}$. One way to study its additive structure is to see how large sets $\mathcal{A}$ and $\mathcal{B}$ exist with $\mathcal{A}+\mathcal{B} \subset \mathcal{S}$.

Perhaps one can even decompose the whole set so that there exist two sets of integers $\mathcal{A}$ and $\mathcal{B}$ with $|\mathcal{A}| \geq 2$ and $|\mathcal{B}| \geq 2$ such that $\mathcal{A}+\mathcal{B} = \mathcal{S}$ holds. A measure theoretic argument due to Wirsing [44] shows that most sets $\mathcal{S}$ cannot be decomposed in that way; (not even in the asymptotic sense of Definition 1.2.2 below). But, for any particular set $\mathcal{S}$ it may be of interest to show that the set can indeed not be decomposed.

The structure of sumsets and product sets is in the focus of current research activity; see, for example, Tao and Vu [41].

Here we are particularly interested if multiplicatively defined sets can be written as sumsets or if a shifted copy of a multiplicatively defined set can also be a product set.

This survey consists of two parts. In the first part we review results which show that if $\mathcal{A}+\mathcal{B}$ or $\mathcal{AB}+1$ is a subset of a multiplicatively defined set, then the counting functions on $\mathcal{A}(N) = \sum_{a \in \mathcal{A}, a \leq N} 1$ and $\mathcal{B}(N)$ cannot be too large. In particular, we give a proof that the set of primes $\mathcal{P}$ cannot be written in the form $\mathcal{AB}+c$, where $|\mathcal{A}|, |\mathcal{B}| \geq 2$, even if finitely many exceptions are allowed.

In the second part we use purely combinatorial counting arguments which show that there do exist sets $\mathcal{A}$ and $\mathcal{B}$ of certain finite sizes, such that $\mathcal{A}+\mathcal{B} \subset \mathcal{P}$, for example.

While the first part introduces sieve methods, the second part is based on pigeonhole-principle-type arguments, or extremal graph theory.

1.2 Part I

1.2.1 Multiplicative decompositions of sumsets

In earlier work such as [10, 11, 13], the author studied the question if multiplicatively defined sets can be additively decomposed. We say a set $\mathcal{S}$ can be additively decomposed if there exist sets $\mathcal{A}, \mathcal{B}$ with at least two elements each, such that $\mathcal{A}+\mathcal{B}=\mathcal{S}$.

Let us study two examples.

Example 1.2.1. Let $a\mathbb{Z} = \{an : n \in \mathbb{Z}\}$. Then $a\mathbb{Z} + b\mathbb{Z} = \gcd(a,b)\mathbb{Z}$. Note that $\mathbb{Z}\mathbb{Z} = \mathbb{Z}$, so that this sumset also has a trivial multiplicative decomposition:

$$a\mathbb{Z} + b\mathbb{Z} = (\gcd(a,b)\mathbb{Z})\,\mathbb{Z}.$$

If one allows positive integers only, i.e., if one studies $a\mathbb{N} + b\mathbb{N}$, then all sufficiently large multiples of $\gcd(a,b)$ can be represented. So, while $a\mathbb{N} + b\mathbb{N} = \gcd(a,b)\mathbb{N}$ does not quite hold, it holds apart from finitely many exemptions. This suggests the study of asymptotic additive decompositions.

Definition 1.2.2 (see [37, vol. 1, p. 5]). Let $\mathcal{S}$ be a set of positive integers. We say that $\mathcal{S}$ is *asymptotically additively irreducible* (or we say that *no asymptotic additive decomposition exists*) if there do not exist two sets of positive integers $\mathcal{A}$ and $\mathcal{B}$, with at least two elements each, such that for a sufficiently large N_0

$$(\mathcal{A}+\mathcal{B}) \cap [N_0, \infty] = \mathcal{S} \cap [N_0, \infty].$$

Another example of a set which has an additive and a multiplicative structure is the following.

Example 1.2.3. Let $\mathcal{A} = \{x^2 : x \in \mathbb{Z}\}$, $\mathcal{B} = \{2y^2 : y \in \mathbb{Z}\}$. It is well known that n can be written in the from $n = x^2 + 2y^2$ if and only if n is zero, or n has a prime factorization of the from

$$n = 2^r \prod_{p_i \text{ prime with } p_i \equiv 1,3 \bmod 8} p_i^{s_i} \prod_{q_i \text{ prime with } q_i \equiv 5,7 \bmod 8} q_i^{2t_i},$$

where the r, s_i, t_i are non-negative integers. In other words with $\mathcal{C} = \{n \in \mathbb{N} : p \mid n \Rightarrow (p = 2 \text{ or } p \equiv 1,3 \bmod 8)\}$ and $\mathcal{D} = \{n^2 : n \in \mathbb{Z}\}$ we find that $\mathcal{A}+\mathcal{B} = \mathcal{CD}$.

Note that such a decomposition is not necessarily unique. Let $\mathcal{D}' = \{n^2 : p \mid n \Rightarrow p \equiv 5, 7 \bmod 8\}$, then $\mathcal{A} + \mathcal{B} = \mathcal{C}\mathcal{D}'$ also holds.

Example 1.2.3 comes from the theory of quadratic forms. Observe that $\mathcal{A}$ and $\mathcal{B}$ occupy modulo primes $\frac{p+1}{2}$ of the residue classes. The sum of two random sets with about half of the residue classes modulo a prime p would usually cover all residue classes modulo p. In this example, modulo many primes this is not the case, namely if $p \equiv 5, 7 \bmod 8$, and $x, y \not\equiv 0 \bmod p$, then the corresponding sets $\mathcal{A}'$ and $\mathcal{B}'$ occupy $\frac{p-1}{2}$ of the residue classes, but $x^2 + 2y^2 \not\equiv 0 \bmod p$. Moreover, the considered sets are very "large". This makes the example interesting.

It may be conjectured that large sumsets that have a multiplicative property must come from related examples based on algebraic polynomials. For example, Croot and myself formulated the following conjecture:

Problem 1.2.1. Let $\mathcal{A} \subset [1, N]$ with $|\mathcal{A}| > N^{0.4}$. Assume that $|\mathcal{A} \mod p| \leq \frac{2}{3}p$, for every prime $p \leq N$. Must any such $\mathcal{A}$ be contained in the set of values of a quadratic polynomial, apart from $N^{1/3+\varepsilon}$ exceptions?

The values of a quadratic polynomial $f(x) = ax^2 + bx + c$ are in about one half the residue classes modulo primes.

The philosophy behind this conjecture is, if one removes from the interval $[1, N]$ a positive proportion of the residue classes modulo a positive proportion of the primes, and one arrives at a set $\mathcal{A}$ with $|\mathcal{A}| > N^{\delta}$, where $\delta > 0$, then there should be an algebraic reason for this, since for a random sieve process one would expect that $|\mathcal{A}|$ is very much smaller.

One of the long-standing open problems about additive decompositions is Ostmann's conjecture.

Conjecture 1.2.2 (Ostmann, [37, vol. 1, p. 13]). The set of primes $\mathcal{P}$ is asymptotically additively irreducible.

As partial results in the direction of this conjecture, the author proved (see Elsholtz [10, 13]) the following theorem.

Theorem 1.2.4. *Suppose that there is an asymptotic additive decomposition of the set of primes, i.e.,* $\mathcal{P} \cap [N_0, \infty] = \mathcal{P}' \cap [N_0, \infty]$, *for some large* N_0, *and* $\mathcal{A} + \mathcal{B} = \mathcal{P}'$, *then*

$$\sqrt{N}(\log N)^{-3} \ll \mathcal{A}(N) \ll \sqrt{N}(\log N)^2.$$

The same bounds hold for $\mathcal{B}(N)$.

Also $\mathcal{A}(N)\mathcal{B}(N) \ll N$ is known, a result that was independently proved by several authors. For a survey on this see [13].

As a consequence of a result such as Theorem 1.2.4, the author proved in [10] the following corollary.

Corollary 1.2.5. *The set of primes is not asymptotically additively decomposable into* three *sets* $\mathcal{A}, \mathcal{B}, \mathcal{C}$ *containing at least two elements each.*

For the additive decomposability of multiplicatively defined sets the following holds (see [13]).

Theorem 1.2.6. *Let $\mathcal{T}$ denote a set of primes with*

$$\left| \sum_{\substack{p\le x\\ p\in\mathcal{T}}} \frac{\log p}{p} - \tau \log x \right| < C.$$

Here $0<\tau<1$ denotes a real constant. Let

$$\mathcal{Q}(\mathcal{T}) = \{n \in \mathbb{N} : p|n \Rightarrow p \in \mathcal{T}\}.$$

Let $\mathcal{A}+\mathcal{B} \subseteq \mathcal{Q}'(\mathcal{T})$, where $\mathcal{Q}(\mathcal{T})\cap[N_0,\infty] = \mathcal{Q}'(\mathcal{T})\cap[N_0,\infty]$, for sufficiently large N_0. Then

$$\mathcal{A}(N)\mathcal{B}(N) \ll_{\tau,C} N(\log N)^{2\tau}.$$

Corollary 1.2.7. *Let $\mathcal{T}$ consist of $p=2$ and all primes $p\equiv 1,3 \bmod 8$. Then in the above theorem $\tau=\frac{1}{2}$ is admissible. Let $\mathcal{A}$ and $\mathcal{B}$ be as above. Then*

$$\mathcal{A}(N)\mathcal{B}(N) \ll N(\log N).$$

This upper bound is quite close to the following constructive example.

Example 1.2.8. Let $\mathcal{A}=\{n^2 : n\in\mathbb{N}\}$ and $\mathcal{B}=\{2n^2 : n\in\mathbb{N} \text{ and } (p \mid n \Rightarrow p\equiv 1,3 \bmod 8)\}$. No element $a_i+b_j\in\mathcal{A}+\mathcal{B}$ contains any prime factor $q\equiv 5,7 \bmod 8$ since for such q one would have

$$m^2+2n^2 \equiv 0 \bmod q, \quad \text{i.e.,} \quad \frac{m^2}{n^2} \equiv -2 \bmod q.$$

Since -2 is a quadratic non-residue modulo primes $p\equiv 5,7 \bmod 8$, this implies that both $m\equiv 0 \bmod q$ and $n\equiv 0 \bmod q$ must hold, which is not the case by construction. Hence $\mathcal{A}+\mathcal{B}\subset\mathcal{Q}(\mathcal{T})$, where $\mathcal{T}=\{p\in\mathcal{P} : p=2 \text{ or } p\equiv 1,3 \bmod 8\}$, and $\tau=\frac{1}{2}$.

Note that $\mathcal{A}(N)\sim N^{1/2}$ and $\mathcal{B}(N)\sim c\frac{N^{1/2}}{(\log N)^{1/2}}$, for some positive constant c. Therefore the general upper bound $N\log N$ on $\mathcal{A}(N)\mathcal{B}(N)$ and the actual value $c\frac{N}{(\log N)^{1/2}}$ in this example is only off by a small logarithmic factor.

1.2.2 Multiplicative decompositions of shifted sets

In [14] the author studied the multiplicative analogues.

Definition 1.2.9. Let $\mathcal{S}$ be a set of positive integers. We say that $\mathcal{S}$ is *asymptotically multiplicatively translation-irreducible* if there is no decomposition of the type $\mathcal{S}'=\mathcal{A}\mathcal{B}+c$, where $\mathcal{A},\mathcal{B}$ are sets of positive integers with at least two elements each, $c\neq 0$ an integer, and where $\mathcal{S}'$ is asymptotically equal to $\mathcal{S}$.

Theorem 1.2.10. *The set of primes is asymptotically multiplicatively translation-irreducible.*

The following theorem proves the corresponding result for sets of integers composed of certain prime factors only.

Theorem 1.2.11. *Let $\mathcal{T} \subset \mathcal{P}$ be a set of primes with the property that*

$$\sum_{p\leq N,\ p\in\mathcal{T}} 1 = \tau\frac{N}{\log N} + O\left(\frac{N}{(\log N)^2}\right)$$

for some constant $0 < \tau < 1$. Let

$$\mathcal{Q}(\mathcal{T}) = \{1\} \cup \{n \in \mathbb{N} : p \mid n \Rightarrow p \in \mathcal{T}\}.$$

Then $\mathcal{Q}(\mathcal{T})$ is asymptotically multiplicatively translation-irreducible.

In this survey we will give a proof of Theorem 1.2.10. The proof of Theorem 1.2.11 uses the same methods.

As a motivation for this type of results we observe that famous open problems are closely related. It is not known if there are infinitely Sophie Germain primes, or infinitely Carmichael numbers with a given number of prime factors. A Sophie Germain prime is a prime p where $2p+1$ is also prime. Let $\mathcal{A} = \{1,2\}$ and $\mathcal{AB}-1 \subset \mathcal{P}$. If p is a Sophie Germain prime, then $p+1 \in \mathcal{B}$. So, the question if an infinite set of shifted primes $\mathcal{P}'+1$ can be multiplicatively decomposed into $\mathcal{AB}$, where $\mathcal{A} = \{1,2\}$, is equivalent to the question whether there are infinitely Sophie Germain primes or not.

A Carmichael number n is a composite number such that $a^n \equiv a \bmod n$. Thus a Carmichael number is a pseudoprime to all bases a. By Korselt's criterion a squarefree number n is a Carmichael number if for all prime factors $p \mid n$: $p-1 \mid n-1$.

Let $\mathcal{A} = \{6,10,12,40\}$ and $\mathcal{AB}+1 \subset \mathcal{P}$. If it is possible to find infinitely many b such that $6b+1, 10b+1, 12b+1$ and $40b+1$ are simultaneously prime, then the product $n = (180b+7)(300b+11)(360b+13)(1200b+41)$ is a Carmichael number. This follows since $n-1$ is divisible by $120(30b+1)$ so that it is divisible by $180b+6, 300b+10, 360b+12, 1200b+40$. The smallest example of this type is $n = 7\cdot 11\cdot 13\cdot 41 = 41041$. For further details on Carmichael numbers see [21].

1.2.3 Some background from sieve methods

The two main ingredients of the proof are the large sieve inequality, in a form due to Montgomery, and Gallagher's larger sieve.

Lemma 1.2.12 (Montgomery [35]). *Let $\mathcal{P}$ denote the set of primes. Let p be a prime. Let $\mathcal{A}$ denote a set of integers which avoids $\omega_{\mathcal{A}}(p)$ residue classes modulo p. Here $\omega_{\mathcal{A}} : \mathcal{P} \to \mathbb{N}$ with $0 \leq \omega_{\mathcal{A}}(p) \leq p-1$. Let $\mathcal{A}(N)$ denote the counting function*

$\mathcal{A}(N) = \sum_{a\le N, a\in\mathcal{A}} 1$. *Let μ denote the Möbius function. Then the following upper bound on the counting function holds:*

$$\mathcal{A}(N) \le \frac{N+Q^2}{L}, \text{ where } L = \sum_{q\le Q} \mu^2(q) \prod_{p\mid q} \frac{\omega_{\mathcal{A}}(p)}{p-\omega_{\mathcal{A}}(p)}.$$

One typically chooses $Q = N^{1/2}$. There are many excellent expositions of a proof of this statement (or variants of it), including those by Montgomery [35], Brüdern [2], Davenport [8], Gallagher [18], and Tenenbaum [42].

Vaughan has found a suitable lower bound of L if $\sum_{p\le y}\frac{\omega(p)}{p}$ is known.

Lemma 1.2.13 (Vaughan [43]). *For sufficiently large Q,*

$$L \ge \sum_{m=1}^{\infty} \exp\left(m \log \left(\frac{1}{m} \sum_{p\le Q^{1/m}} \frac{\omega(p)}{p} \right)\right).$$

The sum $\sum_{m=1}^{\infty}$ is in fact a finite sum only. The parameter m denotes the number of prime factors of q in the definition of L. Hence $1 \le m \le \frac{\log Q}{\log 2}$. As there are at most $O(\log N)$ summands, a lower bound on L can be found by choosing a suitable value of m, and replacing the sum by this one summand. The loss of a factor of size at most $O(\log N)$ is small in typical large sieve applications.

Lemma 1.2.14 (Gallagher's larger sieve, [19]). *Let $\mathcal{S}$ denote a set of primes such that $\mathcal{A}$ lies modulo p (for $p\in\mathcal{S}$) in at most $\nu_{\mathcal{A}}(p)$ residue classes. Then the following bound holds, provided the denominator is positive:*

$$\mathcal{A}(N) \le \frac{-\log N + \sum_{p\in\mathcal{S}} \log p}{-\log N + \sum_{p\in\mathcal{S}} \dfrac{\log p}{\nu_{\mathcal{A}}(p)}}.$$

This sieve has a very elementary proof.

Proof. Let $\mathcal{A} = \{a_1, a_2, \dots, a_{|\mathcal{A}|}\} \subset [1,N]$, where $a_1 < a_2 < \cdots$. We study upper and lower bounds of $\prod_{1\le i<j\le|\mathcal{A}|}(a_j - a_i)$. The following upper bound is trivial, since $a_j - a_i < a_j \le N$:

$$\prod_{1\le i<j\le|\mathcal{A}|} (a_j - a_i) \le N^{\frac{1}{2}|\mathcal{A}|(|\mathcal{A}|-1)}.$$

To provide a lower bound we observe that the product is divisible by many small primes to a high power. Let $p^{s(p)} \parallel \prod_{1\le i<j\le|\mathcal{A}|}(a_j - a_i)$. A factor of p arises whenever a_i and a_j are in the same residue class modulo p. (Some additional factors of p arise modulo prime powers, leading to a slightly sharper sieve inequality, but for simplicity we ignore this here.) Let $t_k = |\{a_i : a_i \equiv k \bmod p\}|$, so that $\sum_k t_k = |\mathcal{A}|$. Then $s(p) \ge \sum_k \frac{1}{2} t_k(t_k - 1)$. The smallest that this latter sum can

be is if the a_i are as equidistributed among the $\nu(p)$ residue classes as possible, i.e., if $t_i \approx \frac{|\mathcal{A}|}{\nu(p)}$. This is a consequence of Cauchy's inequality:

$$\nu(p) \sum_k t_k^2 = \sum_{k:t_k>0} 1^2 \sum_{k:t_k>0} t_k^2 \geq \left(\sum_{k:t_k>0} 1 \cdot t_k \right)^2 = |\mathcal{A}|^2.$$

This implies that

$$\sum_{k=0}^{p-1} t_k(t_k - 1) \geq |\mathcal{A}| \left(\frac{|\mathcal{A}|}{\nu(p)} - 1 \right).$$

Hence $s(p) \geq \frac{|\mathcal{A}|}{2} \left(\frac{|\mathcal{A}|}{\nu(p)} - 1 \right)$.

Combining the upper and lower bounds gives

$$\prod_{p \in \mathcal{S}} p^{\frac{|\mathcal{A}|}{2}\left(\frac{|\mathcal{A}|}{\nu(p)} - 1\right)} \leq N^{\frac{|\mathcal{A}|}{2}(|\mathcal{A}|-1)}.$$

Simplifying and taking logarithms gives

$$|\mathcal{A}| \sum_{p \in \mathcal{S}} \frac{\log p}{\nu(p)} - \sum_{p \in \mathcal{S}} \log p \leq (|\mathcal{A}| - 1) \log N,$$

$$|\mathcal{A}| \left(\sum_{p \in \mathcal{S}} \frac{\log p}{\nu(p)} - \log N \right) \leq -\log N + \sum_{p \in \mathcal{S}} \log p,$$

which proves the larger sieve inequality, provided $\sum_{p \in \mathcal{S}} \frac{\log p}{\nu(p)} > \log N$ holds. □

1.2.4 Proof of Theorem 1.2.10

Proposition 1.2.15. *Let $\mathcal{A}, \mathcal{B}$ be sets of positive integers with at least two elements each. Suppose that $\mathcal{P}' = \mathcal{A}\mathcal{B} + c$, where $\mathcal{P}' \cap [N_0, \infty) = \mathcal{P} \cap [N_0, \infty)$. For sufficiently large N the following holds:*

$$\mathcal{A}(N) \ll N^{\frac{1}{2}+\frac{1}{7}}.$$

The same holds for $\mathcal{B}(N)$.

Proof. Suppose that $\mathcal{P}' = \mathcal{A}\mathcal{B} + c$, where $\mathcal{P}' \cap [N_0, \infty) = \mathcal{P} \cap [N_0, \infty)$. Let $b_1 < b_2$ be the least two elements of $\mathcal{B}$. Without loss of generality we may assume that $N_0 > \max(b_2, c)$; otherwise, we just increase N_0. Let N be a sufficiently large integer. We begin by showing that for any prime $q \in [N_0, N^{1/2}]$, the set $\mathcal{A}_1 = \mathcal{A} \cap [N^{1/2}, N]$ avoids at least two residue classes modulo q. If $a \in \mathcal{A}_1$, then $ab_1 + c = p$ is a prime with $b_1 < N_0 < q < N^{1/2} < p$. Now $b_1 \not\equiv 0 \bmod q$, so $b_1^{-1} \bmod q$ exists and $a \not\equiv -b_1^{-1}c \bmod q$. Similarly $a \not\equiv -b_2^{-1}c \bmod q$. It follows that $a \in \mathcal{A}_1$ avoids the two residue classes $-b_1^{-1}c$ and $-b_2^{-1}c$ modulo q. These are distinct since $0 < b_1 < b_2 < q$ and $0 < c < q$.

Applying Lemma 1.2.12 with $\omega(q) = 2$ gives the upper bound

$$\begin{aligned}\mathcal{A}_1(N) &\leq \frac{2N}{\sum_{q\leq N^{1/2}} \mu^2(q) \prod_{p|q} \frac{2}{p-2}}\\ &\leq \frac{2N}{\sum_{q\leq N^{1/2}} \mu^2(q) \prod_{p|q} \frac{2}{p}}\\ &= \frac{2N}{\sum_{q\leq N^{1/2}} \mu^2(q) \frac{d(q)}{q}}\\ &\ll \frac{N}{(\log N)^2},\end{aligned}$$

where $d(q)$ denotes the number of divisors of q. So $\mathcal{A}(N) \leq \sqrt{N} + \mathcal{A}_1(N) \ll \frac{N}{(\log N)^2}$. In view of $\mathcal{P}(N) \gg \frac{N}{\log N}$ we must have $\mathcal{B}(N) \gg \log N$. This trivially implies that $\mathcal{B}(N) \to \infty$, as $N \to \infty$.

Let $b_1 < b_2 < \cdots < b_k$ be the first $k = 8$ elements of $\mathcal{B}$. We adapt the argument above with the change that $N_0 > b_8$. A sieve with $\omega(q) = 8$, for $q \in [N_0, N^{\frac{1}{2}}]$, implies that $\mathcal{A}(N) \ll \frac{N}{(\log N)^8}$ and therefore $\mathcal{B}(N) \geq c'(\log N)^7$, for some positive constant c'.

To iterate this further we need a lower bound on $\omega(q)$ on average. Since a residue class $b \mod q$ forbids a class $a \mod q$ for $\mathcal{A}$, we actually count those classes modulo primes that occur in $\mathcal{B}$.

Let $\nu_{\mathcal{B}}(p) = |\mathcal{B} \mod p|$; i.e., $\nu_{\mathcal{B}}(p)$ is the number of residue classes modulo p that contain at least one element of $\mathcal{B}$.

Let

$$y = c'(\log N)^8 \tag{1.2.1}$$

and $\mathcal{S} = \{p : N_0 < p \leq y\}$. Assume that $\sum_{p\in\mathcal{S}} \frac{\log p}{\nu_{\mathcal{B}}(p)} > 3\log N$. This means that $\nu_{\mathcal{B}}(p)$ is small, say of size at most $p^{7/8}$ modulo many small primes $p \leq y$. We therefore apply Gallagher's larger sieve. Using (1.2.1) and Chebyshev's bound $\sum_{p\leq y} \log p < 2y$ we obtain a contradiction:

$$\begin{aligned}c'(\log N)^7 \leq \mathcal{B}(N) &\leq \frac{-\log N + \sum_{p\in\mathcal{S}} \log p}{-\log N + \sum_{p\in\mathcal{S}} \frac{\log p}{\nu_{\mathcal{B}}(p)}}\\ &< \frac{2y}{-\log N + \sum_{p\in\mathcal{S}} \frac{\log p}{\nu_{\mathcal{B}}(p)}} < c'(\log N)^7.\end{aligned}$$

Consequently,

$$\sum_{p\in\mathcal{S}} \frac{\log p}{\nu_{\mathcal{B}}(p)} \leq 3\log N.$$

This means that $\nu_{\mathcal{B}}(p)$ is modulo many small primes $p \leq y$ not as small as originally assumed. Using Montgomery's large sieve, this knowledge will give a good upper bound on $\mathcal{A}$, since any class that occurs in $\mathcal{B}$ forbids one in $\mathcal{S}$. But in order to use Montgomery's sieve we first need to transform the information from a measure on $\sum_{p\in\mathcal{S}} \frac{\log p}{\nu_{\mathcal{B}}(p)}$ to a measure on $\sum_{N_0<p\le y} \frac{\nu_{\mathcal{B}}(p)}{p}$.

From the Cauchy–Schwarz inequality and by partial summation from Chebyshev's bound on the number of primes $\pi(y) \gg \frac{y}{\log y}$ we find that

$$\left(\sum_{N_0<p\le y} \frac{\log p}{\nu_{\mathcal{B}}(p)}\right)\left(\sum_{N_0<p\le y} \frac{\nu_{\mathcal{B}}(p)}{p}\right) \ge \left(\sum_{N_0\le p\le y} \left(\frac{\log p}{p}\right)^{1/2}\right)^2 \gg \frac{y}{\log y}.$$

We combine these two inequalities to obtain

$$\sum_{p\in\mathcal{S}} \frac{\nu_{\mathcal{B}}(p)}{p} \gg \frac{y}{(\log y)(3\log N)} \gg \frac{(\log N)^7}{\log\log N}.$$

It then follows by Montgomery's large sieve and by Vaughan's Lemma 1.2.13 that $\mathcal{A}_1(N) \le \frac{2N}{L}$, where

$$\begin{aligned} L &= \sum_{q\le N^{1/2}} \mu^2(q) \prod_{p|q} \frac{\omega_{\mathcal{A}}(p)}{p-\omega_{\mathcal{A}}(p)} \\ &\ge \max_{m\in\mathbb{N}} \exp\left(m \log\left(\frac{1}{m} \sum_{p\le N^{1/(2m)}} \frac{\omega_{\mathcal{A}}(p)}{p}\right)\right). \end{aligned}$$

We choose the integer m such that $N^{1/(2m)}$ is close to y. Hence

$$m = \frac{1}{16}\frac{\log N}{\log\log N} + O(1).$$

We may choose

$$\omega_{\mathcal{A}}(p) = \begin{cases} \nu_{\mathcal{B}}(p) & \text{for } p\in\mathcal{S}, \\ 0 & \text{otherwise.} \end{cases}$$

Therefore we find that for all $\varepsilon' > 0$,

$$\begin{aligned} \log L &\ge \left(\frac{\log N}{16\log\log N} + O(1)\right) \log\left(\frac{16\log\log N}{\log N}\frac{c''(\log N)^7}{\log\log N}\right) \\ &\ge \left(\frac{1}{2} - \frac{1}{8} - \varepsilon'\right)\log N, \end{aligned}$$

$$\mathcal{A}(N) \le \mathcal{A}_1(N) + N^{\frac{1}{2}} \le \frac{2N}{L} + N^{\frac{1}{2}} \ll_{\varepsilon'} N^{\frac{1}{2}+\frac{1}{8}+\varepsilon'} \ll_{\varepsilon} N^{\frac{1}{2}+\frac{1}{7}} = N^{\frac{9}{14}}.$$

The same upper bound holds for $\mathcal{B}(N)$, by symmetry. So, the proposition follows. □

Proof of Theorem 1.2.10. If $ab + c = p \leq N$, then at least one of $a \ll \sqrt{N}$ or $b \ll \sqrt{N}$ must hold. Since all $\pi(N) + O(1) \gg \frac{N}{\log N}$, primes in $[N_0, N]$ have at least one presentation as $ab + c$, we not only have $\mathcal{A}(N)\mathcal{B}(N) \gg \frac{N}{\log N}$ but also have

$$\mathcal{A}(N)\mathcal{B}(\sqrt{N}) + \mathcal{A}(\sqrt{N})\mathcal{B}(N) \gg \frac{N}{\log N}.$$

Since our proposition holds for all sufficiently large N, we can apply it independently several times. Applying it once with $N_1 = \sqrt{N}$, a second time with $N_2 = N$ gives $\mathcal{A}(N) \ll N^{9/14}, \mathcal{B}(N) \ll N^{9/14}$ and $\mathcal{A}(\sqrt{N}) \ll N^{9/28}, \mathcal{B}(\sqrt{N}) \ll N^{9/28}$, whence

$$\frac{N}{\log N} \ll \mathcal{A}(N)\mathcal{B}(\sqrt{N}) + \mathcal{A}(\sqrt{N})\mathcal{B}(N) \ll N^{\frac{9}{14}+\frac{9}{28}} \ll N^{\frac{27}{28}}.$$

This contradiction proves the theorem. □

1.3 Part II

1.3.1 Sumsets

If one works in finite sets $[1, N]$, one can show that there exist two sets $\mathcal{A}, \mathcal{B}$ with at least $\frac{\log N}{\log \log N}$ many elements each, with $\mathcal{A} + \mathcal{B} \subset \mathcal{P}$. (If one chooses one of the sets smaller, then the other set can be taken considerably larger!) Below we will study some combinatorial counting methods like the pigeonhole principle to prove this lower bound. For results of this type we make essentially use of the counting function $\mathcal{P}(N) \gg \frac{N}{\log N}$ of the set of primes only. So, the same type of results would hold for other sets with such a counting function. But in order to strengthen the constants in the results slightly, we take a stronger result on the prime counting function. Recall that by the prime number theorem, $\pi(N) = \mathrm{li(N)} + \mathrm{O}(\frac{\mathrm{N}}{(\log \mathrm{N})^k})$ holds for all k. Here $\mathrm{li(N)} = \int_2^{\mathrm{N}} \frac{\mathrm{dt}}{\log \mathrm{t}}$. In particular (see, for example, Landau [33, p. 47]), $\pi(N) = \frac{N}{\log N} + \frac{N}{(\log N)^2} + 2\frac{N}{(\log N)^3} + O(\frac{N}{(\log N)^4})$.

On the other hand, using the large sieve method one can show that two sets $\mathcal{A}, \mathcal{B} \subset [1, N]$ of the same size $|\mathcal{A}| = |\mathcal{B}|$ with $\mathcal{A} + \mathcal{B} \subset \mathcal{P}$ can at most be of size $|\mathcal{A}| = O(N^{1/2})$ (compare [13]). There is a huge gap between the lower bound and the upper bound, and we conjecture that the upper bound should be $O_\varepsilon(N^\varepsilon)$, for all $\varepsilon > 0$. Observe that Examples 1.2.3 and 1.2.8 show that there are square-like sequences, where the sumset essentially avoids half of all prime factors as divisors. The conjecture says that if the sums are required to be primes, then no such square-like sequences exist.

1.3.2 Counting methods

The following result is due to Erdős, Stewart and Tijdeman [16] and is a very useful counting tool.

Lemma 1.3.1. *Let N be a positive integer and let $\mathcal{S} \subset [1, N]$ be a nonempty set. Let k be an integer with $1 \leq k \leq |\mathcal{S}|$. Then there exist a set $\mathcal{A} \subset \mathcal{S}$ and a set of non-negative integers $\mathcal{B} \subset [0, N-1]$ such that*

$$\mathcal{A} + \mathcal{B} \subset \mathcal{S}, \quad |\mathcal{A}| \geq \frac{\binom{|\mathcal{S}|}{k}}{\binom{N-1}{k-1}}, \quad |\mathcal{B}| = k.$$

Proof. There are $\binom{|\mathcal{S}|}{k}$ subsets of $\mathcal{S}$ containing k elements. To each of these subsets $\{s_1, \ldots, s_k\}$ with $s_1 < \cdots < s_k$ we associate the $(k-1)$-element difference-subset

$$\{s_2 - s_1, \ldots, s_k - s_1\} \subset [1, N-1].$$

By the pigeonhole principle there exists a $(k-1)$-element set $\{h_1, \ldots, h_{k-1}\}$ which is the difference subset of at least

$$t = \frac{\binom{|\mathcal{S}|}{k}}{\binom{N-1}{k-1}}$$

distinct k-element subsets of $\mathcal{S}$. The least elements of these t sets are denoted by $a_1, a_2, \ldots, a_t$. (Note that all a_i are distinct since otherwise two k-element sets would be the same.) Thus the lemma follows with $\mathcal{A} = \{a_1, \ldots, a_t\}$ and $\mathcal{B} = \{0, h_1, \ldots, h_{k-1}\}$. □

1.3.3 A result from extremal graph theory

We shall make use of results from extremal graph theory. The applicability of results from graph theory to number theory has been promoted by Erdős. For an early example see Erdős [15].

More recently it was stated in a paper by Győry, Stewart and Tijdeman [28] that M. Simonovits observed the possibility to apply the Kővari–Sós–Turán theorem to number theory. That theorem was, for example, applied in Gyarmati [26] to squarefree sumsets and by the author in [11] to triples of primes in arithmetic progression.

Let us state the Kővari–Sós–Turán theorem (see [31]). (Compare Theorem IV.10 (page 113) of Bollobás [1].)

Theorem 1.3.2. *Let $G(m, n)$ denote a bipartite graph with m vertices in the first class and n in the second. Let $z(m, n; s, t)$ denote the maximal number of edges of G such that G does not contain a complete bipartite graph $K_{s,t}$ with s vertices in the first class and t in the second. Then, for all natural numbers m, n, s and t we have*

$$z(m, n; s, t) \leq s^{\frac{1}{t}} n m^{1-\frac{1}{t}} + (t-1)m.$$

Proof. Let us consider a bipartite graph $G(V_1 \cup V_2, E)$ with $|V_1| = m, |V_2| = n$ that does not contain a complete bipartite subgraph $K_{s,t}$, with s elements in the first vertex set and t elements in the second. (For convenience, we say that a graph $K_{u,v}$ is oriented if the u vertices are in V_1 and the v vertices are in V_2.) Let us count the number of oriented $K_{1,t}$ subgraphs. On the one side, this number is $\sum_{i=1}^{m} \binom{d_i}{t}$, where d_i denotes the degree of the i-th vertex in V_1. On the other side, there are $\binom{n}{t}$ choices of t out of n elements and each of these choices of t elements is counted at most $s-1$ times, since there is no oriented $K_{s,t}$ subgraph. This implies that

$$\sum_{i=1}^{m} \binom{d_i}{t} \leq (s-1)\binom{n}{t}.$$

Since $f(z) = \binom{z}{t}$ is a convex function, and since $\sum_i d_i = |E|$, it follows by Jensen's inequality (see below) that

$$\sum_{i=1}^{m} \binom{d_i}{t} \geq m\binom{\frac{|E|}{m}}{t}.$$

With $|E| \geq m(t-1)$ (for a graph without $K_{s,t}$ but with a maximal number of edges) this implies that

$$(s-1)n^t \geq m\left(\frac{|E|}{m} - t + 1\right)^t$$

so that we have

$$z(m,n;s,t) \leq (s-1)^{\frac{1}{t}} n m^{1-\frac{1}{t}} + (t-1)m. \qquad \square$$

Let us recall Jensen's inequality for convex functions.

Lemma 1.3.3. *If f is a convex function on the interval $[a,b]$, then*

$$f\left(\sum_{i=1}^{n} \lambda_i x_i\right) \leq \sum_{i=1}^{n} \lambda_i f(x_i),$$

where (for all $1 \leq i \leq n$) $0 \leq \lambda_i \leq 1$, $\sum_{i=1}^{n} \lambda_i = 1$ and $x_i \in [a,b]$.

An important special case is with $\lambda_i = \frac{1}{n}$:

$$f\left(\frac{1}{n}\sum_{i=1}^{n} x_i\right) \leq \frac{1}{n}\sum_{i=1}^{n} f(x_i);$$

that is, the value of the function at the mean of the x_i is less than or equal to the mean of the values of the function at each x_i.

1.3.4 The case of primes

The difference counting approach

Pomerance, Sárközy and Stewart [38] proved the following results.

Theorem 1.3.4 (Pomerance, Sárközy and Stewart). *Let N, k be positive integers with $k < \log N$. There is an effectively computable constant c_1 such that if $N > c_1$, then there exist $\mathcal{A}, \mathcal{B} \subset [1, N]$ with $|\mathcal{B}| = k$:*

$$|\mathcal{A}| > \frac{N}{k(\log N)^k} \text{ and } \mathcal{A} + \mathcal{B} \subset \mathcal{P} \cap [1, N].$$

Proof. Recall that $\pi(N) \geq \frac{N}{\log N} + \frac{N}{(\log N)^2}$, for sufficiently large N.

By Theorem 1.3.1 there exist $\mathcal{A} \subset \mathcal{P}$ and $\mathcal{B}$ with $\mathcal{A} + \mathcal{B} \subset \mathcal{P} \cap [1, N]$ and $|\mathcal{B}| = k$:

$$|\mathcal{A}| \geq \frac{\binom{\pi(N)}{k}}{\binom{N-1}{k-1}} \geq \frac{\frac{(\pi(N)-k)^k}{k!}}{\frac{N^{k-1}}{(k-1)!}} \geq \frac{\left(\frac{N}{\log N}\right)^k}{kN^{k-1}} = \frac{N}{k(\log N)^k}.$$

So far we have $\mathcal{A} \subset [2, N], \mathcal{B} \subset [0, N-2]$. Shifting the set $\mathcal{A}$ down by 1, and shifting $\mathcal{B}$ up by 1, proves the theorem. □

Remark 1.3.5. An application of the purely combinatorial Theorem 1.3.1 allows to have that one of the sets $\mathcal{A}$ or $\mathcal{B}$ is itself a subset of the primes (or a shifted copy thereof). In fact, the second set can also have a prime restriction. In order to see this recall that Chudakov [4], van der Corput [5], Estermann [17] and Chowla [3] proved that almost all even integers are the sum of two primes with about the expected number of representations. The exceptional set has a counting function of at most $O_k\left(\frac{N}{(\log N)^k}\right)$, for all k. In particular, almost all integers of the form $2p$ are of the form $2p = p_1 + p_2$, about the expected number of times. This means that there are infinitely many triples of primes in arithmetic progression; see Chowla [3]. More precisely, the number of solutions of this equation with primes $p, p_1, p_2 \leq N$ is of order of magnitude $\frac{N^2}{(\log N)^3}$. For a closely related problem we refer the reader to Theorem 1.3.8 in Section 1.3.5.

The graph theoretic approach

Theorem 1.3.4 is certainly a good approximation to the prime k-tuple problem. We observe that the lower bound can be refined as follows. Let $y = \frac{1}{2}\log N$. Let $P = \prod_{p \leq y} p$. Define a bipartite graph $G(V_1 \cup V_2, E)$, where $V_1 = \{P, 2P, \ldots, \lfloor \frac{N}{P} \rfloor P\}$ and $V_2 = \{n \leq N : (n, P) = 1\}$. The set of edges is defined by $(v_1, v_2) \in E \Leftrightarrow v_1 + v_2 \in \mathcal{P}$. By this construction $v_1 + v_2$ is not divisible by any prime $p \leq y$. This pre-sieving increases the edge density slightly; for each p by a factor of $\frac{1}{1-\frac{1}{p}}$, and for all $p \le y$ together by a factor of $\prod_{p\le y} \frac{1}{1-\frac{1}{p}} \sim \log y$. For any given $v_1 \in V_1$, the number of $v_2 \in V_2$ such that $v_1 + v_2$ is prime, is $\frac{N}{\log N} \log y$, so that the total number of edges is, with a positive constant c_1,

$$|E| \ge (c_1 + o(1))|V_1|\,|V_2|\frac{\log\log N}{\log N}.$$

A sumset $\mathcal{A}+\mathcal{B} \subset \mathcal{B}$ corresponds to a complete bipartite graph $K_{s,t}$ with $|\mathcal{A}| = s$ and $|\mathcal{B}| = t$. Let us assume that the graph G does not contain any $K_{s,t}$. This gives bounds on the parameters s and t. By Theorem 1.3.2,

$$(c_1 + o(1))|V_1|\,|V_2|\frac{\log\log N}{\log N} \le |E| \le z(m,n,s,t) \le s^{\frac{1}{t}}|V_1||V_2|^{1-\frac{1}{t}} + t|V_2|.$$

This implies that

$$s \ge |V_2| c_2^t \frac{(\log\log N)^t}{(\log N)^t}.$$

With $|V_2| \sim N\prod_{p\le y}\left(1-\frac{1}{p}\right) \ge \frac{1}{2}\frac{N}{(\log\log N)}$ it follows that

$$s \ge \frac{c_3^t N(\log\log N)^{t-1}}{(\log N)^t}.$$

Now, let us assume that s is smaller than this bound. By Theorem 1.3.2 there exists a graph $K_{s,t}$ and hence subsets of the required size.

Corollary 1.3.6. *There exist sets $\mathcal{A}, \mathcal{B}$ with $\mathcal{A}+\mathcal{B} \subset \mathcal{P}$ and with cardinality $|\mathcal{A}|, |\mathcal{B}| \ge \frac{\log N}{\log\log N - \log\log\log N + O(1)} \ge \frac{\log N}{\log\log N}$, for large N.*

This slightly improves upon the bound $(1-\varepsilon)\frac{\log N}{\log\log N}$ proved by Pomerance, Sárközy and Stewart [38].

The lower bound above, $s \ge \frac{c_3^t N(\log\log N)^{t-1}}{(\log N)^t}$, may look somewhat surprising: for the prime k-tuple problem one thinks of the upper bound as $C_k\frac{N}{(\log N)^k}$, whereas here the lower bound has additional $\log\log N$ factors. This is however no contradiction: since V_1 and V_2 depend on P, and so on N the k-tuples considered above are not "constant", but vary as N increases. This example shows that upper bounds of the general prime k-tuple problem, where $n + b_i$, prime, $b_i \in [0,N], i \in \{1,\dots,k-1\}$, i.e., where the coefficients can vary with N, must also include $(\log\log N)$-factors.

1.3.5 Chains of primes in arithmetic progressions

If one studies chains of primes in arithmetic progressions a, $a+d$, $a+2d$, $a+(k-1)d \le N$, then an upper bound sieve shows there are at most $O_k(\frac{N^2}{(\log N)^k})$ of these. Until recently, a corresponding lower bound was only known in the case

$k = 3$ (see the above-mentioned work by Chudakov [4], van der Corput [5], Estermann [17] and Chowla [3]). For a very precise result see Grosswald [24].

Recently, Green and Tao [22] proved the following theorem.

Theorem 1.3.7. *Let*

$$G_k(N) := |\{(p_1 < p_2 < \cdots < p_k) : p_k \le N,\ p_i \in \mathcal{P} \text{ and } p_i = p_1 + (i-1)d\}|.$$

Then the following lower bound holds for some constant $C_k > 0$:

$$G_k(N) \ge (C_k + o(1))\frac{N^2}{(\log N)^k}.$$

An asymptotic was known before for $k = 3$, and Green and Tao [23] more recently established one for $k = 4$.

In this section we combine the Green–Tao result with the graph theoretic counting above and show the following results.

Theorem 1.3.8. *Let*

$$G_k(N) := |\{(p_1 < p_2 < \cdots < p_k) : p_k \le N,\ p_i \in \mathcal{P} \text{ and } p_i = p_1 + (i-1)d\}|.$$

Let $G_k(N) \ge (C_k + o(1))\frac{N^2}{(\log N)^k}$ and $C'_k = \frac{\ln C_k}{k-2}$. Let N be sufficiently large. For $t \ge 2$, $\varepsilon > 0$ and $s \ge (C_k + o(1))^t \frac{N}{(\log N)^{t+1}}$, there exist disjoint sets of primes $\mathcal{A}, \mathcal{B} \subseteq \mathcal{P} \cap [1, N]$ with $|\mathcal{A}| = s$, $|\mathcal{B}| = t$ such that for all $a_i \in \mathcal{A}$, $b_j \in \mathcal{B}$ and all $\lambda_r \in \{0, \frac{1}{k}, \dots, \frac{k-1}{k}, 1\}$: all $\lambda_r a_i + (1 - \lambda_r)b_j$ are prime.

Remark 1.3.9. Let us remark that Green and Tao actually proved a stronger theorem. Let $\mathcal{S}$ be a set of primes with positive upper density, i.e., $\limsup \frac{|\mathcal{S} \cap [1,N]|}{\pi(N)} > 0$, then $\mathcal{S}$ contains such progressions. Our corollary works in the density situation as well.

Corollary 1.3.10. *The same theorem holds for any finite set $L = \{\lambda_1, \dots, \lambda_l\} \subset [0, 1]$ of rational numbers.*

Proof. Let z denote the last common multiple of the denominators of the λ_r. Then $L \subset \{0, \frac{1}{z}, \frac{2}{z}, \dots, \frac{z-1}{z}, 1\}$, and Corollary 1.3.10 follows by an application of Theorem 1.3.8 with $k = z$. □

For some other consequences of the Green–Tao theorem see Granville [20].

For sets $\mathcal{A}, \mathcal{B}$ of equal size $|\mathcal{A}| = |\mathcal{B}|$ the theorem implies the following result.

Theorem 1.3.11. *For large N there exist disjoint sets of primes $\mathcal{A}, \mathcal{B} \subseteq \mathcal{P} \cap [1, N]$ as above with $|\mathcal{A}|, |\mathcal{B}| \ge \frac{\log N}{(k-2)\log\log N} - (C'_k + \varepsilon)\frac{\log N}{(\log\log N)^2}$.*

For the proof of Theorem 1.3.8 we use the Green–Tao theorem combined with the counting argument that we already used in the last section. This shows that this counting method is versatile and can be easily adapted to similar situations.

Proof of Theorem 1.3.8. We define the bipartite graph $G(V_1 \cup V_2, E)$ as follows: The sets of vertices are $V_1 = V_2 = \mathcal{P} \cap [1, N]$, the set of edges is

$$E_k = \Big\{(v_1, v_2) \in V_1 \times V_2 \mid v_1 \neq v_2 \quad \text{and} \quad \lambda v_1 + (1-\lambda)v_2 \in \mathcal{P}, \lambda \in \Big\{0, \frac{1}{k}, \dots, \frac{k-1}{k}, 1\Big\}\Big\}.$$

Here v_1 corresponds to p_1 and v_2 to p_k. An edge between v_1 and v_2 corresponds to a $(k+1)$-tuple of primes in progression. Note that $p_1 = p_2 = \cdots = p_k$ is not allowed. A complete bipartite graph $K_{s,t}$ corresponds to disjoint sets of primes $\mathcal{A}$ and $\mathcal{B}$ of sizes s and t such that all $\lambda a_i + (1-\lambda)b_j$ are also prime. By the Green–Tao theorem this graph G contains $G_{k+1}(N) \geq (C_{k+1} + o(1))\frac{N^2}{(\log N)^{k+1}}$ edges. The Kővari–Sós–Turán theorem (Theorem 1.3.2) guarantees that a bipartite graph with many edges must have a large $K_{s,t}$ as a subgraph.

Suppose that G does not contain a complete bipartite graph $K_{s,t}$ for which the first class of $K_{s,t}$ lies in the first class V_1 of G and the second class in V_2. We then find by Theorems 1.3.2 and 1.3.7 that for sufficiently large N,

$$(C_{k+1} + o(1))\frac{N^2}{(\log N)^{k+1}} \leq |E| \leq s^{\frac{1}{t}}\pi(N)^{2-\frac{1}{t}} + t\pi(N).$$

For large N we have the estimate $\pi(N) \geq \frac{N}{\log N}$. This implies for $t = O(N^{\varepsilon})$ that

$$\begin{aligned}\frac{C_k + o(1)}{(\log N)^{k-2}} &\leq s^{\frac{1}{t}}\frac{(\log N)^{1/t}}{N^{1/t}} + t\frac{\log N}{N}\\ &\leq s^{\frac{1}{t}}\frac{(\log N)^{1/t}}{N^{1/t}}\left(1 + \frac{1}{N^{1/3}}\right),\end{aligned}$$

and therefore

$$s \geq (C_k + o(1))^t \frac{N}{(\log N)^{(k-2)t+1}}.$$

Note that, since we can assume without loss of generality that $s \geq t$, the choice $t \leq O(N^{\varepsilon})$ is not restrictive.

Hence for s smaller than the above bound the graph G contains a complete bipartite graph $K_{s,t}$ which proves Theorem 1.3.8.

For Theorem 1.3.11 we want both sets to be of the same size, i.e., $s = t$. An easy computation shows that

$$t = \left\lfloor \frac{\log N}{(k-2)\log\log N} - \left(\frac{\ln C_k}{k-2} + \varepsilon\right)\frac{\log N}{(\log\log N)^2} \right\rfloor$$

is an admissible value, for sufficiently large N. □

Acknowledgment. The author is grateful to Oriol Serra and Javier Cilleruelo, the organizers of the DocCourse Additive Combinatorics at CRM in Barcelona, for the kind invitation and the interesting discussions. The author also thanks Laurence Rackham for discussions and congratulates him for the excellent results achieved.

Bibliography

[1] B. Bollobás, *Modern Graph Theory*, Graduate Texts in Mathematics, 184, Springer, New York, 1998.

[2] J. Brüdern, *Einführung in die analytische Zahlentheorie*, Springer, Berlin, 1995.

[3] S. Chowla, *There exists an infinity of 3–combinations of primes in A. P.*, Proc. Lahore Philos. Soc. **6** (1944), 15–16.

[4] N. G. Chudakov, *On the Goldbach problem*, Dokl. Akad. Nauk SSSR **17** (1937), 335–338.

[5] J. G. van der Corput, *Sur l'hypothèse de Goldbach pour presque tous les nombres pairs*, Acta Arith. **2** (1937), 266–290.

[6] E. S. Croot and C. Elsholtz, *On variants of the larger sieve*, Acta Math. Hungar. **103** (2004), 243–254.

[7] ———, *On thin sets of primes expressible as sumsets*, Acta Math. Hungar. **106** (2005), 197–226.

[8] H. Davenport, *Multiplicative Number Theory*, 2nd edition. Revised by Hugh L. Montgomery, Graduate Texts in Mathematics, 74, Springer, New York, 1980.

[9] C. Elsholtz, *A Remark on Hofmann and Wolke's Additive Decompositions of the Set of Prime*, Arch. Math. **76** (2001), 30–33.

[10] ———, *The Inverse Goldbach Problem*, Mathematika **48** (2001), 151–158.

[11] ———, *Some remarks on the additive structure of the set of primes*, Number Theory for the Millennium, I (Urbana, IL, 2000), A.K. Peters, 2002, pp. 419–427.

[12] ———, *Triples of primes in arithmetic progressions*, Quart. J. Math. **53** (2002), 393–395.

[13] ———, *Additive decomposability of multiplicatively defined sets*, Funct. Approx. Comment. Math. **35** (2006), 61–77.

[14] ———, *Multiplicative decomposability of shifted sets*, Bull. London Math. Soc. **40** (2008), 97–107.

[15] P. Erdős, *On sequences of integers no one of which divides the product of two others*, Izr. Inst. Math. and Mech. Univ. Tomsk **2** (1938), 74–82.

[16] P. Erdős, C. L. Stewart and R. Tijdeman, *Some Diophantine equations with many solutions*, Compositio Math. **66** (1988), 37–56.

[17] T. Estermann, *On Goldbach's problem: Proof that almost all even positive integers are sums of two primes*, Proc. London Math. Soc. (2) **44** (1938), 307–314.

[18] P. X. Gallagher, *The large sieve*, Mathematika **14** (1967), 14–20.

[19] ———, *A larger sieve*, Acta Arith. **18** (1971), 77–81.

[20] A. Granville, *Prime number patterns*. Amer. Math. Monthly **115** no. 4 (2008), 279–296.

[21] A. Granville and C. Pomerance, *Two contradictory conjectures concerning Carmichael numbers*, Math. Comp. **71** (2002), 883–908.

[22] B. Green and T. Tao, *The primes contain aribtrarily long arithmetic progressions*, Ann. of Math. (2) **167** (2008), 481–547.

[23] ———, *Linear equations in primes*, Ann. of Math. (2), to appear.

[24] E. Grosswald, *Arithmetic progressions that consist only of primes*, J. Number Theory **14** (1982), 9–31.

[25] ———, *On the number of quadruples of primes in arithmetic progression below a given bound*, Libertas Math. **2** (1982), 99–112.

[26] K. Gyarmati, *On divisibility properties of integers of the form* $ab+1$, Period. Math. Hungar. **43** (2001), 71–79.

[27] K. Győry, A. Sárközy and C. L. Stewart, *On the number of prime factors of integers of the form* $ab+1$, Acta Arith. **74** (1996), 365–385.

[28] K. Győry, C. L. Stewart and R. Tijdeman, *On prime factors of sums of integers III*, Acta Arith. **49** (1988), 307–312.

[29] A. Hofmann and D. Wolke, *On additive decompositions of the set of primes*, Arch. Math. **67** (1996), 379–382.

[30] B. Hornfeck, *Ein Satz über die Primzahlmenge*, Math. Z. **60** (1954), 271–273, see also the Zentralblatt review by Erdős and the correction in Math. Z. **62** (1955), page 502.

[31] T. Kővari, V. T. Sós and P. Turán, *On a problem of K. Zarankiewicz*, Colloquium Math. **3** (1954), 50–57.

[32] W. B. Laffer and H. B. Mann, *Decomposition of sets of group elements*, Pacific J. Math. **14** (1964), 547–558.

[33] E. Landau, *Handbuch der Lehre von der Verteilung der Primzahlen*, Leipzig, Berlin, 1909, Chelsea reprint New York, 1953.

[34] H. B. Mann, *Addition theorems: The addition theorems of group theory and number theory*, Wiley Interscience, New York, 1965.

[35] H. Montgomery, *The analytic principle of the large sieve*, Bull. Amer. Math. Soc. **84** (1978), 547–567.

[36] M. Nathanson, *Additive Number Theory, Inverse problems and the Geometry of Sumsets.* Graduate Texts in Mathematics, 165, Springer, New York, 1996.

[37] H.-H. Ostmann, *Additive Zahlentheorie*, 2 volumes, Springer, Berlin, 1956, reprint 1968.

[38] C. Pomerance, C. L. Stewart and A. Sárközy, *On Divisors of Sums of Integers III*, Pacific J. Math. **133** (1988), 363–379.

[39] J.-C. Puchta, *On additive decompositions of the set of primes*, Arch. Math. **78** (2002), 24–25.

[40] P. Ribenboim, *The New Book of Prime Number Records*, 3rd edition, Springer, New York, 1995.

[41] T. Tao and V. Vu, *Additive Combinatorics*, Cambridge University Press, 2006.

[42] G. Tenenbaum, *Introduction to Analytic and Probabilistic Number Theory.* Cambridge Studies in Advanced Mathematics, 46, Cambridge University Press, Cambridge, 1995.

[43] R. C. Vaughan, *Some Applications of Montgomery's Sieve*, J. Number Theory **5** (1973), 64–79.

[44] E. Wirsing, *Ein metrischer Satz über Mengen ganzer Zahlen*, Arch. Math. **4** (1953), 392–398.

[45] ______, *Über die Zahlen, deren Primteiler einer gegebenen Menge angehören*, Arch. Math. **7**, 263–272, 1956.

Christian Elsholtz
Department of Mathematics
Royal Holloway
Egham
Surrey TW20 0EX
UK
e-mail: `christian.elsholtz@rhul.ac.uk`

Chapter 2

On the detailed structure of sets with small additive property[1]

Gregory A. Freiman

It is known that a set of k integers A with small doubling (small $|A+A|$) satisfying the condition $|A+A| = 2k-1+b$, $0 \le b \le k-3$, is a part of arithmetic progression of $k+b$ terms. It appeared that the structure of A may be described in a much more detailed way.

The aim of this talk is to describe the deep structure of sets with small doubling (small $|A+A|$) as well as the intersection of two sets A and B with small $|A+B|$. The former case will be treated in Section 2.1 and it is all described in [3]. The second case will be introduced in Section 2.2.

2.1 Small doubling

The results in this section are a continuation of those from the series of papers on Inverse Additive Number Theory published in 1955–1964 (see references [84]–[92], [98] cited in [6]).

We will work with the set $A \subset \mathbb{Z}$ with $|A| = k \ge 3$. We assume that

$$A = \{a_0 = 0 < a_1 < \cdots < a_{k-1}\}$$

and $\gcd(A) = 1$. Let $T = |2A| = |A+A| = |\{a+b \mid a, b \in A\}|$. Notice that $T \ge 2k-1$.

In [2] there is the following result.

[1]This chapter has been written based on notes taken by Jordi Moragas at the seminar delivered by the author at CRM on February 13, 2008.

Theorem 2.1.1. *For $0 \leq b < k-2$ and $T = 2k-1+b$, the set A is a subset of the set $L = \{0, 1, \ldots, k+b-1\}$.* □

That is, if the set A has *small doubling property*, then it is contained in a small interval. From Theorem 2.1.1 it follows that $a_{k-1} = k + b' - 1$ where $b' \leq b$. It is no hard to see that, in fact, $b/2 \leq b' \leq b$.

If $T = 2k-1$, $A = \{0, 1, 2, \ldots, k-1\}$. In general,

$T = 2k-1+b$	b	$\lvert A\rvert$	A
$2k-1$	0	k	● ● ● ● ● ● ●
$2k$	1	$k+1$	● ● ● ● ○ ● ●
$2k+1$	2	$k+2$	● ● ○ ● ● ○ ●
		$k+1$	● ● ○ ● ● ● ○
$2k-1+b$	b	$\leq k+b$	??

However, if $b = k-2$, and thus $T = 3k-3$, then A may be contained on an arbitrarily large interval. We next give an example:

$$A = \{0, 1, \ldots, k_1 - 1, c, c+1, \ldots c + k_2 - 1\}$$

with c sufficiently large. Then, $|A| = k_1 + k_2 = k$, $|2A| = 2k_1 - 1 + 2k_2 - 1 + k_1 + k_2 - 1 = 3(k_1 + k_2) - 3 = 3k - 3$, and $A \subset L$ with $|L| = c + k_2$.

Remark 2.1.2 (Chang's polynomial bound). In 1964, the author discovered a deep fact about the structure of finite sets of integers with small doubling property. A finite d-dimensional arithmetic progression is a set of the form

$$P = \{a + x_1 q_1 + \cdots + x_d q_d \mid 0 \leq x_i < l_i, i = 1, \ldots, d\} \subset \mathbb{Z}.$$

The length of P is $l(P) = \prod_{i=1}^{d} l_i$. Clearly $|P| \leq l(P)$.

Theorem 2.1.3. *Let $A \subset \mathbb{Z}$ be a finite set such that $|2A| \leq \alpha|A|$. Then A is a subset of a d-dimensional arithmetic progression P, where $d \leq d(\alpha)$ and $|P| \leq C(\alpha)|A|$.* □

The bounds on these constants depending on α, given by Chang in [6], are $d \leq K\alpha^2 \log^3 \alpha$ and $C(\alpha) \leq \exp(K\alpha^2 \log^2 \alpha)$.[2] Chang also showed how one can improve the bound on the dimension to $d(\alpha) \lesssim \lfloor d-1 \rfloor$ with the bound on $|P|$ not substantially worse.

Throughout the presentation, we will use the following set to exemplify most of the situations:

$$\mathcal{A} = \{0, 2, 4, 6, 7, 8, 10, 12\}. \tag{2.1.1}$$

[2]In the notes [5], the explicit value $K = 2^{20}$ is obtained.

For this set we have $k = 8$, $b = k - 3 = 5$ and $T = 3k - 4 = 20$.

In what follows we will analyze the structure of a generic set A with small doubling property. We call a set A *stable* if $2A \cap [0, a_{k-1}] = A$. We can see that the sets $\mathcal{A}_1 = \mathcal{A} \cap [0,6] = \{0,2,4,6\}$ and $\mathcal{A}_2 = 12 - ([8,12] \cap \mathcal{A}) = \{0,2,4\}$ are stable.

We will show how the set A can be partitioned into three parts,

$$A = A_1 \cup I \cup (a_{k-1} - A_2),$$

where A_1 and A_2 are stable and I is an interval.[3] The same will happen with $2A$,

$$2A = A_1 \cup J \cup (2a_{k-1} - A_2),$$

where J is an interval.

For our example (2.1.1) we have $I = \{6,7,8\}$ and $J = \{6,7,\ldots,20\}$.

To describe in such precise way the structure of both A and $2A$ we will take a look at the *holes* of the set A. If we define $M(A) = M = [0, a_{k-1}]$, a *hole* in A will be simply an element of M not belonging to A. In the example, the set of holes is $\{1,3,5,9,11\}$. Now define a set $B = A \cup (a_{k-1} + A) \subset 2A$. In the example,

$$\mathcal{B} = \{\underbrace{0,2,4,6,7,8,10,\overbrace{12}_{A},14,16,18,19,20,22,24}^{a_{k-1}+A}\}. \tag{2.1.2}$$

If a is a hole in B, so is $a + a_{k-1}$. We can partition the set of holes in B into pairs $(a, a + a_{k-1})$ where a is a hole in A and $a + a_{k-1}$ is a hole in $a_{k-1} + A$. For (2.1.2) we have

$$(1,13) \quad (3,15) \quad (5,17) \quad (9,21) \quad (11,23).$$

Lemma 2.1.4. *For each pair $(a, a + a_{k-1})$ of holes in B we have*

$$a \in 2A \qquad or \qquad a + a_{k-1} \in 2A.$$

Proof. Let us look at A as a set of residues modulo a_{k-1}. Since the total number of residues is $k + b \leq 2k - 3$ and the sets $A \pmod{a_{k-1}}$ and $a - A \pmod{a_{k-1}}$ contain $k-1$ residues each, the sets of residues A and $a - A$ have a nonzero intersection. Therefore

$$a \in 2A \pmod{a_{k-1}}, \tag{2.1.3}$$

but in the set of integers the residue a is represented by a or by $a + a_{k-1}$. If both of these numbers do not belong to $2A$, then this contradicts (2.1.3). □

[3]This corresponds to the case when $b' = b$ (the length of the interval containing A is the maximum possible). Otherwise, the set I would be an interval with holes.

This lemma leads us to the following question, for the pair $(a, a + a_{k-1})$, one of the numbers belongs to $2A$, but what about the second one? There can be the two possible situations. If in the pair $(a, a + a_{k-1})$ both numbers of the pair belong to $2A$, we call the pair *unstable*. Otherwise, if one of the numbers of the pair does not belong to $2A$, the pair is called *stable*. In this latter case, the number not belonging to $2A$ will be called a *stable hole*. Moreover, if we have

$$a \notin 2A,$$

the pair will be called *left stable*; if

$$a + a_{k-1} \notin 2A,$$

the pair will be called *right stable*.

The number of stable/unstable pairs is described in the following lemma.

Lemma 2.1.5 (see [4]). *The set of b' pairs of holes in B consists of $2b' - b$ stable pairs and $b - b'$ unstable pairs.* □

Finally, let $e = \max\{a \mid (a, a_{k-1}) \text{ is left stable}\}$ and $c = \min\{a \mid (a, a_{k-1})$ is right stable$\}$.

Let us now take a look at the example (2.1.1), $\mathcal{A} = \{0, 2, 4, 6, 7, 8, 10, 12\}$. In this example, $b' = b$ and thus there are no unstable pairs. The following table describes the situation of the pairs.

Pairs	**Left**	**Right**
$(1, 13)$	1 (left stable)	13=6+7
$(3, 15)$	3 (left stable)	15=7+8
$(5, 17)$	5 (left stable) $= e$	17=7+10
$(9, 21)$	9=2+7$= c$	21 (right stable)
$(11, 23)$	11=4+7	23 (right stable)

Lemma 2.1.6 (see [4]). *We have*

$$e < c.$$ □

That is, all left-stable points occur before any right-stable point and all points in between are unstable. This fact ensures the existence of a long interval in $2A$.

We are now prepared to describe accurately the structure of the set A. Let d be the number of unstable pairs (a, a_{k-1}) for which $e < a < c$. The next theorem estimates the length of the interval which belongs to $2A$.

Theorem 2.1.7 (see [4]). *We have $J = [e + 1, c + a_{k-1} - 1] \subset 2A$ and $|J| \geq 2k - 1 + 2d$.* □

We will explain now how we construct a set $A'' \supset A$ (a completion of A) which helps to understand and describe the properties of A.

Take the set A' of numbers p which are holes in A, where p belongs to an unstable pair $(p, p + a_{k-1})$ for which $e < p < c$. Define the set

$$A'' = A \cup A'.$$

Theorem 2.1.8. *The set A'' has the following properties:*

- $2A'' = 2A$,
- *the interval $J' = [e+1, c-1]$ is included in A'' and*
- $|A + A| = 2k - 1 + b' = |A'' + A''| = 2|A''| - 1 + b' - 2d$. □

The most interesting results are those when $a_{k-1} = k - 1 + b$, that is, when the interval containing A has the maximum length for a given T. The following theorem is a corollary of Theorem 2.1.7 for this particular case and sheds light on what we were searching, the structure of A.

Theorem 2.1.9. *Let $T = 2k - 1 + b$ for A, where $0 \leq b < k - 2$, and let the length of the interval M be maximal, i.e., $|M| = k + b$. Then*

- $A_1 = A \cap [0, e+1]$ *is stable (i.e.* $A \cap [0, e+1] = 2A \cap [0, e+1]$*),*
- $A_2 = a_{k-1} - ([c-1, a_{k-1}] \cap A)$ *is stable,*
- $J = [e+1, c + a_{k-1} - 1] \subset 2A$ *and*
- $I = [e+1, c-1] \subset A$. □

2.2 Sums of different sets

In this part we are mainly interested in the structure of $A = \{a_0 = 0 < a_1 < \cdots < a_{k-1}\}$ if $|A - A| = |\{a_1 - a_2 \mid a_1, a_2 \in A\}|$ is small. The sets A and $a_{k-1} - A$ are symmetric with center $a_{k-1}/2$, and we have $|A - A| = |A + (a_{k-1} - A)|$.

To study this situation, we can go to the more general case where we have two sets $A = \{a_0 = 0 < a_1 < \cdots < a_{k-1}\}$ and $B = \{b_0 = 0 < b_1 < \cdots < b_{k-1}\}$ with $a_{k-1} = b_{k-1}$ and *small additive property*; that is, $|A + B| = 2k - 1 + b$ and $b < k - 2$. An approach to this topic is the following theorem due to Y. Stanchescu.

Theorem 2.2.1 (see [7]). *If $|A + B| \leq 2k - 1 + b$ and $b < k - 2$, then*

$$M(A) = M(B) = [0, k - 1 + b'],$$

where $b' \leq b$. □

For our purposes, this is a weak result in the sense that we only know that the sets are contained in a short interval, but we do not know anything about their structure. We will go further.

Take $A \cap B = C_1$ and define $n = |C_1|$. If we could show that $n \simeq k$, we would conclude that A and B are almost the same set.

Define

- $C_2 = A \setminus (A \cap B)$, $|C_2| = k - n = s$,
- $C_3 = B \setminus (A \cap B)$, $|C_3| = k - n = s$,
- $C_4 = A \cup B \subset [0, a_{k-1}]$,
- $C_5 = C_4 \cup (C_4 + a_{k-1})$, $|C_5| = 2k + 2s - 1$.

The number of holes in C_4 is $b' - s$ (and so is the number of pairs of holes in C_5). Moreover, for each pair of holes (a, a_{k-1}) in C_5,

$$a \in A + B,$$

or

$$a + a_{k-1} \in A + B.$$

From these last two facts we obtain the following inequality:

$$|A + B| \geq 2k + 2s - 1 + b' - s = 2k - 1 + b' + s. \tag{2.2.1}$$

Comparing (2.2.1) with $|A + B| = 2k - 1 + b$ we can conclude that $b \geq b' + s = b' + k - n$, and finally

$$n \geq k - (b - b'). \tag{2.2.2}$$

Then, if $b = b'$, that is, when the interval given by Theorem 2.2.1 is the maximum possible for a fixed $|A + B|$, we have that $n = k$ and thus $A = B$.

The conclusion is that, two *different* sets with small additive property are in fact *not* so different.

Bibliography

[1] M. Chang, *A polynomial bound in Freiman's theorem*, Duke Math. J. **113** no. 3 (2002), 399-419.

[2] G. A. Freiman, *The addition of finite sets I*, Izv. Vyss. Ucebn. Zaved. Matematica **6** no. 13 (1959), 202–213 (in Russian).

[3] ———, *Inverse additive number theory. XI. On the detailed structure of sets with small additive property*, preprint.

[4] ———, *Inverse additive number theory XII. Long arithmetic progressions in sets with small sumsets*, Acta Arithmetica, to appear.

[5] B. J. Green, *Edinburgh Lecture Notes on Freiman's Theorem*, unpublished.

[6] M. B. Nathanson, *Additive Number Theory. Inverse Problems and the Geometry of Sumsets*, Graduate Texts in Mathematics, 165, Springer, New York, 1996.

[7] Y. Stanchescu, *On addition of two distinct sets of integers*, Acta Arith. **75** no. 2 (1996), 191-194.

Gregory A. Freiman
The Raymond and Beverly Sackler Faculty of Exact Sciences
School of Mathematical Sciences
Tel Aviv University
Ramat Aviv, Tel Aviv 69978
Israel
e-mail: grisha@post.tau.ac.il

Chapter 3

The isoperimetric method[1] *Yahya O. Hamidoune*

Yahya O. Hamidoune

These notes consist of six sections. In the first three we introduce some classical results in several directions. In fourth section we develop isoperimetric tools for torsion-free groups and provide several very short proofs using the new technique. In fifth section we extend notions and main lemmas for general groups and provide proofs for theorems stated in the first sections. In last section we provide several observations, conjectures and recent results in the area.

3.1 Universal bounds of set products

Let G be a group and let A, B be nonempty finite subsets of G. Define their product as

$$AB = \{xy : x \in A \text{ and } y \in B\}.$$

We shall write

$$A^n = \underbrace{A \cdots A}_{n \text{ times}}.$$

When G is additive, we write $A + B$ and nA, respectively.

In the famous proof that every natural number is the sum of four squares Legendre used a lemma, saying that every integer is the sum of two squares modulo prime p. If we set $Q = \{x^2 : x \in \mathbb{Z}_p\}$, then it can be stated in a following way.

Lemma 3.1.1 (Legendre). $Q + Q = \mathbb{Z}_p$.

This lemma can be generalized as follows.

[1]This chapter has been written based on notes taken by Benjamin Girard at the seminar delivered by the author at CRM on February 6, 2008

Lemma 3.1.2 (Folklore). *Let G be a finite group, and let X, Y be nonempty subsets of G. If $XY \neq G$, then $|X| + |Y| \leq |G|$.*

Proof. Take any $a \in G \setminus (XY)$. Then we have $aX^{-1} \cap Y = \emptyset$. Therefore,

$$|G| \geq |aX^{-1}| + |Y| = |X^{-1}| + |Y| = |X| + |Y|. \qquad \square$$

Some easy bounds of a similar nature can be obtained. Let $X, Y \in G$ be nonempty finite subsets with $1 \in X$. We have

$$|XY| \geq \max(|X|, |Y|).$$

Equality is reached only when X is a subgroup and $Y = \bigcup y \in YXy$, as one can show using the next lemma.

Fix the notations throughout the section. Let G be a group, $T \subset G$ a finite nonempty subset, and let $S \subset G$ be a finite generating subset with $1 \in S$.

Lemma 3.1.3 (Folklore). *If $|TS| \leq |T|$, then $T = G$.*

Proof. Since $T \subset TS$, we have $TS = T$ and therefore $Ty = T$ for all $y \in S$. Since S generates G, we have $Tg = T$ for all $g \in G$. This gives

$$T = \bigcup_{g \in G} Tg = TG = G. \qquad \square$$

This lemma gives the following immediate corollary.

Corollary 3.1.4. *Suppose that G is finite. Set $|G| = n$ and $|S| = k$. Then $S^{n-2k+3} = G$.*

Proof. Suppose that $|S^j| < n$. By Legendre's lemma,

$$n \geq |S^{j-1}| + |S|.$$

Using Lemma 3.1.3 we obtain

$$|S^{j-1}| \geq |S^{j-2}| + 1 \geq |S^{j-3}| + 2 \geq \cdots \geq |S| + j - 2 = k + j - 2.$$

Therefore $n \geq k + k + j - 2$ and hence $j \leq n - 2k + 2$. $\square$

A sharp bound can be obtained from the following result.

Theorem 3.1.5 (Olson (1984) [17], Hamidoune (1981)). *Let $|S| = k$. If $TS \neq G$, then*

$$|TS| \geq |T| + \frac{k}{2}.$$

This bound is the best possible. For example, take a subgroup A and $a \notin A$. Put $B = A \cup Aa$. Then $|AB| = |B| = 2|A| = |A| + \frac{|B|}{2}$.

Olson's proof uses the Kemperman's generalization of the Cauchy–Davenport theorem and requires around 10 pages. However, using isoperimetric methods we can get a very short proof.

Using this sharper bound, we get more precise result than in Corollary 3.1.4.

Corollary 3.1.6 (Hamidoune–Rødseth [13]). *Suppose G is finite. Set $|G| = n$ and $|S| = k$. Assume also that $2k \leq n$. Then*

$$S^{\lfloor \frac{2n}{k} \rfloor - 1} = G.$$

Proof. Suppose that $|S^j| < n$. By Legendre's lemma,

$$n \geq |S^{j-1}| + |S|.$$

Using Theorem 3.1.5 we obtain

$$|S^{j-1}| \geq |S^{j-2}| + \frac{|S|}{2} \geq |S^{j-3}| + 2\frac{|S|}{2} \geq \cdots \geq |S| + (j-2)\frac{|S|}{2} = j\frac{k}{2}.$$

Therefore $n \geq k + j\frac{k}{2}$ and hence $j \leq \frac{2n}{k} - 2$. □

3.2 The Frobenius problem

Addition theorems have several applications. In this section we will see how each of the above bounds apply to the Frobenius problem. First let us define Frobenius semigroup.

Definition 3.2.1. Let $A \subset \mathbb{N}$ be such that $\gcd(A) = 1$. The semigroup

$$F(A) = \bigcup_{n \geq 1} nA$$

is called Frobenius semigroup.

Also set $m = \max(A)$ and $|A| = k$. Let $\phi : \mathbb{Z} \mapsto \mathbb{Z}_m$ be the canonical morphism. We write $\phi(A) = \bar{A}$.

Legendre's lemma implies the following.

Lemma 3.2.2 (Folklore). *If $|A| \geq m/2$, then $[m-1, \infty[\subset F(A)$.*

Proof. Take $x \geq m-1$ such that $x \notin F(A)$. By Legendre's lemma, $2(A \bar{\cup \{0\}} = \mathbb{Z}_m$. In particular $x = a_1 + a_2 + sm$, for some $a_1, a_2 \in A$, $a_1 \leq a_2$.

Since $x \notin F(A)$, we have $s = -1$. Since $x \geq m-1$,we have $2m-1 \leq a_1+a_2 \leq 2m$, but $x \neq m$, so $a_1 + a_2 = 2m-1$. This gives $a_1 = m-1$, $a_2 = m$, $x = m-1$ and therefore $x \in A \subset F(A)$, a contradiction. □

Applying Corollary 3.1.4 we get a bound for a little more general setting.

Lemma 3.2.3 (Folklore). *$[m(m-k)+1, \infty[\subset F(A)$.*

Proof. By Corollary 3.1.4,

$$(m - 2k + 3)A \bar{\cup \{0\}} = \mathbb{Z}_m.$$

Take $x \geq m(m-2k+2)+1$. Then $x = a_1 + \cdots + a_{m-2k+3} + sm$ for some $s \in \mathbb{Z}$. We have

$$m(m-2k+2)+1 \leq x = a_1 + \cdots + a_{m-2k+3} + sm \leq (m-2k+3)m + sm.$$

Hence $s \geq 0$ which, together with $m \in A$, implies $x \in F(A)$. □

Corollary 3.1.6 gives an even better result.

Lemma 3.2.4 (Rødseth, conjectured by Graham and Erdős). $\mathbb{N} \cap [\frac{(2m-k)m}{k}, \infty[\subset F(A)$.

Proof. By Corollary 3.1.6,

$$qA \,\bar{\cup}\, \{0\} = \mathbb{Z}_m,$$

where $q = \lfloor \frac{2m}{k} \rfloor - 1$. Take $x \geq \frac{(2m-k)m}{k}$. Then $x = a_1 + \cdots + a_q + sm$ for some $s \in \mathbb{Z}$. We have

$$\frac{(2m-k)m}{k} \leq x = a_1 + \cdots + a_q + sm \leq mq + ms \leq m(\frac{2m}{k} - 1) + sm.$$

Hence $s \geq 0$ which, together with $m \in A$, implies $x \in F(A)$. □

Further improvements in this direction were made by Dixmier.

Theorem 3.2.5 (Dixmier [6], conjectured by Graham and Erdős). $[m(m-1)/(k-1), \infty[\subset F(A)$.

The proof of Dixmier is ingenious and uses Kneser's theorem. A more precise bound conjectured by Lewin states that for sufficiently large m, $[(m-2)(m-k+1)/k+1, \infty[\subset F(A)$. Lewin's conjecture was proved by Dixmer in some special case and by Hamidoune in the general case (1997).

3.3 The $\alpha + \beta$-theorems

3.3.1 Direct theorems

Cauchy obtained the following generalization of Legendre's lemma.

Theorem 3.3.1 (Cauchy–Davenport [4], [5]). *Let p be a prime and let A and B be nonempty subsets of $\mathbb{Z}_p$ with $A + B \neq \mathbb{Z}_p$. Then*

$$|A+B| \geq |A| + |B| - 1.$$

The result of Cauchy was rediscovered by Davenport in 1935. Cauchy proved using his theorem that

$$|\{x_1^k + \cdots + x_t^k\}| \geq \min(p, 1 + t(p-1)/k),$$

where $1 \leq k \leq p-2$.

Until recently, the following Kemperman's generalization of the Cauchy–Davenport theorem was the main tool in the investigation of $|AB|$ in the non-abelian case.

Theorem 3.3.2 (Kemperman). *Let A and B be finite subsets of a multiplicative group G and let $a \in A$, $b \in B$. Then there exists a finite subgroup H with $aHb \subset AB$ and*

$$|AB| \geq |A| + |B| - |H|.$$

The proof of this result requires a beautiful transform. Notice that the same transform is recently used by Dicks and Ivanov in connection with a deep conjecture in group theory.

Because of unusual requirement $aHb \subset AB$ we do not know any isoperimetric proof of Kemperman's theorem, but it is suspected that such a proof exists (possibly very short). However, later on we will give a simple proof for the following corollary.

Corollary 3.3.3 (Kemperman). *Let A and B be finite subsets of a multiplicative torsion-free group G. Then*

$$|AB| \geq |A| + |B| - 1.$$

3.3.2 Inverse theorems

Inverse theorems or critical pair theorems are results describing sets reaching the bounds. The critical pair for the Cauchy–Davenport theorem is due to Vosper.

Theorem 3.3.4 (Vosper [20]). *Let A, B be subsets of $\mathbb{Z}_p$ such that $|A|,|B| \geq 2$ and $|A+B| = |A| + |B| - 1 \leq p - 2$. Then A, B are arithmetic progressions with the same difference.*

The critical pair for Vosper's theorem is as follows.

Theorem 3.3.5 (Hamidoune–Rødseth [14]). *Let A, B be subsets of $\mathbb{Z}_p$ such that $|A| \geq 3$, $|B| \geq 4$ and $|A+B| = |A| + |B| \leq p - 4$. Then there is d such that A (resp., B) is contained in arithmetic progression with difference d and cardinality $|A|+1$ (resp., $|B|+1$).*

Analogues for these two theorems for torsion-free groups are also known.

Theorem 3.3.6 (Brailowski–Freiman [1]). *Let A, B be finite subsets of a torsion-free group G such that $|A|,|B| \geq 2$ and $|A+B| = |A| + |B| - 1$. Then there are $a, b \in G$ such that $aA = \{1, r, \ldots, r^{|A|-1}\}$ and $Bb = \{1, r, \ldots, r^{|B|-1}\}$.*

The original proof based on Kemperman transform requires 18 pages. We shall give a very simple proof later. The critical pair theorem is as follows.

Theorem 3.3.7 (Hamidoune–Lladó–Serra [12]). *Let A, B be finite subsets of a torsion-free group G such that $|A| \geq 3$ and $|B| \geq 4$ and $|AB| = |A| + |B|$. Then there are $a, b \in G$ such that $aA \subset \{1, r, \ldots, r^{|A|}\}$ and $Bb \subset \{1, r, \ldots, r^{|B|}\}$.*

3.4 Torsion-free groups: The isoperimetric approach

For this section, G is a torsion-free group, $T \subset G$ is a finite nonempty subset and $S \subset G$ is a finite generating subset with $1 \in S$. By $\mathcal{F}(G)$ we shall mean the set of finite nonempty subsets of G.

Consider the objective function

$$\partial : \mathcal{F}(G) \mapsto \mathcal{F}(\mathcal{G}), \text{ where } \partial(\mathcal{F}) = (\mathcal{F}\mathcal{S}) \setminus \mathcal{F}.$$

Notice that $\partial(F)$ is the boundary of F in the Cayley graph on G defined by S.

Recall some more definitions.

Definitions. k^{th}-*connectivity* of S (or k^{th}-*isoperimetric number*) is defined as

$$\kappa_k(S) = \min\{|\partial(X)| : k \leq |X| < \infty\}.$$

A finite subset X of G such that $|X| \geq k$ and $|\partial(X)| = \kappa_k(S)$ is called a k-*fragment* of S.

A k-fragment with minimal cardinality is called a k-*atom*.

Note that if A is a k-atom, then for every x, xA is also a k-atom. In particular, there exists a k-atom containing 1.

Let us settle a very useful lemma.

Lemma 3.4.1 (Hamidoune (1996)). *Let X, Y be two distinct k-atoms of S. Then $|X \cap Y| \leq k - 1$.*

Proof. One can check the inequality $|\partial(X \cap Y)| + |\partial(X \cup Y)| \leq |\partial(X)| + |\partial(Y)|$ as an easy exercise. Assume that $|X \cap Y| \geq k$. Then by definition of $\kappa_k(S)$ we get

$$\begin{aligned} 2\kappa_k(S) &\leq |\partial(X \cap Y)| + |\partial(X \cup Y)| \\ &\leq |\partial(X)| + |\partial(Y)| \\ &= 2\kappa_k(S). \end{aligned}$$

Therefore $\kappa_k(S) = |\partial(X \cap Y)|$, and since $|X \cap Y| < |X| = |Y|$ this contradicts that X, Y are k-atoms. □

This lemma allows us to prove following statement.

Corollary 3.4.2 (Hamidoune (1986)). *If A is 1-atom of S, then $|A| = 1$.*

Proof. By translating, if necessary, we may assume that $1 \in A$. Now for every $x \in A$, xA and A are two 1-atoms containing x. Then $xA = A$, otherwise it would contradict Lemma 3.4.1. Since equality holds for all $x \in A$ we get $A = A^2$. Therefore A is a finite subgroup of a torsion-free group. Thus $A = \{1\}$. □

Now we are ready to give a short proof of the Cauchy–Davenport theorem for torsion-free groups.

Proof of Corollary 3.3.3. We have $|TS| - |T| \geq \kappa_1(S) = |\partial(\{1\})| = |1 \cdot S| - 1 = |S| - 1$. □

Note that $\kappa_2(S) \geq |S| - 1$. We will show that equality is possible, only when S is an arithmetic progression. We will need the following lemma.

Lemma 3.4.3. *Let A be a 2-atom. If $|A| = 2$ and $\kappa_2 = |S| - 1$, then S is an arithmetic progression.*

Proof. Put $A = \{1, r\}$. Then $\kappa_2 = |S| - 1 = |AS \setminus A| = |AS| - 2 \Rightarrow |AS| = |S| + 1$. Since $AS = S \cup Sr$, we see that S can only be arithmetic progression. □

Proposition 3.4.4 (Hamidoune (1996)). *If $\kappa_2(S) = |S| - 1$, then S is an arithmetic progression.*

Proof. Suppose the contrary and take a counterexample S having a minimal length. Let A be a 2-atom with $1 \in A$. By the previous lemma $|A| \geq 3$.

From the assumption on $\kappa_2(S)$ it follows that $|AS| - |A| = |S| - 1$, therefore

$$|S^{-1}A^{-1}| = |S| + |A| - 1.$$

This shows that $\kappa_2(A^{-1}) \leq |A| - 1$, hence

$$\kappa_2(A^{-1}) = \kappa_2(A) = |A| - 1.$$

By minimality assumption, either A has to be arithmetic progression or $|A| \geq |S|$. We show that both are impossible.

The case A is an arithmetic progression. Set $A = r^j\{1, r, r^2, \ldots\}$. Then $|rA \cap A| \geq 2$ which contradicts Lemma 3.4.1.

The case $|A| \geq |S|$. Let us begin with the following observation. For every $x \in A$ there is an element $s_x \in S \setminus \{1\}$ such that $x(s_x)^{-1} \in A$, since otherwise $A \setminus \{x\}$ would be a 2-fragment.

Since $|A| \geq |S|$, there are $x \neq y$ with $s_x = s_y = s$. Put $x(s_x)^{-1} = a \in A$ and $y(s_y)^{-1} = b \in A$. Then

$$s \in a^{-1}A \cap b^{-1}A.$$

But

$$1 \in a^{-1}A \cap b^{-1}A,$$

therefore $a^{-1}A = b^{-1}B$ and then $a = b$ and hance $x = y$, a contradiction. □

We now give an isoperimetric proof of the Brailowski–Freiman theorem.

Proof of Theorem 3.3.6. Suppose $|TS| = |T| + |S| - 1$ with $|T|,|S| \geq 2$. Without loss of generality, we may assume that $1 \in S \cap T$. Let K be the subgroup generated by S. We have $T \subset S$, otherwise there is a partition $T = T_1 \cup T_2$ with $1 \in T_1 \subset K$ and $T_2 \cap K = \emptyset$ which gives a contradiction:

$$|TS| = |T_1S| + |T_2S| \geq |T_1| + |S| - 1 + |T_2| + |S| - 1 \geq |T| + |S|.$$

Now the assumption $|TS| = |T|+|S|-1$ gives us $\kappa_2(S) \leq |S|-1 \Rightarrow \kappa_2(S) = |S|-1$, and by a previous corollary S is an arithmetic progression. After repeating the same argument for T we are done. □

3.5 The isoperimetric formalism

In this section we again consider arbitrary group G ($1 \in S \subset G$ — finite generating subset, T — finite nonempty subset).

Definition 3.5.1. We shall say that S is *k-separable* if there is a subset X with $|X| \geq k$ and $|XS| + k \leq |G|$.

Note that in the infinite group every subset is k-separable. Analogously we define k^{th}-*connectivity* of a k-separable subset S (or k^{th}-*isoperimetric number*) as

$$\kappa_k(S) = \min\{|\partial(X)| : k \leq |X| < \infty \text{ and } |XS| + k \leq |G|\}.$$

A finite subset X of G such that $|X| \geq k$, $|XS| + k \leq |G|$ and $|\partial(X)| = \kappa_k(S)$ is called a *k-fragment* of S. A k-fragment with minimal cardinality is called a *k-atom*. The cardinality of k-atom of S will be denoted by $\alpha_k(S)$.

A k-fragment of S^{-1} will be called a *negative k-fragment*. We use the following notations:

- $\alpha_{-k}(S) = \alpha_k(S^{-1})$,
- $\kappa_{-k}(S) = \kappa_k(S^{-1})$.

For a subset $X \subset G$ we shall write $X^S = G \setminus (XS)$.

The following lemma may be used as definition of $\kappa_k(S)$.

Lemma 3.5.2. *$\kappa_k(S)$ is the maximal integer j such that for every finite subset $X \subset G$ with $|X| \geq 4$,*

$$|XS| \geq \min(|G| - k + 1, |X| + j).$$

As one could expect there is a close isoperimetric relation between S and S^{-1}.

Lemma 3.5.3. *Suppose that G is finite and that S is k-separable. Then S^{-1} is k-separable. Moreover, $\kappa_k(S) = \kappa_{-k}(S)$ and $(X^S)^{S^{-1}} = X$.*

Now we can generalize Lemma 3.4.1 to this more general setting.

Theorem 3.5.4 (Hamidoune 1977 for $k = 1$ [7], 1996 for arbitrary k). *One of the following holds.*

1. *Two distinct k-atoms intersect in at most $k - 1$ elements.*
2. *Two distinct negative k-atoms intersect in at most $k - 1$ elements.*

Even though we do not know which of the two holds, the theorem is still very useful. Let us take a look at the following corollary.

Corollary 3.5.5 (Hamidoune (1981)). *Suppose $S \neq G$. Let H be a 1-atom of S with $1 \in H$ and let K be a 1-atom of S^{-1} with $1 \in K$. Then H is a subgroup or K is a subgroup.*

Proof. By the previous theorem one of the following reasonings is true.

For every $x \in H$, we have $x \in xH \cap H$. Then $H^2 = H$ and therefore H is a subgroup.

For every $x \in K$, we have $x \in xK \cap K$. Then $K^2 = K$ and therefore K is a subgroup. □

Sometimes we know that both cases in Theorem 3.5.4 hold simultaneously. Namely,

- $S = S^{-1}$. In this case k-atoms are also negative k-atoms.
- G is abelian. In this case $-A$ is a negative k-atom if A is a k-atom.
- G is infinite.

Let us now prove the Cauchy–Davenport theorem in a general case.

Proof of Theorem 3.3.1. We can take $0 \in B$. By the previous corollary there is a subgroup H with $\kappa_1(B) = |H + B| - |H| = |B| - 1$. Therefore,

$$|A + B| - A \geq \kappa_1(B) = |B| - 1.$$ □

Finally we proof a sharp bound for $|AB|$ stated in Section 3.1.

Proof of Theorem 3.1.5. First we prove that $\kappa_1(S) \geq \frac{|S|}{2}$. Let H be a 1-atom of S with $1 \in H$ and let K be 1-atom of S^{-1} with $1 \in K$.

If H is a subgroup, then using the fact that S generates G we have $|HS| \geq 2|H|$. Hence

$$\kappa_1(S) = |HS| - |H| \geq \frac{|HS|}{2} \geq \frac{|S|}{2}.$$

If H is not a subgroup, then K is a subgroup and G is finite. We obtain

$$\kappa_{-1}(S) \geq \frac{|S|}{2}.$$

But by Lemma 3.5.3, $\kappa_{-1}(S) = \kappa_1(S)$ and

$$|TS| - |T| \geq \kappa_1(S) \geq \frac{|S|}{2}.$$ □

3.6 The structure of k-atoms

In the previous section we saw that the structure of 1-atoms is known for one of the sets S or S^{-1}. However, the structure of 2-atoms in torsion-free groups is very mysterious. The following conjecture could be true.

Conjecture. Suppose that G is torsion-free and let $1 \in S \subset G$ be a subset. Let A be a k-atom of S. Then $|A| = k$.

The answer for this conjecture is not known even for $k = 2$.

In a private discussion Freiman formulated the following conjecture.

Conjecture. If $1 \in A$ and $|A^2| \leq 3|A| - 4$, then $\langle A \rangle$ is abelian.

It is suspected that this conjecture could be reduced from the previous one.

On the other hand, in $\mathbb{Z}_p$ the situation is a bit clearer. The following theorem holds.

Theorem 3.6.1 (Serra–Zèmor [19]). *Let S be a subset of $\mathbb{Z}_p$ and $0 \in S$. Let A be a 2-atom of S. Put $\mu = \kappa_2(S) - |S|$. If $|S| < p - \binom{\mu+4}{2}$, then $|A| = 2$. This bound is the best possible.*

In the case of a general abelian group we have the following result.

Theorem 3.6.2 (Hamidoune–Serra–Zèmor (2006)). *Let G be a finite abelian group and let $0 \in S$ be a generating 2-separable subset of G such that $|S| \geq 3$ and*

$$\kappa_2(S) - |S| = \mu \leq 4.$$

Let A be a 2-atom of S containing 0. If $|S| < |G| - \binom{\mu+4}{2}$, then either $|A| = 2$ or A is a subgroup of G.

In the end we formulate several very recent results. For the sake of brevity let us give a very vague formulation of the following theorem, which extends Kemperman's theorem.

Theorem 3.6.3 (Grynkiewicz [2]). *Let A and B be finite, nonempty subsets of an abelian group G with $|A+B| = |A| + |B|$. If $A+B$ is aperiodic, then there exist $\alpha, \beta \in G$ such that $|A \cup \{\alpha\} + B \cup \{\beta\}| = |A \cup \{\alpha\}| + |B \cup \{\beta\}| - 1$, or otherwise sets A and B fall into certain categories, which are listed in [2].*

A recent improvement of Kneser's theorem is due to Balandraud.

Theorem 3.6.4 (Balandraud (2008) [3]). *Let G be an abelian group and $B \subset G$ a finite subset, such that there exists a finite set $A \in G$ satisfying*

$$|A+B| < |A| + |B| - 1.$$

Then there exists a finite subgroup $H_B \neq \{0\}$ such that for all A satisfying $|A+B| < |A| + |B| - 1$ we have

$$A + B + H_B = A + B.$$

Moreover, $|A+B| < |A+H_B| + |B+H_B| - |H_B|$.

Bibliography

[1] L.V. Brailowski, G.A. Freiman, *On a product of finite subsets in a torsion free group*, J. Algebra **130** (1990), 462–476.

[2] D. Grynkiewicz, *A step beyond Kempermann structure theorem*, preprint, May 2006.

[3] E. Balandraud, *Un nouveau point de vue isopérimetrique appliqué au théorème de Kneser*, Ann. Inst. Fourier (Grenoble), to appear.

[4] A. Cauchy, *Recherches sur les nombres*, J. École polytechnique **9** (1813), 99–116.

[5] H. Davenport, *On the addition of residue classes*, J. London Math. Soc. **10** (1935), 30–32.

[6] J. Dixmier, *Proof of a conjecture by Erdös, Graham concerning the problem of Frobenius*, J. Number Theory **34** (1990), 198–209.

[7] Y.O. Hamidoune, *Sur les atomes d'un graphe orienté*, C. R. Acad. Sc. Paris Sér. A **284** (1977), 1253–1256.

[8] ———, *Quelques problémes de connexité dans les graphes orientés*, J. Combin. Theory Sér. B **30** (1981), 1–10.

[9] ———, *An application of connectivity theory in graphs to factorizations of elements in groups*, European J. Combin. **2** (1981), 349–355.

[10] ———, *Additive group theory applied to network topology.* Combinatorial Network Theory, Appl. Optim., vol. 1, Kluwer Academic Publishers, Dordrecht 1996, 1–39,

[11] ———, *An isoperimetric method in additive theory.* J. Algebra **179** (1996), no. 2, 622–630.

[12] Y.O. Hamidoune, A. Llàdo and O. Serra, *On subsets with a small product in torsion-free groups.* Combinatorica **18** (1998), no. 4, 513–549.

[13] Y.O. Hamidoune and Ø.J. Rødseth, *On bases for σ-finite groups.* Math. Scand. **78** (1996), no. 2, 246–254.

[14] ———, *An inverse theorem modulo p*, Acta Arith. **92** (2000) 251–262.

[15] Y.O. Hamidoune, O. Serra and G. Zémor, *On the critical pair theory in* $\mathbb{Z}_p$, Acta Arith. **121** (2006), no. 2, 99–115.

[16] ———, *On the critical pair theory in Abelian groups: Beyond Chowla's theorem*, Combinatorica **28** (2008), no. 4, 441–467.

[17] J.E. Olson, *On the sum of two sets in a group*, J. Number Theory **18** (1984), 110–120.

[18] Ø.J. Rødseth, *Two remarks on linear forms in non-negative integers, Math. Scand.* **51** (1982), 193–198.

[19] O. Serra and G. Zémor, *On a generalization of a theorem by Vosper*, Integers **0**, Paper A10, 2000.

[20] G. Vosper, *The critical pairs of subsets of a group of prime order*, J. London Math. Soc. **31** (1956), 200–205.

Yahya O. Hamidoune
Université Pierre et Marie Curie (Paris 6)
E.Combinatoire – case 189
4, place Jussieu
75252 Paris Cedex 05
France
e-mail: `hamidoune@math.jussieu.fr`

Chapter 4

Additive structure of difference sets

Norbert Hegyvári

Additive structure of difference and iterated difference sets are investigated. In this survey we collect results and some applications of theorems of Bogolyubov and Følner. Some exercises are also included.

4.1 Introduction

Let A be a subset of integers with positive upper density,

$$\overline{d}(A) := \limsup_{n\to\infty} \frac{A(n)}{n} > 0,$$

where $A(n) := \sum_{\substack{a\in A\\ 1\le a\le n}} 1$. The difference set is defined as follows:

$$D(A) = \{a - a' : a, a' \in A\}.$$

The iteration of this operation (i.e., the second difference set) is

$$D_2 = D(D(A)) = A - A + A - A,$$

and generally for $k > 1$,

$$D_{k+1} = D(D_k(A)).$$

Define the time of stability of A by $T(A) = \min\{k \mid D^+_{k+1}(A) = D^+_k(A)\}$, where the operation $D^+(\cdot)$ takes just the positive part of $D(\cdot)$.

Stewart and Tijdeman and later Ruzsa proved that

$$T(A) \leq 2 + \log_2(\overline{d}(A)^{-1} - 1).$$

Let G be a countable torsion group and let $H_1 \subseteq H_2 \subseteq \cdots \subseteq H_n \subseteq \cdots$ be a sequence of finite subgroups of G. Then G is said to be σ-finite with respect to $\{H_n\}$ if $G = \bigcup_{n=1}^{\infty} H_n$.

Let $A \subseteq G$. The asymptotic upper density of A is defined by

$$\overline{d}(A) = \limsup_{n\to\infty} \frac{|A \cap H_n|}{|H_n|}. \tag{4.1.1}$$

We can introduce the time of stability in groups as well.

Assume that the sequence $\{D_k(A);\ k \geq 0\}$ is stable (i.e., for some k, $D_{k+1}(A) = D_k(A)$). Let $T(A, G)$ be the time of stability defined by

$$T(A,G) = \min\{k \mid D_{k+1}(A) = D_k(A)\}.$$

Theorem 4.1.1. [5] *Let G be a σ-finite abelian group with respect to $\{H_n\}$ and let A be a nonempty subset of G. Let $\overline{d}(A)$ be the upper density of A defined by* (4.1.1). *If $\overline{d}(A) > 0$, then*

$$T(A,G) \leq \log_2(\overline{d}(A)^{-1}) + 2.$$

F. Hennecart and the author extended this theorem to finite groups (Theorem 4.1.2) and σ-finite groups (not necessary abelian, Theorem 4.1.3), see [9].

Theorem 4.1.2. [8] *Let A be a generating subset of a finite group G such that $1 \in A$. Let k_0 be defined by*

$$k_0 = \begin{cases} 1 & \text{if } |G|/2 < |A| \leq |G|, \\ 2 & \text{if } |G|/3 < |A| \leq |G|/2, \\ \left\lfloor \log_2\left(\frac{2|G|}{3|A|} - 1\right)\right\rfloor + 3 & \text{if } |A| \leq |G|/3. \end{cases}$$

Then, for any integer $k \geq k_0$

$$D_k(A) = G. \tag{4.1.2}$$

Theorem 4.1.3. [8] *Let G be a σ-finite group with respect to $\{H_n\}$ and let A be a nonempty subset of G. Assume that A has a positive upper density and $\alpha := \overline{d}(A)^{-1} \geq 3$. Then*

$$T(A,G) \leq \lfloor \log_2(2\alpha/3 - 1)\rfloor + 3, \tag{4.1.3}$$

where $\lfloor u \rfloor$ denotes the greatest integer less than or equal to the real number u.

Exercises.

1. For $A \subseteq \mathbb{Z}$, $\overline{d}(A) := \limsup_{n\to\infty} \frac{A(n)}{n} > 0$, we have that $D(A)$ has bounded gaps; i.e., writing $D(A) = \{d_1 < d_2 < \cdots < d_k < \cdots\}$ there exists a $K > 0$, for which $d_{k+1} - d_k \leq K$.

   The proof will be given in a more general structure (see the proof of Lemma 4.2.2).

2. Prove that for $A \subseteq \mathbb{Z}$, $\overline{d}(A) := \limsup_{n\to\infty} \frac{A(n)}{n} > 0$ and that for every $k \in \mathbb{N}$ the sequence $D(A)$ contains a k-terms arithmetic progression.

3. Nevertheless the size of the gaps is uncertain; for every $\delta > 0$ and K there exists a set $A \subseteq \mathbb{Z}$, $\overline{d}(A) := \limsup_{n\to\infty} \frac{A(n)}{n} = \delta$, and $\max_k d_{k+1} - d_k \geq K$.

4. Let G be an Abelian group and assume that the sequence $\{D_k(A)\}$ is stable, i.e., there exists a k_0, such that $k \geq k_0$,

$$D_{k+1} = D_k(A).$$

   Show that $D_k(A)$ is a subgroup in G.

5. Let $m \in \mathbb{N}$, $A = \{x : x \equiv 0, 1 (\mod m)\} \subseteq \mathbb{N}$. Prove that $d(A)$ exists where $d(A)$ is the density of a sequence defined by $\lim_{n\to\infty} \frac{A(n)}{n}$ and find its value. Determine $T(A)$, the time of the stability as well.

4.2 The case D(A)

V. Bergelson [1] investigated the additive structure of $D(A)$.

Let us define first the Banach density of a set of integers as follows:

$$d^*(A) := \sup\Big\{L : \forall m,\ \exists(a_m, b_m)\ |b_m - a_m| \geq m, and\ \frac{|A \cap (a_m, b_m)|}{|b_m - a_m|} \geq L\Big\}.$$

(We shall define $d^*(A)$ in higher dimension as well.)

Exercise. Prove that $\overline{d}(A) \leq d^*(A)$.

Bergelson proved the following theorem.

Theorem 4.2.1. [1] *There exists an infinite set B of integers for which*

$$A - A \supseteq B + B + \cdots + B = B \cdot k,$$

provided A has positive Banach density.

The proof of this theorem is based on an ergodic theorem (Fürstenberg Correspondence Principle).

Theorem 4.2.2. [?] *Given a set $E \subseteq \mathbb{Z}$, with positive upper density, there exist a probability measure preserving system $(X, \mathcal{B}, \mu, T)$ and a set $A \in \mathcal{B}$, $\mu(A) = \overline{d}(E)$, such that for any $k \in \mathbb{N}$, $n_1, n_2, \dots, n_k \in \mathbb{Z}$, one has*

$$d^*(E \cap (E-n_1) \cap (E-n_2) \cap \cdots \cap (E-n_k)) = \mu(A \cap T^{-n_1}A \cap T^{-n_2}A \cap \cdots \cap T^{-n_k}A).$$

$(X, \mathcal{B}, \mu, T)$ is a probability measure preserving system, where
X is a set,
$\mathcal{B}$ is a σ-algebra over X,
$\mu(\cdot)$ is a probability measure,
$T\colon X \mapsto X$ is a measurable transformation; i.e., for all $A \in \mathbf{B}$, $\mu(T^{-1}A) = \mu(A)$.

Now we give a pure combinatorial proof for Bergelson's theorem in higher dimension as well.

Definition 4.2.1. Let $A \subseteq \mathbb{Z}^n$, the counting function is defined by

$$A(x) = \sum_{a \in A; |a| \le x} 1,$$

where $|a|$ is the length of a (the distance from the origin).

Define the *discrete* rectangle of $\mathbb{Z}^n$ by

$$R = [a_1, b_1] \times [a_2, b_2] \times \cdots [a_n, b_n] \cap \mathbb{Z}^n.$$

The volume of R is $|R| = \prod_i (b_i - a_i + 1)$.

An upper Banach density of A now is

$$d^*(A) := \sup\Big\{L : \forall m,\ \exists R_m,\ \min_i |b_i - a_i| \ge m, s.t.\ \frac{|A \cap R_m|}{|R_m|} \ge L\Big\}.$$

We formulate the theorem now as follows.

Theorem 4.2.3. [6] *Let $A \subseteq \mathbb{Z}^n$, with $d^*(A) = \gamma > 0$. For every integer M there is an infinite set $B \subseteq \mathbb{Z}^n$ such that*

$$D(A) \supseteq = B \cdot M := B + B + \cdots + B \ (M \text{ times}).$$

Proof. $A - A$ is symmetric to the origin. Without loss of generality, assume $M > 0$. Consider the lattice points of the cube $\{\mathbf{x}_i\}_{i=1}^{M^n}$; $\mathbf{x}_i = (x_{i_1}, x_{i_2}, \dots, x_{i_n})$; $0 \le x_{i_j} \le M - 1$. Write

$$\mathbf{u} = (u_1, u_2, \dots, u_n) \equiv \mathbf{v} = (v_1, v_2, \dots, v_n) \pmod{M}$$
$$\Leftrightarrow u_i \equiv v_i \pmod{M} \text{ for all } i,\ 1 \le i \le n.$$

Let

$$A_i = \{\mathbf{a} \in A : \mathbf{a} \equiv \mathbf{x}_i \pmod{M}\},$$
$$d^*(A) = \gamma > 0 \ \Rightarrow\ d^*(A_i) = \rho > 0$$

for some i.

Let

$$A' = A_i - \mathbf{x}_i \subseteq L := \{\mathbf{u} \equiv \mathbf{0} \pmod{M}\}.$$

Surely

$$A' - A' = A_i - A_i \subseteq A - A.$$

□

Lemma 4.2.2. *There exists a finite set U such that $A' - A' + U = L$.*

Proof. Let $U = \{\mathbf{u}_1, \mathbf{u}_2, \dots, \mathbf{u}_r, \dots\}$ be the maximal subset of $\mathbb{Z}^n$, such that the sets

$$\mathbf{u}_1 + A', \mathbf{u}_2 + A', \dots, \mathbf{u}_r + A', \dots$$

are pairwise disjoint.

Claim: $r \leq 4/\rho$.

Since $d^*(A') = d^*(A_i) = \rho > 0$, there is a rectangle R such that $|R \cap A'| \geq \frac{\rho|R|}{2}$. Assume that the minimal length of edge of R is large enough, we get

$$\begin{aligned}|R| &\geq |R \cap \{(\mathbf{u}_1 + A') \cup \dots \cup (\mathbf{u}_r + A')\}| \\ &= |R \cap (\mathbf{u}_1 + A')| + \dots + |R \cap (\mathbf{u}_r + A')| \geq r\frac{|R \cap A'|}{2} \geq r\frac{\rho|R|}{4}\end{aligned}$$

which gives $r \leq 4/\rho$.

If U would not fulfil the statement of the lemma, then there will be an $x \in L \setminus (A' - A' + U)$, for which

$$x \notin A' - A' + \mathbf{u}_i$$

for $i = 1, 2, \dots r$, or, or equivalently,

$$x + A' \cap \mathbf{u}_i + A' = \emptyset.$$

It contradicts the fact that U is the maximal subset of $\mathbb{Z}^n$.

We introduce an r-coloring

$$\chi(\mathbf{x}_1, \dots, \mathbf{x}_M) \mapsto \{1, 2, \dots, r\}$$

of all M element subsets of L as follows: for an M-tuple $\mathbf{x}_1, \dots, \mathbf{x}_M$, the color of it is determined by

$$\chi(\mathbf{x}_1, \dots, \mathbf{x}_M) \in \{i : \mathbf{x}_1 + \dots + \mathbf{x}_M \in A' - A' + \mathbf{u}_i\}.$$

(Note the coloring is not necessary unique; if it is not, then use an arbitrary color.) □

Lemma 4.2.3 (Ramsey). *Let X be a countable set and color all of M-tuples of X by r colors (color the M-uniform graph of X).*

There exists an infinite set B' which is monochromatic.

Now by the lemma, there is an infinite set B' which is monochromatic; i.e., for every M-tuple $\mathbf{x}_1, \ldots, \mathbf{x}_M$ in B',

$$\mathbf{x}_1 + \cdots + \mathbf{x}_M \in A' - A' + \mathbf{u}_s$$

for some fixed s.

Finally let

$$B := B' - \frac{\mathbf{u}_s}{M},$$

$$\mathbf{u}_s \in L \Rightarrow \frac{\mathbf{u}_s}{M} \in \mathbb{Z}^n.$$

Then

$$\begin{aligned} A' - A' + \mathbf{u}_s &\supseteq B' + B' + \cdots + B' \ \ (M \text{ times}) \\ &= \left(B + \frac{\mathbf{u}_s}{M}\right) + \cdots + \left(B + \frac{\mathbf{u}_s}{M}\right) \ \ (M \text{ times}) \\ &= B + B + \cdots + B + M\frac{\mathbf{u}_s}{M}, \end{aligned}$$

which implies

$$A - A \supseteq A' - A' \supseteq B + B + \cdots + B \ \ (M \text{ times}).$$

4.3 An intermezzo; D(D(A)) is highly well structured

By a Bohr set we mean the set

$$B(S, \varepsilon) = \{m \in \mathbb{Z} : \max_{s \in S} \|sm\| < \varepsilon\},$$

where S is a finite subset of the set of real numbers and $\|x\| = \min_{k\in\mathbb{Z}} |x - k|$.

One of the central observations in Additive Combinatorics is the following.

Theorem 4.3.1. [2] *Let A be a subset of integers, $\overline{d}(A) := \gamma > 0$.*

There exists a Bohr set $B(S, \varepsilon)$ contained in $D_2(A)$, in fact S, and ε can be chosen to

$$|S| \leq \frac{1}{\gamma^2}; \quad \varepsilon = \frac{1}{4}.$$

One can ask:

What about the first difference sets?

4.4 First difference sets and Bohr sets I

In this section we investigate two questions.

1. *Is there a Bohr set contained in $A-A$ provided A has positive upper density?*
2. *Can one cover $A-A$ by a "small" Bohr set?*

For the first question the answer is negative.

Theorem 4.4.1. *There exists a set A of positive integers with positive density for which the first difference set $A-A$ does not contain any Bohr set.*

Proof. For the proof we need some notion from the graph theory.

A graph G is defined on the set $\mathbb{Z}$ of integers; i.e., $G = G(V, E)$, where the set of vertices V is the set of integers, $E \subseteq \mathbb{Z} \times \mathbb{Z}$.

• G is said to be *shift invariant* if

$$(x, y) \in E \quad \Leftrightarrow \quad (x+n, y+n) \in E$$

for every $n \in \mathbb{Z}$.

• The chromatic number $\chi(G) = r$ of a graph G is the smallest number r for which there exists a coloring of V by r colors such that for every $x, y \in V$, $\chi(x) \neq \chi(y)$, if $(x, y) \in E$. If there is no such a number, then $\chi(G) = \infty$.

• A set $S \subseteq V$ is said to be *independent* if

$$(S \times S) \cap E = \emptyset,$$

i.e., there is no edge in the set S. □

We need the following result of *Křiž*.

Lemma 4.4.1. *For every $\varepsilon > 0$ there exists a shift-invariant graph G on $\mathbb{Z}$ with chromatic number $\chi(G) = \infty$ and containing an independent set A of dense $>$ $1/2 - \varepsilon$.*

From this lemma we derive that for this given independent set A, $A-A$ does not contain any Bohr set.

Indeed, since A is an independent set and the graph is shift invariant, we conclude for every u, w,

$$u - w \in A - A \quad \Rightarrow \quad (u, w) \notin E(G). \tag{$*$}$$

(There exists an integer $x \in \mathbb{Z}$ such that $(u - x, w - x) \in A$.)

On the contrary assume there exists a Bohr set

$$B(S, \varepsilon) \subseteq A - A.$$

Let $|S| = m$, and split the m-dimensional cube $[0, 1]^m$ into subcubes $C_1, C_2, \dots, C_s$ such that the sides of $C_i < \varepsilon/2$.

Color $\mathbb{Z}$ by the following way:

$$Z_i := \{n : \|sn\| \in C_i; \forall s \in S\}.$$

It implies that for every i,

$$\bigcup_{i=1}^{s} Z_i = \mathbb{Z} \quad \text{and} \quad Z_i - Z_i \subseteq B(S,\varepsilon);$$

i.e., we have an s coloring of $\mathbb{Z}$ for which $Z_i - Z_i \subseteq B(S,\varepsilon) \subseteq A - A$. By $(*)$ we conclude that for every $u, w \in Z_i$, $(u,v) \notin E(G)$. Hence

$$s \geq \chi(G) = \infty,$$

a contradiction.

For the second question we investigate just the modular version. (The case of integers comes is similarly as in the proof of Bogolyubov's theorem.) The following theorem is a version of a theorem of Tao and Vu.

Theorem 4.4.2. [11] *Let $\gamma > 0$, $A \subseteq \mathbb{Z}_N$, $|A| > \gamma N$. There exists a Bohr set $B(S,\varepsilon)$, which covers $A - A$, $|S| \leq (1-\alpha)^{-2}$, and*

$$\varepsilon = \sqrt{\frac{\alpha}{2\gamma}}.$$

Proof. Let as usual $\widehat{B}(r) = \sum_x B(x)e(rx)$, where $B(x)$ is the indicator of the set B. Denote $D = A - A$.

Let

$$S = \{r : |\widehat{D}(r)| > (1-\alpha)N\},$$

where $\alpha > 0$.

By the Parseval formula we conclude that $|S| \leq (1-\alpha)^{-2}$.

Let $r \in S$. Now there exists a $u \in \mathbb{Z}_N$ such that

$$\operatorname{Re} \sum_{z \in D} e(rz + u) \geq (1-\alpha)N,$$

thus

$$\sum_{z \in D} (1 - \operatorname{Re} e(rz+u)) \leq \alpha N.$$

Fix $x, y \in A$, and since the terms are non-negative we can write

$$\sum_{a \in A} |1 - \operatorname{Re} e(r(x-a)+u)| \leq \alpha N$$

and

$$\sum_{a \in A} |1 - \operatorname{Re} e(r(y-a)+u)| \leq \alpha N.$$

Using Cauchy–Schwarz inequality, we get

$$\sum_{a\in A}|1-\operatorname{Re} e(r(x-a)+u)|^{1/2}\le\sqrt{|A|}\sqrt{\sum_{a\in A}|1-\operatorname{Re} e(r(x-a)+u)|}$$
$$\le\sqrt{|A|}\sqrt{\alpha}\sqrt{N}=\sqrt{\alpha\gamma}N$$

and a similar bound for y.

Exercise. Prove

$$|1-e(\alpha)|\le\sqrt{2}|1-\operatorname{Re} e(\alpha)|^{1/2}.$$

By this exercise we obtain

$$\sum_{a\in A}|1-e(r(x-a)+u)|\le\sqrt{2\alpha\gamma}N$$

and

$$\sum_{a\in A}|1-e(r(y-a)+u)|\le\sqrt{2\alpha\gamma}N.$$

Now by the triangle inequality we have

$$\sum_{a}|e(r(x-a)+u)-e(r(y-a)+u)|\le 2\sqrt{2\alpha\gamma}N.$$

The quantity $|e(r(x-a)+u)-e(r(y-a)+u)|$ is the same as $|1-e(r(x-y))|$, and using $4\|z\|\le|1-e(z)|$, finally we obtain

$$4\gamma N\|r(x-y)\|\le|A|\|r(x-y)\|=\sum_{a}|1-e(r(x-y))|\le 2\sqrt{2\alpha\gamma}N,$$

rearranging we have

$$\|r(x-y)\|\le\frac{\sqrt{2}}{2}\sqrt{\frac{\alpha}{\gamma}};$$

i.e., for every $x,y\in A$, $x-y\in B(S,\varepsilon)$, where

$$\varepsilon=\sqrt{\frac{\alpha}{2\gamma}}.$$

□

4.5 Raimi's theorem; difference set of partitions

A branch of combinatorial analysis – called Ramsey theory – investigates partitions of certain structures.

In this section we investigate a partition question which is related to difference sets. We will generalize a theorem of Raimi which sounds as follows:

Theorem 4.5.1. *There exists $E \subseteq \mathbb{N}$ such that, whenever $r \in \mathbb{N}$ and $\mathbb{N} = \bigcup_{i=1}^{r} D_i$, there exist $i \in \{1, 2, \ldots, r\}$ and $k \in \mathbb{N}$ such that $(D_i + k) \cap E$ is infinite and $(D_i + k) \setminus E$ is infinite.*

One can read this theorem as follows.

Theorem 4.5.2. *There exists a partition of $\mathbb{N} = E_1 \cup E_2$ such that, whenever $r \in \mathbb{N}$ and $\mathbb{N} = \bigcup_{i=1}^{r} D_i$, there exist $i \in \{1, 2, \ldots, r\}$ and $k \in \mathbb{N}$ such that both $k \in E_1 - D_i$ and $k \in E_2 - D_i$ hold infinitely many times.*

A generalization of it is the following.

Theorem 4.5.3. [7] *Let $r \in \mathbb{N}$ and let $\alpha_1, \alpha_2, \ldots, \alpha_r$ be positive real numbers such that $\sum_{i=1}^{r} \alpha_i = 1$. There exists a disjoint partition $\mathbb{N} = \bigcup_{i=1}^{r} E_i$ such that*

(1) *for every $i \in \{1, 2, \ldots, r\}$, $d(E_i) = \alpha_i$ and*

(2) *for each $t \in N$ and each partition $\mathbb{N} = \bigcup_{j=1}^{t} F_j$, there exist $m \in \{1, 2, \ldots, t\}$ and a sequence $\{x_n\}_{n=1}^{\infty}$ in $\mathbb{N}$ such that for every $h \in FS(\{x_n\}_{n=1}^{\infty})$ and every $i \in \{1, 2, \ldots, r\}$, $(F_m + h) \cap E_i$ is infinite.*

Sketch of the proof. We prove the simplest case when $r = 2$ and $\alpha_1 = \alpha_2 = 1/2$. The proof of the general case is similar just more technical.

Now color the unit interval $[0, 1)$ as follows: the first half by two colors: $[0, 1/4)$ is red and $[1/4, 1/2)$ is blue. Color the half of the rest again by two colors; i.e., $[1/2, 5/8)$ is red and $[5/8, 3/4)$ is blue. Again color the half of the rest by two colors etc. So we have an infinite set of red intervals with total length

$$\frac{1}{4} + \frac{1}{8} + \frac{1}{16} + \cdots = \frac{1}{2},$$

and blue intervals with the same length.

Now our partition will be the following.

Let γ be a positive irrational. Let

$$E_1 = \{x \in \mathbb{N} : \langle \gamma x \rangle \in \text{some red interval}\}$$

and

$$E_2 = \{x \in \mathbb{N} : \langle \gamma x \rangle \in \text{some blue interval}\}.$$

Let $t \in N$ and let $\mathbb{N} = \bigcup_{j=1}^{t} F_j$.

Exercise.

1. Prove that $d(E_i) = \alpha_i = 1/2$.

2. For any c, d with $0 \leq c < d \leq 1$ there exists $m \in \{1, 2, \ldots, t\}$ and there exist a, b, with $c \leq a < b \leq d$, such that $\{\langle \gamma x \rangle : x \in F_m\}$ is dense in (a, b).

The sequence $\{x_n\}_{n=1}^{\infty}$ will be defined inductively.

Since the length of the red-blue pairs of intervals tends to zero and since the sequence $\{\langle \gamma m \rangle\}$ $(m \in \mathbb{N})$ is uniformly distributed, we get that there exists an integer x_1, such that $\{\langle \gamma(x + x_1)\rangle : x \in F_m\} \cap (a, b)$ covers a red-blue pair of intervals (there are intervals with length $1/2^{k+1} < b - a$). It means that there are infinitely many coincidence of the sets both of $x_1 + F_m$, E_1, and $x_1 + F_m$, E_2.

Assume that a sequence $\{x_n\}_{n=1}^{N}$ has been defined. Since

$$FS(\{x_n\}_{n=1}^{N+1} = FS(\{x_n\}_{n=1}^{N}) + \{0, x_{N+1}\},$$

and again since $\{\langle \gamma m \rangle\}$ $(m \in \mathbb{N})$ is uniformly distributed, we obtain that there exists an integer x_{N+1} such that $\{\langle \gamma(\sum_{i=1}^{N+1} x_i + x)\rangle : x \in F_m\} \cap (a, b)$ covers the given red-blue pairs of intervals. □

4.6 First difference sets and Bohr sets II; Følner's theorem

Recall that by a Bohr set we mean the set

$$B(S, \varepsilon) = \{m \in \mathbb{Z} : \max_{s \in S} \|sm\| < \varepsilon\},$$

where S is a finite subset of the set of real numbers.

Følner proved the following theorem.

Theorem 4.6.1. *Assume that $A \subseteq \mathbb{N}$ and $\overline{d}(A) > 0$. There exists a Bohr set $B(S, \varepsilon)$ such that*

$$B(S, \varepsilon) \setminus (A - A)$$

has density 0.

Exercises.

1. Prove that $d(B(S, \varepsilon)) \geq \varepsilon^{|S|}$.
2. Prove that a Bohr set $B(S, \varepsilon)$ has bounded gaps.

4.7 Applications of Følner theorem

(Joint work with Imre Ruzsa. The results will be published in another place [10].)

As in the introduction we mentioned that $D(D(A))$ always contains a Bohr set, while the set $D(A)$ not necessary contains a Bohr set. Now we investigate the threefold sum-differences of A.

Theorem 4.7.1. *There is a symmetric set A of integers such that $0 \in A$, the positive elements of A form a set of positive density and the set $A + A + A$ does not contain a Bohr set.*

On the other hand, we prove that $A+A-A$ is always a Bohr neighborhood of some $a \in A$.

Theorem 4.7.2. *Assume that* $\overline{d}(A) > 0$. *There exists a subset* A' *of* A, $\overline{d}(A') > 0$, *such that for every* $a' \in A'$, *the set* $A+A-A-a'$ *contains a Bohr set.*

Corollary 1. *Følner theorem implies Bogolyubov theorem.*

Proof. Since for $a' \in A' \subseteq A$, the set $A+A-A-a'$ contains a Bohr set and by $A+A-A-a' \subseteq A-A+A-A$, we get Bogolyubov's result. □

By the Følner's theorem, Bergelson's theorem can be proved.

Theorem 4.7.3. *Assume* $\overline{d}(A) > 0$. *There exists an infinite set* C *such that*

$$A-A \supseteq FS(C) \cup FP(C),$$

where

$$FS(C) = \Big\{ \sum_{x \in X} : X \subseteq C;\ X \text{ is finite} \Big\},$$

$$FP(C) = \Big\{ \prod_{x \in X} : X \subseteq C;\ X \text{is finite} \Big\}.$$

Bibliography

[1] V. Bergelson, *Sets of reccurence of $\mathbb{Z}^m$-actions and properties of sets of differences*, J. London Math. Soc. (2) **31** (1985), 295–304.

[2] N. N. Bogolyubov, *Some algebraic properties of almost periods*, Zapiski katedry matematichnoi fiziji (Kiev) **4** (1939), 185-194 (in Russian).

[3] E. Følner, *Generalization of a theorem of Bogoliuboff to topological Abelian groups. With an appendix on Banach mean values in non-Abelian groups*, Math. Scand. **2** (1954), 5–18.

[4] ______, *Note on a generalization of a theorem of Bogoliuboff*, Math. Scand. **2** (1954), 224–226.

[5] N. Hegyvári, *On iterated difference sets in groups*, Period. Math. Hungar. **43** (2001), no. 1-2, 105–110.

[6] ______, *Note on difference sets in $\mathbb{Z}^n$*, Period. Math. Hung. **44** (2002), no. 2, 183–185.

[7] ______, *On intersecting properties of partitions of integers*, Combin. Probab. Comput. **14** (2005), no. 3, 319–323.

[8] N. Hegyvári and F. Hennecart, *Iterated Difference Sets in Sigma-finite Groups*, to appear.

[9] N. Hegyvári and F. Hennecart, *Iterated difference sets in sigma-finite groups*, Annales Univ. Sci. Budapest, to appear.

[10] N. Hegyvári and I. Z. Ruzsa, *Additive Structure of Difference Sets and a Theorem of Følner*, available at http://hegyvari.web.elte.hu/folner(RH)3.pdf

[11] T. Tao and V. Vu, *Additive Cominatorics*, Cambridge Stud. in Adv. Math., vol. 105, Cambridge University Press, Cambridge, 2006.

Norbert Hegyvári
Institute of Mathematics, Eötvös University, Pázmány st. 1, H-1117 Budapest
Hungary
e-mail: `hegyvari@caesar.elte.hu`

Chapter 5

The polynomial method in additive combinatorics[1]

Gyula Károlyi

5.1 Introduction

First of all we are going to introduce what we call the polynomial lemma, a simple but very powerful result.

Given a field F and a polynomial $f \in F[x]$ (if $f = 0$ we will say that $\deg f = -\infty$), it is well known from algebra that if $\deg f = t \geq 0$, then f has at most t roots. So, we have the following observation.

Observation 1. If $f \in F[x]$, $\deg f \leq t$ and we have $a_1, \ldots, a_{t+1}$ such that $f(a_i) = 0$ for $1 \leq i \leq t+1$, then $f = 0$.

Following the same argument, we make another observation.

Observation 2. If $f \in F[x,y]$, $\deg f$ in x is $\leq t_1$ and $\deg f$ in y is $\leq t_2$ and we have $a_1, \ldots, a_{t_1+1}, b_1, \ldots, b_{t_2+1}$ such that $f(a_i, b_j) = 0$ for every $1 \leq i \leq t_1 + 1$ and $1 \leq j \leq t_2 + 1$, then $f = 0$.

Indeed, we can write

$$f(x,y) = f_{t_2}(x)y^{t_2} + \cdots + f_0(x)$$

with $f_i(x) \in F[x]$ and $\deg f_i \leq t_1$. Then $f(a_i, y) \in F[y]$ $(1 \leq i \leq t_1 + 1)$ has $b_1, \ldots, b_{t_2+1}$ as roots and has degree $\leq t_2$. According to Observation 1, $f(a_i, y) = 0$ $(1 \leq i \leq t_1 + 1)$.

[1]This chapter has been written based on notes taken by Carlos Vinuesa at the seminar delivered by the author at CRM on January 16, 2008.

This means that, for each $1 \le i \le t_1 + 1$ and $1 \le j \le t_2 + 1$, we have $f_j(a_i) = 0$. But then every $f_j \in F[x]$ has a degree at most t_1 and at least $t_1 + 1$ different roots, so we get $f_j = 0$ for $1 \le j \le t_2 + 1$, and finally $f = 0$.

It is clear how to extend the above observations to several variables. More appropriate is, however, to consider the following version, which proved very useful in different areas of combinatorics.

Lemma 5.1.1 (Polynomial Lemma, Alon [2]). *If $f \in F[x_1, \ldots, x_n]$, $\deg f = t_1 + t_2 + \cdots + t_n$, and $cx_1^{t_1}x_2^{t_2}\cdots x_n^{t_n}$ is a leading term[2] of f with some $c \neq 0$, then f cannot vanish on $S_1 \times S_2 \times \cdots \times S_n$ whenever $|S_i| > t_i$ for $i = 1, 2, \ldots, n$.*

For example, if $f(x, y) = x^3y^3 + 2x^5y + 3y^5$, then $\deg f = 6$ and according to the Polynomial Lemma f cannot vanish on a 4×4 Cartesian product, neither on a 6×2 Cartesian product, whereas Observation 2 only implies that f cannot vanish on a 6×6 Cartesian product. This strength of the Polynomial Lemma over Observation 2 is what can be fully exploited in various situations, and what we are going to demonstrate now through some examples.

Example 5.1.2 (Komjáth's conjecture). Let $C_d = \{0, 1\}^d$ be the vertex set of the unit cube in d-space. We are interested in covering $C_d \setminus \{0\}$ with hyperplanes, but with the condition that none of them passes through 0.

In other words, we are looking for a collection of hyperplanes $H_1, H_2, \ldots, H_m$ such that $C_d \setminus \{0\} \subseteq \cup H_i$ and $0 \notin \cup H_i$.

Claim (Alon–Füredi, [5]). If the hyperplanes $H_1, H_2, \ldots, H_m$ have the above property, then $m \ge d$.

The following examples show how to make it with d hyperplanes.

1. $H_i : x_i = 1$ $(i = 1, \ldots, d)$,
2. $H_i : x_1 + x_2 + \cdots + x_d = i \Longleftrightarrow \frac{1}{i}x_1 + \frac{1}{i}x_2 + \cdots + \frac{1}{i}x_d = 1$ $(i = 1, \ldots, d)$.

The matrices of the "normalized" equations are

1.
$$\begin{pmatrix} 1 & & 0 \\ & \ddots & \\ 0 & & 1 \end{pmatrix},$$

2.
$$\begin{pmatrix} 1 & \cdots & 1 \\ \vdots & \vdots & \vdots \\ \frac{1}{d} & \cdots & \frac{1}{d} \end{pmatrix}.$$

[2]A leading term is a term of the polynomial which has the highest degree.

Notice that the permanent[3] of each matrix is 1.

To prove the claim, choose $S_1 = S_2 = \cdots = S_d = \{0,1\}$ so that $C_d = S_1 \times S_2 \times \cdots \times S_d$.

Assume for contradiction, that $m < d$. The hyperplanes can be defined by m linear equations of the form

$$H_i : \sum_{j=1}^{d} a_{ij}x_j - 1 = 0.$$

The equations have this shape because the independent term cannot be 0 by the condition $0 \notin H_i$. Thus, were this term different from 1, dividing the equation through it we would arrive at an equation of the desired form.

Consider the polynomial $f \in \mathbb{R}[x_1, \ldots, x_d]$ defined by

$$f(x_1, \ldots, x_d) = \prod_{i=1}^{m} \left(\sum_{j=1}^{d} a_{ij}x_j - 1 \right) - (-1)^{m-d} \prod_{j=1}^{d} (x_j - 1).$$

The first term on the right-hand side vanishes at every point of $C_d \setminus \{0\}$ because all these points satisfy the equation of at least one hyperplane. Its value at 0 is $(-1)^m$.

The second term is also 0 at every point of $C_d \setminus \{0\}$ for all of them have at least one coordinate equal to 1. Its value at 0 is also $(-1)^m$. Therefore f, which is the difference of these two terms, vanishes at every point of C_d.

So $f|_{S_1 \times S_2 \times \cdots \times S_d} = 0$. Also, since we have assumed $m < d$, we have $\deg f = d$. Moreover, the coefficient of the leading term $x_1 x_2 \ldots x_d$ is $-(-1)^{m-d}$ which is different from 0. But then the Polynomial Lemma claims that f cannot vanish on $S_1 \times S_2 \times \cdots \times S_d$, which is a contradiction. Moreover, a closer inspection also reveals that if $m = d$ and the hyperplanes H_i do the job, then the permanent of the corresponding matrix must be 1.

Example 5.1.3 (Jamison's theorem [8], Brouwer–Schrijver [6]). Now we work in the affine geometry $AG(2,p)$, which means the set of points with two coordinates in the Galois field $GF(p)$ of p elements, p a prime. We ask: What is the minimum number of points such that each line in $AG(2,p)$ contains at least of them?

It is clear that $2p-1$ points suffice, since the union of the two coordinate axes $x = 0$ and $y = 0$ consists of $2p-1$ points. Each line of the affine plane intersects at least one of these two lines.

---

[3]The permanent of a matrix $M = (a_{ij})_{1 \le i,j \le n}$ is $Per\ M = \sum_{\pi \in S_n} \prod_{i=1}^{n} a_{i\pi(i)}$. We consider the product of n elements, one of each row and each column, and add up all such possible products. Thus, the permanent of a matrix looks similar to its determinant, but "without signs". It may look a simpler expression, but in fact it is a lot more complicated object from the computational point of view.

Theorem. *The minimum number of points such that each line in* $AG(2,p)$ *contains one or more of them is* $2p-1$.

Since each translation leaves the set of lines invariant, we may assume that one of the points is $(0,0)$. The lines that do not contain $(0,0)$ have an equation of the form

$$ax + by - 1 = 0$$

for again, the independent term is not 0, and thus we can divide the equation through it.

If the points $(0,0)$, (a_1,b_1), (a_2,b_2), ..., (a_m,b_m) cover each line, then the relation

$$\prod_{i=1}^{m}(aa_i + bb_i - 1) = 0$$

holds for any pair $(a,b) \neq (0,0)$. The left-hand side is a polynomial in the variables a, b which vanishes everywhere except at the origin. Its value at the origin is $(-1)^m$. Now we to introduce a "modification term" as in Example 5.1.2 to obtain a polynomial $f(a,b) \in GF(p)[a,b]$:

$$f(a,b) = \prod_{i=1}^{m}(aa_i + bb_i - 1) - (-1)^m(1 - a^{p-1})(1 - b^{p-1}),$$

which vanishes everywhere because the second term vanishes at every point except from the origin by the Fermat–Euler Theorem[4] and its value at the origin is $(-1)^m$, the same value the first term has, so when we subtract the two terms we have 0.

For a contradiction, suppose that $m < 2p-2$. Then $\deg f = 2p-2$ and the coefficient of the leading term $a^{p-1}b^{p-1}$ is $-(-1)^m$, which is different from 0, so the Polynomial Lemma says that f cannot vanish on a $p \times p$ Cartesian product, that is, f cannot vanish on the whole plane $AG(2,p)$, which is a contradiction.

To extend the result to $AG(2,q)$, where $q = p^\alpha$ for some positive integer α and some prime number p, the role of the Fermat–Euler Theorem must be transferred to the fact that the multiplicative group of any finite field is cyclic. The proof easily generalizes to higher dimensions, yielding that the minimum number of points one needs to cover each affine hyperplane in $AG(d,q)$ equals $d(q-1)+1$.

Example 5.1.4. The following result, with a complicated proof, answered a long-standing open quiestion in graph theory.

Theorem (Tashkinov [11]). *Every 4-regular simple graph has a 3-regular subgraph.*[5]

Not every vertex has to be included in the 3-regular subgraph (if this was the case, then the other edges would give us a perfect matching and this is not

---

[4] **Fermat–Euler Theorem.** *If* $(a,n) = 1$, *then* $a^{\varphi(n)} \equiv 1 \pmod{n}$, *where* φ *is Euler's totient function.* Remember that for a prime number p, $\varphi(p) = p-1$.

[5] A k-regular graph is a graph in which every vertex has degree k.

always possible, for example it is not possible for a graph with an odd number of vertices).

With the Polynomial Lemma it is possible to prove a slightly weaker version.

Theorem (Alon–Friedland–Kalai [3]). *If G is a 4-regular simple graph plus 1 edge, then G contains a 3-regular subgraph.*

In fact, they proved the following more general version. Taking $p = 3$ one immediately obtains the above result.

Theorem. *If p is a prime and G a simple graph of average degree $> 2p-2$ and maximum degree $\leq 2p-1$, then G contains a p-regular subgraph.*

To prove it, the goal is to assign to each edge of G 0 or 1 such a way, that there is at least one edge assigned with 1, and for every vertex, the number of edges starting at that vertex and having the value 1 is either 0 or p. In view of the assumption on the maximum degree we can rephrase the second condition: the number of such edges must be divisible by p for every vertex. In such an assignment, the edges assigned with 1 form a p-regular subgraph.

If $G = G(V, E)$, where V is the vertex set and E is the set of edges of G, the incidence matrix of G is the 0–1 matrix $(a_{ve})_{1\leq v\leq |V|,\ 1\leq e\leq |E|}$, where $a_{ve} = 1$ if and only if the vertex v is incident to the edge e. Thus, if we introduce a 0–1 variable X_e for every edge $e \in E$, the second condition can be formulated as

$$\sum_{e\in E} a_{ve}X_e \equiv 0 \pmod{p}. \tag{5.1.1}$$

Accordingly, we consider the following polynomial $f \in GF(p)[X_e|e \in E]$:

$$f(X_e|e \in E) = \prod_{v\in V}\left(1 - \left(\sum_{e\in E} a_{ve}X_e\right)^{p-1}\right) - \prod_{e\in E}(1 - X_e).$$

Since the average degree of G is $> 2p-2$, that is, $(2p-2)|V| < 2|E|$, we have $(p-1)|V| < |E|$ and then $\deg f = |E|$. Then $\prod_{e\in E} X_e$ is a leading monomial with coefficient $-(-1)^{|E|} \neq 0$. By the Polynomial Lemma, f cannot vanish on $\{0,1\}^{|E|}$. This means that we have a choice for each X_e (0 or 1) such that $f(X_e|e \in E) \neq 0$. Were all the X_e's 0, the value of the polynomial would also be 0. Then, for this choice, $\prod_{e\in E}(1-X_e)$ is 0 and so $\prod_{v\in V}\left(1 - \left(\sum_{e\in E} a_{ve}X_e\right)^{p-1}\right)$ must be different from 0, which means, again by the Fermat–Euler Theorem, that for every vertex v, condition (5.1.1) is satisfied.

5.2 Applications of the polynomial method to additive problems

The Problem of Snevily. Consider a finite abelian group G of odd order, that is, $2 \nmid |G|$.

Conjecture (Snevily [10]). If $a_1, a_2, \ldots, a_k \in G$ are different and $b_1, b_2, \ldots, b_k \in G$ are different, then there exists a permutation $\pi \in \mathcal{S}_k$ such that the elements $a_1 + b_{\pi(1)}, a_2 + b_{\pi(2)}, \ldots, a_k + b_{\pi(k)}$ are also different.

If $2 \mid |G|$, this is clearly false. Indeed, it follows from Cauchy's Theorem[6] that in such a case there is an element of the group $g \neq 0$ such that $g + g = 0$, and taking $a_1 = b_1 = 0$ and $a_2 = b_2 = g$ one can check that the conjecture fails.

a) If $G = \mathbb{Z}/p\mathbb{Z}$, $p > 2$ a prime, we can identify G with the additive group of the field $GF(p)$. The following idea is due to Alon [1].

First of all, if $k = p$, then we take $b_{\pi(i)} = a_i$. The sums $a_i + b_{\pi(i)} = 2a_i$ are all different. Accordingly, we can assume that $k < p$. In fact, in this case the assumption that $a_1, a_2, \ldots, a_k$ are different will not be necessary.

Introduce, for each $1 \leq i \leq k$, a variable x_i. Our goal is to assign to each variable a value in the set $\{b_1, b_2, \ldots, b_k\}$ such that the next two conditions are satisfied for each pair of indices $0 \leq j < i \leq k$:

$$\begin{cases} a_i + x_i - a_j - x_j \neq 0, \\ x_i - x_j \neq 0. \end{cases}$$

Consider the polynomial $f \in GF(p)[x_1, x_2, \ldots, x_k]$ defined by

$$f(x_1, x_2, \ldots, x_k) = \prod_{1 \leq j < i \leq k} (x_i - x_j)(x_i - x_j + a_i - a_j).$$

Choosing $S_i = \{b_1, b_2, \ldots, b_k\}$ for $i = 1, 2, \ldots, k$, we only need to see that $f|_{S_1 \times \cdots \times S_k} \neq 0$, in other words, that there exist $s_1 \in S_1, s_2 \in S_2, \ldots, s_k \in S_k$ such that $f(s_1, s_2, \ldots, s_k) \neq 0$. Indeed, in this case there is an evaluation of the variables $x_i = s_i$ for $i = 1, \ldots, k$ with the following properties:

- $s_i \in \{b_1, b_2, \ldots, b_k\}$ for $1 \leq i \leq k$,
- $s_i \neq s_j$ for $i \neq j$,
- $a_i + s_i \neq a_j + s_j$ for $i \neq j$.

Then there is a unique permutation π with $s_i = b_{\pi(i)}$, which does the job.

We have that $\deg f = 2\binom{k}{2} = k(k-1)$ and $|S_i| > k - 1$ for $1 \leq i \leq k$. If we can see that the coefficient of the monomial $x_1^{k-1} x_2^{k-1} \cdots x_k^{k-1}$ in f is different from 0 in $GF(p)$, then the existence of such an evaluation follows immediately from the Polynomial Lemma.

[6] **Cauchy's Theorem.** *If G is a finite group and p is a prime number dividing the order of G, then G contains an element of order p.*

To this end we make use Lemma 5.3.1 of the appendix. In our case, the coefficient of the leading term $x_1^{k-1}x_2^{k-1}\cdots x_k^{k-1}$ comes from the product

$$\prod_{1\le j<i\le k}(x_i - x_j)^2,$$

because the other terms have lower degree. According to the lemma, the coefficient can be calculated as

$$(-1)^{\binom{k}{2}} Per\ V(1,1,\dots,1) = (-1)^{\binom{k}{2}} k!$$

Since we assumed that $k < p$, we have that the integer $k!$ is not divisible by p, thus the coefficient in $GF(p)$ is indeed not 0.

b) When G is an arbitrary cyclic group of odd order, we try the following multiplicative analogue of the previous method, see [7]. Assume that G can be embedded into the multiplicative group $F^\times$ of a suitable field. That is, we can identify G with a subgroup of $F^\times$. In this case we want to find numbers $s_i \in \{b_1, b_2, \dots, b_k\}$ such that the the next two conditions to are satisfied:

$$\begin{cases} a_i s_i - a_j s_j \neq 0, \\ s_i - s_j \neq 0. \end{cases}$$

Accordingly, we consider the polynomial $f \in F[x_1, x_2, \dots, x_k]$ given by

$$f(x_1, x_2, \dots, x_k) = \prod_{1\le j<i\le k}\Big((x_i - x_j)(a_i x_i - a_j x_j)\Big).$$

In this case an application of Lemma 5.3.1 of the appendix gives that the coefficient of the leading term $x_1^{k-1}x_2^{k-1}\cdots x_k^{k-1}$ is $(-1)^{\binom{k}{2}} Per\ V(a_1, a_2, \dots, a_k)$. Assume that $Per\ V(a_1, a_2, \dots, a_k)$ is not 0 in F. Then one can apply the Polynomial Lemma as before to find $\pi \in S_k$ such that all the products $a_1 b_{\pi(1)}$, $a_2 b_{\pi(2)}$, …, $a_k b_{\pi(k)}$ are different. But, as we have already mentioned, permanents are not so easy to handle.

This difficulty can be circumvented the following way. If the characteristic of the field is 2, say $F = GF(q)$ with $q = 2^\alpha$, then $+1 = -1$ in F, which means that "Permanent = Determinant". And $Det\ V(a_1, a_2, \dots, a_k) \neq 0$ if a_1, a_2, …, a_k are different.[7]

This yields to the following question. Does there always exist a field F of characteristic 2 so that G can be identified with a subgroup of $F^\times$?

The answer is yes. If $F = GF(2^\alpha)$, then $F^\times$ is a cyclic group of $2^\alpha - 1$ elements. This group has a (unique) subgroup of order n, where $n = |G|$, if and only if $n \mid (2^\alpha - 1)$, that is, if and only if $2^\alpha \equiv 1 \pmod{n}$. Of course, if

[7] Remember that $Det\ V(a_1, a_2, \dots, a_k) = \prod_{1\le j<i\le k}(a_i - a_j)$.

this is possible, this subgroup must be isomorphic to G because both groups are cyclic. And the Fermat–Euler theorem asserts that this is possible taking $\alpha = \varphi(n)$. This completes the proof of Snevily's conjecture for arbitrary cyclic groups of odd order.

The Cauchy–Davenport Theorem. In an abelian group, G, we are given two sets of elements of G: $A = \{a_1, \dots, a_k\}$ and $B = \{b_1, \dots, b_l\}$. We define the sumset

$$A + B = \{a_i + b_j \mid 1 \le i \le k, 1 \le j \le l\}.$$

Assume that G is torsion-free. Then there exists a linear order $<$ on G such that $a < b \Longrightarrow a + c < b + c$. Thus, if $a_1 < \cdots < a_n$ and $b_1 < \cdots < b_l$ then

$$a_1 + b_1 < a_2 + b_1 < \ldots < a_k + b_1 < a_k + b_2 < \ldots < a_k + b_l,$$

which is a chain of length $k + l - 1$.

So in any torsion-free abelian group G we have the following.

Proposition. *If A and B are subsets of the torsion-free abelian group G, $|A| = k$ and $|B| = l$, then $|A + B| \ge k + l - 1$.*

What happens in $\mathbb{Z}/p\mathbb{Z}$? The answer is given by the following theorem.

Theorem (Cauchy–Davenport). *If A and B are subsets of $\mathbb{Z}/p\mathbb{Z}$, with $|A| = k$ and $|B| = l$, and if $p \ge k + l - 1$, then $|A + B| \ge k + l - 1$.*

The polynomial method supports a proof of this result, too, once again looking at $\mathbb{Z}/p\mathbb{Z}$ as the additive group of $GF(p)$. Assume for a contradiction, that $A + B \subseteq C$ where $|C| = k + l - 2 = (k - 1) + (l - 1)$. Consider the polynomial

$$f(x, y) = \prod_{c \in C} (x + y - c) \in GF(p)[x, y].$$

The degree of this polynomial is $(k - 1) + (l - 1)$ and it vanishes on $A \times B$, which is a $k \times l$ Cartesian product. But the coefficient of the leading term $x^{k-1}y^{l-1}$ is $\binom{k+l-2}{k-1}$, which is different from 0 in $GF(p)$ given that $k+l-2 < p$. This contradicts the Polynomial Lemma.

The Erdős–Heilbronn Conjecture. Introduce the restricted sumset

$$A \stackrel{\bullet}{+} B = \{a + b \mid a \in A, b \in B, a \ne b\}.$$

Observe that in the case $A = B$ this is a natural notion, we add up the elements of 2-element subsets of A.

The next conjecture of Erdős and Heilbronn was proved in 1994 by Dias da Silva and Hamidoune.

Conjecture. If $A \subseteq \mathbb{Z}/p\mathbb{Z}$, where $p \ge 2k - 3$ is a prime, then $|A \stackrel{\bullet}{+} A| \ge 2k - 3$.

The complicated proof can be simplified with the polynomial method of Alon via the next result.

Theorem (Alon–Nathanson–Ruzsa [4]). *If $A, B \subseteq \mathbb{Z}/p\mathbb{Z}$, p a prime, are such that $|A| = k$ and $|B| = l$ with $k \neq l$ and $p \geq k+l-2$, then $|A \dot{+} B| \geq k+l-2$.*

This theorem immediately implies the conjecture, taking $B = A \setminus \{a\}$ for any $a \in A$. Observe also that the above theorem cannot be proved for $A = B$, take for example $A = \{0, 1, \ldots, k-1\}$, then $|A \dot{+} A| = \{1, 2, \ldots, 2k-3\}$.

The polynomial proof, very similar to the previous one, goes as follows.

Assume, for contradiction, that $A \dot{+} B \subseteq C$ where $|C| = k+l-3$. Introduce the polynomial

$$f(x,y) = (x-y) \prod_{c \in C} (x+y-c) \in GF(p)[x,y].$$

It has degree $k+l-2$ and it vanishes on the $k \times l$ cartesian product $A \times B$ (if $(x,y) \in A \times B$ with $x \neq y$, then the product on the right vanishes, and if $x = y$, then $x - y$ vanishes). The coefficient of the leading term $x^{k-1}y^{l-1}$ is

$$\binom{k+l-3}{k-2} - \binom{k+l-3}{k-1} = \frac{(k-l)(k+l-3)!}{(k-1)!(l-1)!}$$

which is different from 0 since $0 < |k-l| < p$ and $k+l-3 < p$. This contradicts the Polynomial Lemma and completes the proof.

5.3 Appendix: The Vandermonde matrix

Here we give the proof of the following technical lemma.

Lemma 5.3.1. *Let F be an arbitrary field and $a_1, a_2, \ldots, a_k \in F$. Then the coefficient of the leading term $x_1^{k-1}x_2^{k-1}\cdots x_k^{k-1}$ in the polynomial*

$$f = \prod_{1 \leq j < i \leq k} \Big((x_i - x_j)(a_i x_i - a_j x_j) \Big)$$

is

$$(-1)^{\binom{k}{2}} \mathit{Per}\ V(a_1, a_2, \ldots, a_k),$$

where $V(a_1, a_2, \ldots, a_k)$ is the Vandermonde matrix

$$\begin{pmatrix} 1 & a_1 & \ldots & a_1^{k-1} \\ 1 & a_2 & \ldots & a_2^{k-1} \\ \vdots & \vdots & & \vdots \\ 1 & a_k & \ldots & a_k^{k-1} \end{pmatrix}.$$

Proof. Recall that $\prod_{1\le j<i\le k}(y_i-y_j)$ equals the determinant of the Vandermonde matrix $V(y_1,y_2,\dots,y_k)$. That is,

$$f(x_1,x_2,\dots,x_k)=\mathit{Det}\ V(x_1,x_2,\dots,x_k)\cdot \mathit{Det}\ V(a_1x_1,a_2x_2,\dots,a_kx_k),$$

which can be rewritten as[8]

$$\left(\sum_{\pi\in\mathcal{S}_k}\operatorname{sgn}(\pi)\prod_{i=1}^{k}x_{\pi(i)}^{i-1}\right)\left(\sum_{\tau\in\mathcal{S}_k}\operatorname{sgn}(\tau)\prod_{i=1}^{k}(a_{\tau(i)}x_{\tau(i)})^{i-1}\right).$$

Let us express τ in terms of an other permutation π satisfying $\pi(1)=\tau(k)$, $\pi(2)=\tau(k-1)$, …, $\pi(k)=\tau(1)$. That is, $\pi(k-i+1)=\tau(i)$ for $1\le i\le k$. One can see[9] that $\operatorname{sgn}(\tau)=\operatorname{sgn}(\pi)\cdot(-1)^{\binom{k}{2}}$. Thus, $f(x_1,x_2,\dots,x_k)$ can be written in the form

$$\left(\sum_{\pi\in\mathcal{S}_k}\operatorname{sgn}(\pi)\prod_{i=1}^{k}x_{\pi(i)}^{i-1}\right)\left(\sum_{\pi\in\mathcal{S}_k}\left(\operatorname{sgn}(\pi)\cdot(-1)^{\binom{k}{2}}\right)\prod_{i=1}^{k}(a_{\pi(k-i+1)}x_{\pi(k-i+1)})^{i-1}\right),$$

and after rearranging the indices in the last product, also as

$$\left(\sum_{\pi\in\mathcal{S}_k}\operatorname{sgn}(\pi)\prod_{i=1}^{k}x_{\pi(i)}^{i-1}\right)\left(\sum_{\pi\in\mathcal{S}_k}\left(\operatorname{sgn}(\pi)\cdot(-1)^{\binom{k}{2}}\right)\prod_{i=1}^{k}(a_{\pi(i)}x_{\pi(i)})^{k-i}\right).$$

The coefficient of the monomial $x_1^{k-1}x_2^{k-1}\cdots x_k^{k-1}$ in this expression is

$$(-1)^{\binom{k}{2}}\sum_{\pi\in\mathcal{S}_k}\prod_{i=1}^{k}a_{\pi(i)}^{k-i}$$

which, renaming the index in the product, becomes

$$(-1)^{\binom{k}{2}}\sum_{\pi\in\mathcal{S}_k}\prod_{i=1}^{k}a_{\pi(k-i+1)}^{i-1}$$

and, using the same notation as before, turns into

$$(-1)^{\binom{k}{2}}\sum_{\tau\in\mathcal{S}_k}\prod_{i=1}^{k}a_{\tau(i)}^{i-1}.$$

Applying the definition of the permanent to the Vandermonde matrix $V(a_1,a_2,\dots,a_k)$, the last line reads as $(-1)^{\binom{k}{2}}\mathit{Per}\ V(a_1,a_2,\dots,a_k)$. □

---

[8]The determinant of a matrix $M=(a_{ij})_{1\le i,j\le n}$ is $\mathit{Det}\ M=\sum_{\pi\in\mathcal{S}_n}\operatorname{sgn}(\pi)\prod_{i=1}^{n}a_{i\pi(i)}$.

[9]One permutation can be transformed into the other by $(k-1)+(k-2)+\cdots+2+1$ transpositions.

Bibliography

[1] N. Alon, *Additive Latin transversals*, Israel J. Math. **117** (2000), 125–130.

[2] ———, *Combinatorial Nullstellensatz*, Combin. Probab. Comput. **8** (1999), 7–29.

[3] N. Alon, S. Friedland, and G. Kalai, *Every 4-regular graph plus an edge contains a 3-regular subgraph*, J. Combin. Theory Ser. B **37** (1984), 92–93.

[4] N. Alon, M. B. Nathanson, and I. Z. Ruzsa, *Adding distinct congruence classes modulo a prime*, Amer. Math. Monthly **102** (1995), 250–255.

[5] N. Alon and Z. Füredi, *Covering the cube by affine hyperplanes*, European J. Combin. **14** (1993), 79–83.

[6] A. E. Brouwer and A. Schrijver, *The blocking number of an affine space*, J. Combin. Theory Ser. A **24** (1978), 251–253.

[7] S. Dasgupta, G. Károlyi, O. Serra, B. Szegedy, *Transversals of additive Latin squares*, Israel J. Math. **126** (2001), 17–28.

[8] R. Jamison, *Covering finite fields with cosets of subspaces*, J. Combin. Theory Ser. A **22** (1977), 253–266.

[9] G. Károlyi, *A compactness argument in the additive theory and the polynomial method*, Discrete Math. **302** (2005), 124–144.

[10] H. Snevily, *The Cayley addition table of Z_n*, Amer. Math. Monthly **106** (1999) 584–585.

[11] V. A. Tashkinov, *3-regular subgraphs of 4-regular graphs*, Math. Notes **36** (1984), 612–623.

Gyula Károlyi
Institute of Mathematics, Eötvös University, Pázmány st. 1, H-1117 Budapest
Hungary
e-mail: karolyi@cs.elte.hu

Chapter 6

Problems in additive number theory, III

Melvyn B. Nathanson

6.1 What sets are sumsets?

Let $\mathbb{N}$, $\mathbb{N}_0$, $\mathbb{Z}$ and $\mathbb{Z}^d$ denote, respectively, the sets of positive integers, non-negative integers, integers and d-dimensional integral lattice points. Let $\mathcal{G}$ denote an arbitrary abelian group and let $\mathbf{X}$ denote an arbitrary abelian semigroup, written additively. Let $|S|$ denote the cardinality of the set S. For any sets A and B, we write $A \sim B$ if their symmetric difference is finite, that is, if $|(A \setminus B) \cup (B \setminus A)| < \infty$.

Definition. Let A and B be nonempty subsets of an additive abelian semigroup $\mathbf{X}$. The most important definition in additive number theory is the *sumset* $A+B$ *of the sets* A *and* B:

$$A + B = \{a + b : a \in A, b \in B\}.$$

Let $A + B = \emptyset$ if $A = \emptyset$ or $B = \emptyset$. If $h \geq 3$ and $A_1, \ldots, A_h$ are subsets of $\mathbf{X}$, then we construct the sumset $A_1 + \cdots + A_h$ inductively as follows:

$$\begin{aligned} A_1 + \cdots + A_{h-1} + A_h &= (A_1 + \cdots + A_{h-1}) + A_h \\ &= \{a_1 + \cdots + a_h : a_i \in A_i \text{ for all } i = 1, \ldots, h\}. \end{aligned}$$

If $A_1 = A_2 = \cdots = A_h = A$, then

$$hA = \underbrace{A + \cdots + A}_{h \text{ times}}$$

is called the *h-fold sumset of* A.

Definition. Let A be a nonempty subset of an additive abelian semigroup $\mathbf{X}$. The set A is called a *basis of order h for* $\mathbf{X}$ if $hA = \mathbf{X}$. The set A is called an *asymptotic basis of order h for* $\mathbf{X}$ if $hA \sim \mathbf{X}$.

A basic problem is, What sets are sumsets? More precisely,

Problem 1. Given a set $S \subseteq \mathbf{X}$, do there exist sets $A_1, \ldots, A_h$ of integers such that $A_1 + \cdots + A_h = S$ or $A_1 + \cdots + A_h \sim S$?

Problem 2. Given a set $S \subseteq \mathbf{X}$, does there exist a set A of integers such that $hA = S$ or $hA \sim S$?

These problems are particularly important in the classical cases $\mathbf{X} = \mathbb{N}_0$, $\mathbf{X} = \mathbb{Z}$ and $\mathbf{X} = \mathbb{Z}^d$.

6.2 Describing the structure of hA as $h \to \infty$

Definition. Let A be a set of non-negative integers. The *counting function* $A(x)$ of the set A counts the number of positive elements of A not exceeding x, that is, $A(x) = |A \cap [1, x]|$. The *lower asymptotic density* of A is

$$d_L(A) = \liminf_{x \to \infty} \frac{A(x)}{x}.$$

Theorem 6.2.1 (Nash–Nathanson [20]). *If A is a set of non-negative integers with* $\gcd(A) = 1$ *such that the sumset $h_0 A$ has positive lower asymptotic density for some positive integer h_0, then there exists a number $h \in \mathbb{N}$ such that $hA \sim \mathbb{N}_0$.*

Equivalently, a set A of non-negative integers is an asymptotic basis of finite order if and only if $\gcd(A) = 1$ and $d_L(h_0 A) > 0$ for some positive integer h_0.

Let A be a set of non-negative integers with $0 \in A$ and $\gcd(A) = 1$. We have the increasing sequence of sets

$$A \subseteq 2A \subseteq \cdots \subseteq hA \subseteq (h+1)A \subseteq \cdots.$$

If some sumset has positive lower asymptotic density, then this sequence becomes eventually constant and equal to $A \setminus \mathcal{F}$ for some finite set $\mathcal{F}$ of integers. An important unsolved question is the following.

Problem 3. Suppose that $d_L(hA) = 0$ for all $h \geq 1$. Describe the evolution of the structure of the sumset hA as $h \to \infty$.

6.3 Representation functions

Let $\mathcal{A} = (A_1, \ldots, A_h)$ be an h-tuple of subsets of an additive abelian semigroup $\mathbf{X}$. We want to count the number $R_{\mathcal{A}}(x)$ of representations of an element $x \in \mathbf{X}$ in the form $x = a_1 + \cdots + a_h$ with $a_i \in A_i$ for $i = 1, \ldots, h$. We discuss here only the special case when $A_i = A$ for $i = a, \ldots, h$. We shall consider two different representation functions.

Definition. The *ordered representation function of order* h for the set A is the function $R_{A,h} : \mathbf{X} \to \mathbb{N}_0 \cup \{\infty\}$ defined by

$$R_{A,h}(x) = \left|\left\{(a_1, \ldots, a_h) \in A^h : x = a_1 + \cdots + a_h\right\}\right|.$$

Definition. Let $(a_1, \ldots, a_h) \in A^h$ and $(a'_1, \ldots, a'_h) \in A^h$ be h-tuples that represent x; that is, $x = a_1 + \cdots + a_h = a'_1 + \cdots + a'_h$. These representations are called *equivalent* if there is a permutation σ of the set $\{1, 2, \ldots, h\}$ such that $a'_i = a_{\sigma(i)}$ for $i = 1, 2, \ldots, h$. The *unordered representation function* $r_{A,h}(x)$ *of order* h counts the number of equivalence classes of representations of x.

If $\mathbf{X}$ is a linearly ordered semigroup such as $\mathbb{N}_0$ or $\mathbb{Z}$, then we can write

$$r_{A,h}(x) = \left|\left\{(a_1, \ldots, a_h) \in A^h : x = a_1 + \cdots + a_h \text{ and } a_1 \le a_2 \le \cdots \le a_h\right\}\right|.$$

Let $\mathcal{F}(\mathbf{X})$ denote the set of all functions $f : \mathbf{X} \to \mathbb{N}_0 \cup \{\infty\}$, and let

$$\mathcal{R}_h(\mathbf{X}) = \{r_{A,h} : A \subseteq \mathbf{X}\}$$

denote the set of all unordered representation functions of order h of subsets of $\mathbf{X}$. Then $\mathcal{R}_h(\mathbf{X}) \subseteq \mathcal{F}(\mathbf{X})$. A simple question is the following.

Problem 4. What functions are representation functions?

This problem seems hopelessly difficult at this time. We consider the special case of representation functions of asymptotic bases. Define the function space

$$\mathcal{F}^{(0)}(\mathbf{X}) = \left\{f : \mathbf{X} \to \mathbb{N}_0 \cup \{\infty\} : \left|f^{-1}(0)\right| < \infty\right\}.$$

This is the space of functions on $\mathbf{X}$ with only finitely many zeros. Let

$$\mathcal{R}_h^{(0)}(\mathbf{X}) = \{r_{A,h} : A \subseteq \mathbf{X} \text{ and } hA \sim \mathbf{X}\}$$

be the set of representation functions of asymptotic bases of order h for $\mathbf{X}$. Then

$$\mathcal{R}_h^{(0)}(\mathbf{X}) \subseteq \mathcal{F}^{(0)}(\mathbf{X}).$$

Problem 5. What functions in $\mathcal{F}_0(\mathbf{X})$ are representation functions of asymptotic bases? Equivalently, what are necessary and sufficient conditions for a function on $\mathbf{X}$ with only finitely many zeros to be the representation function of an additive basis?

For the group $\mathbb{Z}$ of integers there is the following amazing result.

Theorem 6.3.1 (Nathanson [24, 25, 26]). *For every integer $h \ge 2$ and for every function $\mathcal{F}_0(\mathbb{Z})$ there exists a set $A \subseteq \mathbb{Z}$ such that $f = r_{A,h}$. Equivalently,*

$$\mathcal{F}_0(\mathbb{Z}) = \mathcal{R}_2^{(0)}(\mathbb{Z}) = \mathcal{R}_3^{(0)}(\mathbb{Z}) = \cdots = \mathcal{R}_h^{(0)}(\mathbb{Z}) = \cdots.$$

The classification of arbitrary representation functions for the integers is still open.

Problem 6. What functions in $\mathcal{F}(\mathbb{Z})$ are representation functions of sumsets of order h?

Very little is known about representation functions of sums of sets of non-negative integers, even for sets that are asymptotic bases. G. A. Dirac gave an elegant proof of the following beautiful result.

Theorem 6.3.2 (Dirac [4]). *If A is an asymptotic basis of order 2 for $\mathbf{N}_0$, then the unordered representation function $r_{A,2}$ is not eventually constant.*

Proof. The generating function $G_A(z) = \sum_{a\in A} z^a$ converges in the open unit disc and diverges as $z \to 1^-$. We have $G_A^2(z) = \sum_{n=0}^{\infty} R_A(n)z^n$ and

$$\frac{G_A^2(z) + G_A(z^2)}{2} = \sum_{n=0}^{\infty} r_{A,2}(n)z^n.$$

If $r_{A,2}(n) = c$ for all $n \geq n_0$, then there is a polynomial $P(z)$ such that

$$\frac{G_A^2(z) + G_A(z^2)}{2} = P(z) + \frac{c}{1-z}.$$

Let $0 < x < 1$ and $z = -x$. Then

$$G_A(x^2) \leq G_A^2(-x) + G_A(x^2) = 2P(-x) + \frac{2c}{1+x}.$$

As $x \to 1^-$, the right side approaches $2P(-1) + c$ while the left side diverges to ∞, which is absurd. □

One of the most famous and tantalizing unsolved problems in additive number theory is the following.

Problem 7 (Erdős–Turán [8]). Let $h \geq 2$. Prove that if A is an asymptotic basis of order h for the non-negative integers, then the representation function $r_{A,h}$ must be unbounded.

6.4 Sets with more sums than differences

Let A be a finite set of integers. We define the sumset

$$A + A = \{a + a' : a, a' \in A\}$$

and the difference set

$$A - A = \{a - a' : a, a' \in A\}.$$

Since $2+3=3+2$ but $2-3\neq 3-2$, that is, since addition is commutative but subtraction is not commutative, it would be reasonable to conjecture that $|A+A|\leq|A-A|$ for every finite set A of integers. In some special cases, for example, if A is an arithmetic progression or if A is symmetric (that is, if $A=c-A$ for some $c\in\mathbb{Z}$), then $|A+A|=|A-A|$. The "conjecture," however, is false. The simplest counterexample is the set

$$A^*=\{0,2,3,4,7,11,12,14\}$$

for which $|A^*+A^*|=26>25=|A^*-A^*|$. This set is "almost" symmetric. We have $A^*=A'\cup\{4\}$ where $A'=\{0,2\}\cup\{3,7,11\}\cup\{12,14\}$ satisfies $A'=14-A'$. A set with more sums than differences is called an *MSTD set.* Nathanson [28] showed that if $k\geq 3$ and $A'=\{0,2\}\cup\{3,7,11,15,19,23,\ldots,4k-1\}\cup\{4k,4k+2\}$ and $A^*=A'\cup\{4\}$, then A^* is an MSTD set. Many other examples of MSTD sets have been constructed by Hegarty [12].

Finite sets A of integers with the property that $|A+A|>|A-A|$ are extremely interesting, since a sumset really should have more elements than the corresponding difference set.

Conjecture 1. $|A+A|\leq|A-A|$ for almost all finite sets A.

Martin and O'Bryant [19] studied the uniform probability measure on the set of all subsets of $\{1,\ldots,N\}$; that is, they assigned to each subset the probability 2^{-N}. Counting sets in this way, they calculated that the average cardinality of a sumset was

$$\frac{1}{2^N}\sum_{A\subseteq\{1,\ldots,N\}}|A+A|=2N-11$$

and the average cardinality of a difference set was

$$\frac{1}{2^N}\sum_{A\subseteq\{1,\ldots,N\}}|A-A|=2N-7.$$

Thus, on average, a difference set contains four more elements than the sumset. However, they also proved the following result.

Theorem 6.4.1 (Martin–O'Bryant [19]). *With the uniform probability measure, there exists a $\delta>0$ such that*

$$Prob\,(|A+A|>|A-A|:A\subseteq\{1,\ldots,N\})>\delta$$

for all $N\geq N_0$

Thus, choosing a uniform probability measure, it is not true that almost all sets have more differences than sums. Of course, with the uniform probability distribution most subsets of the interval $\{1,\ldots,N\}$ are large and satisfy $|A+A|=|A-A|=2N-1$. This skews the calculation.

Using a binomial probability distribution, Hegarty and Miller obtained a very different result.

Theorem 6.4.2 (Hegarty–Miller [13]). *Let $p : \mathbb{N} \to (0,1)$ be a function such that $\lim_{N\to\infty} p(N) = 0$ and $\lim_{N\to\infty} Np(N) = \infty$. Define the function $q : \mathbb{N} \to (0,1)$ by $q(N) = 1 - p(N)$. Consider the binomial probability distribution with parameter $p(N)$ on the space of all subsets of $\{1,\ldots,N\}$, so that a subset of size k has probability $p(N)^k q(N)^{N-k}$. Then*

$$\lim_{N\to\infty} Prob\Big(\,|A+A| > |A-A| : A \subseteq \{1,\ldots,N\}\Big) = 0.$$

These theorems seem to contradict each other, but they do not because they use different probability measures.

Problem 8. A difficult and subtle problem is to decide what is the appropriate method of counting (or, equivalently, the appropriate probability measure) to apply to MSTD sets.

6.5 Comparative theory of linear forms

Let $f(x_1,\ldots,x_m)$ be an integer-valued function on the integers and let A be a set of integers. We define the set

$$f(A) = \{f(a_1,\ldots,a_m) : a_1,\ldots,a_m \in A\}.$$

In particular, if $f(x_1,x_2) = u_1x_1 + u_2x_2$ is a linear form with nonzero integer coefficients, then

$$f(A) = \{u_1a_1 + u_2a_2 : a_1,a_2 \in A\}.$$

For example, if $s(x_1,x_2) = x_1 + x_2$, then $s(A)$ is the sumset $A+A$. If $d(x_1,x_2) = x_1 - x_2$, then $d(A)$ is the difference set $A - A$.

The binary linear forms $f(x_1,x_2) = u_1x_1 + u_2x_2$ and $g(x_1,x_2) = v_1x_1 + v_2x_2$ are *related* if

$(v_1,v_2) = (u_2,u_1)$, or
$(v_1,v_2) = (du_1,du_2)$ for some integer d, or
$(v_1,v_2) = (u_1/d,u_2/d)$ for some integer d that divides u_1 and u_2.

The binary linear forms $f(x_1,x_2) = u_1x_1 + u_2x_2$ and $g(x_1,x_2) = v_1x_1 + v_2x_2$ are *equivalent* if there is a finite sequence of binary linear forms $f_0,f_1,\ldots,f_k$ such that $f = f_0$, $g = f_k$, and f_{i-1} is related to f_i for all $i = 1,\ldots,k$. If f and g are equivalent forms, then $|f(A)| = |g(A)|$ for every finite set A.

The binary linear form $f(x_1,x_2) = u_1x_1 + u_2x_2$ is *normalized* if $\gcd(u_1,u_2) = 1$ and $u_1 \geq |u_2| \geq 1$. Every binary linear form is equivalent to a unique normalized form.

The following is the basic result in the comparative theory of linear forms.

Theorem 6.5.1 (Nathanson–O'Bryant–Orosz–Ruzsa–Silva [33]). *Let* $f(x_1,x_2) = u_1x_1+u_2x_2$ *and* $g(x_1,x_2) = v_1x_1+v_2x_2$ *be distinct normalized linear forms. There exist sets* A *and* A' *of integers such that* $|f(A)| < |g(A)|$ *and* $|f(A')| > |g(A')|$.

Consider now linear forms in more than two variables.

Problem 9. Let $f(x_1,\ldots,x_m) = u_1x_1+\cdots+u_mx_m$ and $g(x_1,\ldots,x_m) = v_1x_1+\cdots+v_mx_m$ be linear forms with nonzero integer coefficients in $m \geq 3$ variables. Suppose that $\gcd(u_1,\ldots,u_m) = \gcd(v_1,\ldots,v_m) = 1$ and that g cannot be obtained from f by some permutation of the coefficients or by multiplication by -1. Does there exist a finite set A of integers such that $|f(A)| > |g(A)|$?

Problem 10. Let $f(x_1,\ldots,x_m)$ and $g(x_1,\ldots,x_m)$ be polynomials with integer coefficients in $m \geq 2$ variables. Under what conditions does there exist a finite set A of integers such that $|f(A)| > |g(A)|$?

Definition. Let f be an integer-valued function defined on $\mathbb{Z}$. Define $N_f(k) = \min\{|f(A)| : A \subseteq \mathbb{Z} \text{ and } |A| = k\}$.

Problem 11. Let f be an integer-valued function defined on $\mathbb{Z}$, for example, a linear form or a polynomial with integer coefficients. Determine $N_f(k)$ and describe the structure of the minimizing sets.

Theorem 6.5.2 (Bukh [2]). *Let* $f(x_1,\ldots,x_m) = u_1x_1+\cdots+u_mx_m$ *be a linear form with nonzero integer coefficients in* $m \geq 2$ *variables. If* $\gcd(u_1,\ldots,u_m) = 1$, *then* $N_f(k) = \Big(|u_1|+\cdots+|u_2|\Big)k - o(k)$.

Theorem 6.5.3. *If* $f(x_1,x_2) = x_1+x_2$, *then* $N_f(k) = 2k-1$ *and the minimizing sets are finite arithmetic progressions. Equivalently, if* A *is a finite set of integers, then* $|A+A| \geq 2|A|-1$ *and* $|A+A| = 2|A|-1$ *if and only if* A *is an arithmetic progression.*

An *affine transform* of a set A of real numbers is a set obtained from A by a sequence of translations and dilations.

Theorem 6.5.4 (Cilleruelo–Silva–Vinuesa [15]). *If* $f(x_1,x_2) = x_1+2x_2$, *then* $N_f(k) = 3k-2$. *Moreover,* $|f(A)| = 3|A|-2$ *if and only if* A *is an arithmetic progression.*

If $f(x_1,x_2) = x_1+3x_2$, *then* $N_f(k) = 4k-4$. *Moreover,* $|f(A)| = 4|A|-4$ *if and only if* A *is* $\{0,1,3\}$ *or* $\{0,1,4\}$ *or* $\{0,3,6,\ldots,3\ell-3\} \cup \{1,4,7,\ldots,3\ell-2\}$, *or an affine transform of one of these sets.*

Definition. Let $\mathcal{U} = (u_1,\ldots,u_m)$ be a sequence of positive integers. A *subsequence sum* of $\mathcal{U}$ is a non-negative integer of the form $\sum_{i\in I} u_i$, where I is a subset of $\{1,\ldots,m\}$. Let $S(\mathcal{U}) = \left\{\sum_{i\in I} u_i : I \subseteq \{1,\ldots,m\}\right\}$ denote the set of all subsequence sums of the sequence $\mathcal{U}$.

A subsequence sum is 0 if and only if $I = \emptyset$. If $\mathrm{U} = u_1+\cdots+u_m$, then $S(\mathcal{U}) \subseteq [0,\mathrm{U}]$.

Definition. The sequence $\mathcal{U}$ is called *complete* if $S(\mathcal{U}) = [0, \mathrm{U}]$.

For example, the sequences $(1,1)$ and $(1,2)$ are complete but $(1,3)$ is not complete.

Theorem 6.5.5 (Nathanson [29]). *Let $\mathcal{U} = (u_1, \ldots, u_m)$ be a complete sequence of positive integers, and let $f(x_1, \ldots, x_m) = u_1x_1 + \cdots + u_mx_m$ be the associated linear form. If $U = u_1 + \cdots + u_m$, then $N_f(k) = Uk - U + 1$ for all positive integers k. Moreover, $|A| = k$ and $|f(A)| = N_f(k)$ if and only if A is an arithmetic progression of length k.*

There is also the dual problem of describing the finite sets of integers whose images under linear maps are large.

Definition. Let f be an integer-valued function defined on $\mathbb{Z}$. Define $M_f(k) = \max\{|f(A)| : A \subseteq \mathbb{Z},\ |A| = k\}$.

Problem 12. Let f be an integer-valued function in m variables defined on $\mathbb{Z}$. Determine $M_f(k)$ and describe the structure of the maximizing sets. For what functions f is $M_f(k) < k^m$?

6.6 The fundamental theorem of additive number theory

Let $A = \{a_0 < a_1 < \cdots < a_{k-1}\}$ be a finite set of integers. Consider the shifted set $A' = A - \{a_0\} = \{0 < a_1 - a_0 < \cdots < a_{k-1} - a_0\}$. Let $d = \gcd\{a_i - a_0 : i = 1, \ldots, k-1\}$, and construct the set

$$A^{(N)} = \frac{1}{d} * A' = \left\{0 < \frac{a_1 - a_0}{d} < \cdots < \frac{a_{k-1} - a_0}{d}\right\}.$$

If hA is the h-fold sumset of A, then $hA' = hA - \{ha_0\}$ and

$$hA^{(N)} = \frac{1}{d} * hA' = \frac{1}{d} * (hA - \{ha_0\}).$$

In particular, $|hA| = |hA^{(N)}|$.

The set $A^{(N)}$ is called the *normalized* form of the set A. In general, a finite set A of integers is normalized if $A = \{0\}$, or if $|A| \geq 2$, $\min(A) = 0$, and $\gcd(A) = 1$.

The following result is often called the Fundamental Theorem of Additive Number Theory.

Theorem 6.6.1 (Nathanson [21]). *Let $A = \{0 < a_1 < \cdots < a_{k-1}\}$ be a normalized finite set of integers. There exist a positive integer h_0, non-negative integers C and D, and finite sets $\mathcal{C} \subseteq [0, C-2]$ and $\mathcal{D} \subseteq [0, D-2]$ such that*

$$hA = \mathcal{C} \cup [C, ha_{k-1} - D] \cup (\{ha_{k-1}\} - \mathcal{D})$$

for all $h \geq h_0$.

If A is a set of non-negative integers that contains 0, then

$$A \subseteq 2A \subseteq \cdots \subseteq hA \subseteq (h+1)A \subseteq \cdots$$

for all $h \geq 1$, and the set

$$\Sigma(A) = \bigcup_{h=1}^{\infty} hA$$

is the additive subsemigroup of the non-negative integers generated by the set A. The fundamental theorem implies that if $\gcd(A) = 1$, then

$$\Sigma(A) = \mathcal{C} \cup [C, \infty).$$

The number $C - 1$ is the largest integer that cannot be represented as a non-negative integral linear combination of $a_1, \ldots, a_{k-1}$. This is called the *Frobenius number* of the set A. Note that $D - 1$ is the Frobenius number of the symmetric normalized set

$$A^\sharp = \{a_{k-1}\} - A = \{0 < a_{k-1} - a_{k-2} < \cdots < a_{k-1} - a_1 < a_{k-1}\}.$$

Also, since $\Sigma(A)$ is a semigroup, it follows that if u and v are non-negative integers with $u + v = C - 1$, then either $u \notin \Sigma(A)$ or $v \notin \Sigma(A)$. Therefore,

$$|\mathbf{N}_0 \setminus \Sigma(A)| = |[0, C-1] \setminus \mathcal{C}| \geq \frac{C}{2}.$$

Problem 13. Let A be a normalized finite set of non-negative integers. Compute $\mathbf{N}_0 \setminus \Sigma(A)$, that is, the set of numbers that cannot be represented as non-negative integral linear combinations of the elements of A.

6.7 Thin asymptotic bases

Let A be an infinite set of non-negative integers that is an asymptotic basis of order h. There is a non-negative integer n_0 such that, if $n_0 \leq n \leq x$, then there exist $a_1, \ldots, a_h \in A$ with $n = a_1 + \cdots + a_h$ and $0 \leq a_i \leq n \leq x$ for $i = 1, \ldots, h$. Denote by $A(x)$ the counting function of the set A. Since the interval $[n_0, x]$ contains at least $x - n_0$ non-negative integers, it follows that $(A(x)+1)^h \geq x - n_0$ and so

$$A(x) \gg x^{1/h}$$

for every asymptotic basis A of order h.

Definition. An asymptotic basis A of order h is called *thin* if $A(x) \ll x^{1/h}$.

Thin asymptotic bases exist, and the first explicit examples were constructed independently by Chartrovsky, Raikov and Stöhr in the 1930s. Thin bases of order h have the property that their counting functions have order of magnitude $x^{1/h}$. Cassels constructed a family of bases of order h whose counting functions are asymptotic to $\lambda x^{1/h}$ for some positive real number λ.

Theorem 6.7.1 (Cassels [3]). *For every* $h \geq 2$, *there exist strictly increasing sequences* $A = \{a_n\}_{n=1}^{\infty}$ *of non-negative integers such that* $hA = \mathbb{N}_0$ *and* $a_n = \lambda n^h + O\left(n^{h-1}\right)$ *for some* $\lambda > 0$.

Problem 14 (Cassels [3]). Let $h \geq 2$. Does there exist an asymptotic basis $A = \{a_n\}_{n=1}^{\infty}$ of order h such that $a_n = \lambda n^h + o\left(n^{h-1}\right)$ for some $\lambda > 0$?

Definition. A positive real number λ will be called an *additive eigenvalue of order* h if there exists an asymptotic basis $A = \{a_n\}_{n=1}^{\infty}$ of order h such that

$$a_n \sim \lambda n^h.$$

We define the *additive spectrum* Λ_h as the set of all additive eigenvalues of order h.

Theorem 6.7.2 (Nathanson [32]). *For every integer* $h \geq 2$, *there is a number* λ_h^* *such that* $\Lambda_h = (0, \lambda_h^*)$ *or* $\Lambda_h = (0, \lambda_h^*]$.

The idea of the proof is to show that if A is an asymptotic basis of order h with eigenvalue λ, and if $0 < \lambda' < \lambda$, then one can adjoin non-negative integers to the set A to obtain an asymptotic basis of order h with eigenvalue λ'. Thus, Λ_h is an interval. Combinatorial and geometric arguments show that the additive spectrum is bounded above.

Problem 15. Compute the upper bound λ_h^* of the additive spectrum Λ_h. Is this upper bound an eigenvalue?

6.8 Minimal asymptotic bases

The set A of non-negative integers is an asymptotic basis of order h if every sufficiently large integer is the sum of exactly h elements of A.

Definition. An asymptotic basis A of order h is *minimal* if, for every element $a^* \in A$, the set $A \setminus \{a^*\}$ is not an asymptotic basis of order h.

Thus, if A is a minimal asymptotic basis of order h, then for every integer $a^* \in A$ there are infinitely many positive integers n that cannot be represented as the sum of h elements of the set $A \setminus \{a^*\}$. Equivalently, every element of A is somehow "responsible" for the representation of infinitely many integers.

Theorem 6.8.1 (Härtter [11], Nathanson [22]). *For every* $h \geq 2$ *there exist minimal asymptotic bases of order* h.

On the other hand, it is not true that every asymptotic basis of order h contains a subset that is a minimal asymptotic basis of order h. In particular, we have the following result.

Theorem 6.8.2 (Erdős–Nathanson [5]). *There exists an asymptotic basis* A *of order* 2 *such that, for every subset* $S \subseteq A$, *the set* $A \setminus S$ *is an asymptotic basis of order* 2 *if and only if* S *is finite.*

Since there is no maximal finite subset of an infinite set, it follows that there exists an asymptotic basis of order 2 that contains no minimal asymptotic basis of order 2.

Problem 16. Let $h \geq 3$. Construct an asymptotic basis A of order h such that, for every subset $S \subseteq A$, the set $A \setminus S$ is an asymptotic basis of order h if and only if S is finite.

Problem 17. Find necessary and sufficient conditions to determine if an asymptotic basis A of order h contains a minimal asymptotic basis of order h.

In a minimal asymptotic basis every element in the basis is responsible for the representation of infinitely many numbers. In particular, if A is a minimal asymptotic basis of order 2, then there must be infinitely many positive integers with a unique representation as the sum of two elements of A.

Let A be a set of integers. Let $r_{A,2}(n)$ denote the *unordered representation function* of the set A; that is,

$$r_{A,2}(n) = |\{\{a_i, a_j\} \subseteq A : n = a_i + a_j\}|.$$

Theorem 6.8.3 (Erdős–Nathanson [7]). *Let A be a set of non-negative integers. If $r_{A,2}(n) > c \log n$ for some $c > 1/\log(4/3)$ and all $n \geq n_0$, then A contains a minimal asymptotic basis of order* 2.

Problem 18. Is this true if $r_{A,2}(n) > c \log n$ for some $c > 0$?

Problem 19. Let A be a set of non-negative integers. If $r_{A,2}(n) \to \infty$ as $n \to \infty$, does A contain a minimal asymptotic basis of order 2?

The idea of minimal asymptotic basis can be generalized in the following way.

Definition. Let $r \geq 1$. The set A is an *r-minimal asymptotic basis of order h* if, for every $S \subseteq A$, the set $A \setminus S$ is an asymptotic basis of order h if and only if $|S| < r$.

Theorem 6.8.4 (Erdős–Nathanson [5]). *For every $r \geq 1$, there exist r-minimal asymptotic bases of order* 2.

Problem 20. Let $h \geq 3$ and $r \geq 2$. Construct an r-minimal asymptotic basis A of order h.

6.9 Maximal asymptotic non-bases

Maximal asymptotic non-bases are the natural dual to minimal asymptotic bases.

Definition. The set A of non-negative integers is an *asymptotic non-basis of order h* if it is not an asymptotic basis of order h, that is, if hA omits infinitely many non-negative integers, that is, the set $\mathbb{N} \setminus hA$ is infinite.

Definition. The set A of non-negative integers is a *maximal asymptotic non-basis of order* h if A is an asymptotic non-basis of order h such that, for every integer $a^* \in \mathbb{N} \setminus A$, the set $A \cup \{a^*\}$ is an asymptotic basis of order h.

The construction of minimal asymptotic bases is difficult, but it is easy to find simple examples of maximal non-bases. For example, the set of all non-negative even integers is a maximal asymptotic non-basis of order h for all $h \geq 2$. The set of all non-negative multiples of a fixed prime number is a maximal asymptotic non-basis of order p. Other examples can be constructed by taking appropriate unions of congruence classes.

Theorem 6.9.1 (Nathanson [23]). *There exist maximal asymptotic non-bases of zero asymptotic density.*

The follow result implies that there exist asymptotic non-bases that cannot be embedded in maximal asymptotic non-bases.

Theorem 6.9.2 (Hennefeld [14]). *There exists an asymptotic non-basis of order* 2 *such that, for every set* $S \subseteq \mathbb{N} \setminus A$, *the set* $A \cup S$ *is an asymptotic non-basis of order* 2 *if and only if* $|\mathbb{N} \setminus (A \cup S)|$ *is infinite.*

There are many beautiful results on minimal asymptotic bases and maximal asymptotic non-bases. Here are two of my favorites.

Theorem 6.9.3 (Erdős–Nathanson [6]). *There exists a partition of* $\mathbb{N}$ *into two sets* A *and* B *such that* A *is a minimal asymptotic basis of order* 2 *and* B *is a maximal asymptotic non-basis of order* 2.

Theorem 6.9.4 (Erdős–Nathanson [6]). *There exists a partition of* $\mathbb{N}$ *into two sets* A *and* B *such that, for any finite subset* F *of* A *and any finite subset* G *of* B, *the partition of* $\mathbb{N}$ *into the sets*

$$(A \setminus F) \cup G$$

and

$$(B \setminus G) \cup F$$

has the following property:

(i) *If* $|F| = |G|$, *then* $(A \setminus F) \cup G$ *is a minimal asymptotic basis of order* 2 *and* $(B \setminus G) \cup F$ *is a maximal asymptotic non-basis of order* 2.

(ii) *If* $|F| = |G| + 1$, *then* $(A \setminus F) \cup G$ *is a maximal asymptotic non-basis of order* 2 *and* $(B \setminus G) \cup F$ *is a minimal asymptotic basis of order* 2.

6.10 Complementing sets of integers

Let A and B be sets of integers, or, more generally, subsets of any additive abelian semigroup $\mathbf{X}$, and let $A + B = \{a + b : a \in A, b \in B\} = C$. If every element of the sumset C has a *unique* representation as the sum of an element of A and an element of B, then we write $A \oplus B = C$. We say that the set A *tessellates* the semigroup $\mathbf{X}$ if there exists a set B such that $A \oplus B = \mathbf{X}$, and that A and B are *complementing subsets of* $\mathbf{X}$.

In this section we consider complementing sets of integers. We examine the special case when A is a finite set of integers, and we want to determine if there exists an infinite set B of integers such that $A \oplus B = \mathbb{Z}$. By translation, we can always assume that A is a finite set of integers with $0 \in A$, and that 0 also belongs to B.

We call a set B *periodic* with *period* m if $b \in B$ implies that $b \pm m \in B$.

Theorem 6.10.1 (Newmann [34]). *Let A be a finite set of integers. If there exists a set B such that $A \oplus B = \mathbb{Z}$, then B is periodic with period*

$$m \leq 2^{diam(A)}$$

where $diam(A) = \max(A) - \min(A)$.

It follows that if $A \oplus B = \mathbb{Z}$, then B is a union of congruence classes modulo m. Defining $\overline{A} = \{a + m\mathbb{Z} : a \in A\}$ and $\overline{B} = \{b + m\mathbb{Z} : b \in B\}$, we obtain a complementing pair $\overline{A} \oplus \overline{B} = \mathbb{Z}/m\mathbb{Z}$. Conversely, suppose that $\overline{A}$ and $\overline{B}$ are sets of congruence classes modulo m such that $\overline{A} \oplus \overline{B} = \mathbb{Z}/m\mathbb{Z}$. Let A be a set of representatives of the congruence classes in $\overline{A}$ and let B be the union of the congruence classes in $\overline{B}$. Then $A \oplus B = \mathbb{Z}$.

Theorem 6.10.2 (Kolountzakis [17], Ruzsa [35], Tijdeman [36]). *Let A and B be sets of integers such that A is finite, $A \oplus B = \mathbb{Z}$ and B has minimal period m. Then*

$$m \ll e^{c\sqrt{diam(A)}}.$$

Theorem 6.10.3 (Biró [1]). *Let A and B be sets of integers such that A is finite, $A \oplus B = \mathbb{Z}$, and B has minimal period m. Then*

$$m \ll e^{c\sqrt[3]{diam(A)}}.$$

Problem 21. Find the least upper bound for the period of a set B of integers that is complementary to a finite set A of diameter d.

We can generalize the problem of complementing sets of integers to higher dimensions. Let $d \geq 2$ and let A be a finite set of lattice points in $\mathbb{Z}^d$. Suppose there exists a set $B \subseteq \mathbb{Z}^d$ such that $A \oplus B = \mathbb{Z}^d$. The following problem is well known.

Problem 22. Is the set B periodic even in one direction? Equivalently, does there exist a lattice point $b_0 \in \mathbb{Z}^d \setminus \{0\}$ such that $B + \{b_0\} = B$?

We can also generalize the problem of complementing sets of integers to linear forms. Rewrite the original question as follows: Let $\varphi(x, y) = x + y$. For sets $A, B \subseteq \mathbb{Z}$, we define the set

$$\varphi(A, B) = \{\varphi(a, b) : a \in A, \, b \in B\}$$

and the representation function

$$r_{A,B}^{(\varphi)}(n) = \{(a,b) \in A \times B : \varphi(a,b) = n\}.$$

Given a finite set A, does there exist a set B such that $\varphi(A, B) = \mathbb{Z}$ and $r_{A,B}^{(\varphi)}(n) = 1$ for all integers n? Now consider the linear forms

$$\psi(x_1, \ldots, x_h) = u_1x_1 + \cdots + u_hx_h$$

and

$$\varphi(x_1, \ldots, x_h, y) = \psi(x_1, \ldots, x_h) + vy = u_1x_1 + \cdots + u_hx_h + vy$$

with nonzero integer coefficients $u_1, \ldots, u_h, v$. Given an h-tuple $\mathcal{A} = (A_1, \ldots A_h)$ of finite sets of integers, and a set B of integers, we define the set

$$\varphi(\mathcal{A}, B) = \{u_1a_1 + \cdots + u_ha_h + vb : a_i \in A_i \text{ for } i = 1, \ldots, h \text{ and } b \in B\}$$

and the representation function

$$r_{\mathcal{A},B}^{(\varphi)}(n) = \{(a_1, \ldots, a_h, b) \in A_1 \times \cdots \times A_h \times B : \varphi(a_1, \ldots, a_h, b) = n\}.$$

Problem 23. Given an h-tuple $\mathcal{A} = (A_1, \ldots A_h)$ of finite sets of integers, determine if there exists a set B such that $\varphi(\mathcal{A}, B) = \mathbb{Z}$ and $r_{\mathcal{A},B}^{(\varphi)}(n) = 1$ for all integers n.

In this case, we say that $\mathcal{A}$ and B are complementing sets of integers with respect to the linear form φ.

Theorem 6.10.4 (Nathanson [31]). *If $\mathcal{A}$ and B are complementing sets of integers with respect to the linear form φ, then B is periodic with period*

$$m \leq 2^{\frac{diam(\psi(A_1,\ldots,A_h))}{|v|}}.$$

Ljujic and Nathanson [18] have extended Biro's cube root upper bound for the period of m to complementing sets of integers with respect to a linear form.

Instead of considering only sets that produce a unique representation for every integer, we can ask for any prescribed number of representations. This suggests the following inverse problem for representation functions associated with linear forms:

Problem 24. Let $\varphi(x_1, \ldots, x_h, y) = u_1x_1 + \cdots + u_hx_h + vy$ be a linear form with integer coefficients, and let $\mathcal{A} = (A_1, \ldots A_h)$ be an h-tuple of finite sets of integers. Given a function $f : \mathbb{Z} \to \mathbb{N}$, does there exist a set $B \subseteq \mathbb{Z}$ such that $r_{\mathcal{A},B}^{(\varphi)}(n) = f(n)$ for all integers n?

We have the following compactness theorem.

Theorem 6.10.5 (Nathanson [31]). *Let $\varphi(x_1, \ldots, x_h, y) = u_1x_1 + \cdots + u_hx_h + vy$ be a linear form with integer coefficients, and let $\mathcal{A} = (A_1, \ldots A_h)$ be an h-tuple of finite sets of integers. Consider a function $f : \mathbb{Z} \to \mathbb{N}$. Suppose there exists a strictly increasing sequence $\{k_i\}_{i=1}^{\infty}$ of positive integers and a sequence $\{B_i\}_{i=1}^{\infty}$ of (not necessarily increasing) sets of integers such that $r^{(\varphi)}_{\mathcal{A},B_i}(n) = f(n)$ for integers n satisfying $|n| \leq k_i$. Then there exists an infinite set B such that $r^{(\varphi)}_{\mathcal{A},B}(n) = f(n)$ for all integers n.*

6.11 The Caccetta–Häggkvist conjecture

Let $G = G(V, E)$ be a finite directed graph with vertex set V and edge set E. Let $n = |V|$. Every edge $e \in E$ is an ordered pair (v, v') of vertices. The vertex v is called the *tail* of e and the vertex v' is called the *head* of e. An edge of the form (v, v) is called a *loop*. We consider only graphs that may have loops, but that do not have multiple edges. A *path of length r* in the graph G is a finite sequence of edges $e_1, e_2, \ldots, e_r$, where $e_i = (v_i, v'_i)$ for $i = 1, \ldots, r$, and $v'_i = v_{i+1}$ for $i = 1, \ldots, r-1$. The path is called a *circuit* if $v'_r = v_1$. A circuit of length 1 is a loop, a circuit of length 2 is called a *digon* and a circuit of length 3 is called a *triangle*.

The *outdegree* of a vertex v, denoted outdegree(v), is the number of edges $e \in E$ whose tail is v. If $|V| = n$ and outdegree$(v) \geq 1$ for every vertex $v \in V$, then the graph G contains a circuit of length at most n. If $|V| = n$ and outdegree$(v) \geq 2$ for every vertex $v \in V$, then it is known that the graph G contains a circuit of length at most $n/2$.

Conjecture 2 (Caccetta–Häggkvist). Let $k \geq 3$. If outdegree$(v) \geq k$ for every vertex $v \in V$, then the graph G contains a circuit of length at most n/k. Equivalently, if *outdegree*$(v) \geq n/k$ for every vertex $v \in V$, then the graph G contains a circuit of length at most k.

Even the case $k = 3$ of the Caccetta–Häggkvist is open: If G is a graph with n vertices and if every vertex is the tail of at least $n/3$ edges, prove that the graph contains a loop, a digon, or a directed triangle. This is a fundamental unsolved problem in graph theory.

Definition. Let Γ be a finite group and let $X \subseteq \Gamma$. The Cayley graph $G(V, E)$ is the graph with vertex set $V = \Gamma$ and edge set $E := \{(\gamma, \gamma x) : \gamma \in \Gamma, x \in \mathbf{X}\}$.

Theorem 6.11.1 (Hamidoune [9, 10]). *The Caccetta–Häggkvist conjecture is true for all Cayley graphs and for all vertex-transitive graphs.*

One proof of this result uses a theorem of Kemperman [16] in additive number theory. An exposition of this and other related results appears in Nathanson [27].

Acknowledgment

This chapter has been written based on notes taken by Fancsali Szabolcs and Itziar Bardaji at the seminar delivered by the author at CRM on January 23 and 25, 2008.

Bibliography

[1] A. Biró, *Divisibility of integer polynomials and tilings of the integers*, Acta Arith. **118** (2005), no. 2, 117–127.

[2] B. Bukh, *Sums of dilates*, arXiv preprint 0711.1610, 2007.

[3] J. W. S. Cassels, *Über Basen der natürlichen Zahlenreihe*, Abh. Math. Sem. Univ. Hamburg **21** (1975), 247–257.

[4] G. A. Dirac, *Note on a problem in additive number theory*, J. London Math. Soc. **26** (1951), 312–313.

[5] P. Erdős and M. B. Nathanson, *Oscillations of bases for the natural numbers*, Proc. Amer. Math. Soc. **53** (1975), no. 2, 253–258.

[6] ______, *Partitions of the natural numbers into infinitely oscillating bases and non-bases*, Comment. Math. Helv. **51** (1976), no. 2, 171–182.

[7] ______, *Systems of distinct representatives and minimal bases in additive number theory*, Number Theory, Carbondale 1979 (Proc. Southern Illinois Conf., Southern Illinois Univ., Carbondale, Ill., 1979), Lecture Notes in Math., vol. 751, Springer, Berlin, 1979, pp. 89–107.

[8] P. Erdős and P. Turán, *On a problem of Sidon in additive number theory, and on some related problems*, J. London Math. Soc. **16** (1941), 212–215.

[9] Y. O. Hamidoune, *An application of connectivity theory in graphs to factorizations of elements in groups*, European J. Combin. **2** (1981), no. 4, 349–355.

[10] ______, *A note on minimal directed graphs with given girth*, J. Combin. Theory Ser. B **43** (1987), no. 3, 343–348.

[11] E. Härtter, *Ein Beitrag zur Theorie der Minimalbasen*, J. Reine Angew. Math. **196** (1956), 170–204.

[12] P. V. Hegarty, *Some explicit constructions of sets with more sums than differences*, Acta Arith. **130** (2007), 61–77.

[13] P. V. Hegarty and S. J. Miller, *When almost all sets are difference dominated*, arXiv preprint 0707.3417, 2007.

[14] J. Hennefeld, *Asymptotic non-bases which are not subsets of maximal aymptotic non-bases*, Proc. Amer. Math. Soc. **62** (1977), 23–24.

[15] C. Vinuesa J. Cilleruelo and M. Silva, *A sumset problem*, preprint, 2008.

[16] J. H. B. Kemperman, *On complexes in a semigroup*, Indag. Math. **18** (1956), 247–254.

[17] M. N. Kolountzakis, *Translational tilings of the integers with long periods*, Electron. J. Combin. **10** (2003), Research Paper 22, 9 pp.

[18] Z. Ljujic and M. B. Nathanson, *Complementing sets of integers with respect to a multiset*, preprint, 2008.

[19] G. Martin and K. O'Bryant, *Many sets have more sums than differences*, Additive Combinatorics, CRM Proc. Lecture Notes, vol. 43, Amer. Math. Soc., Providence, RI, 2007, pp. 287–305.

[20] J. C. M. Nash and M. B. Nathanson, *Cofinite subsets of asymptotic bases for the positive integers*, J. Number Theory **20** (1985), no. 3, 363–372.

[21] M. B. Nathanson, *Sums of finite sets of integers*, Amer. Math. Monthly **79** (1972), 1010–1012.

[22] ______, *Minimal bases and maximal non-bases in additive number theory*, J. Number Theory **6** (1974), 324–333.

[23] ______, *s-maximal non-bases of density zero*, J. London Math. Soc. (2) **15** (1977), no. 1, 29–34. MR MR0435021 (55 #7983)

[24] ______, *Unique representation bases for the integers*, Acta Arith. **108** (2003), no. 1, 1–8.

[25] ______, *The inverse problem for representation functions of additive bases*, Number Theory (New York, 2003), Springer, New York, 2004, pp. 253–262.

[26] ______, *Every function is the representation function of an additive basis for the integers*, Port. Math. (N.S.) **62** (2005), no. 1, 55–72.

[27] ______, *The Caccetta-Häggkvist conjecture and additive number theory*, arXiv: math.CO/0603469, 2006.

[28] ______, *Sets with more sums than differences*, Integers **7** (2007), paper A5, 24 pp.

[29] ______, *Inverse problems for linear forms over finite sets of integers*, J. Ramanujan Math. Soc. **23** (2008), no. 2, 1–15.

[30] ______, *Problems in additive number theory, I*, Additive Combinatorics, CRM Proc. Lecture Notes, vol. 43, Amer. Math. Soc., Providence, RI, 2007, pp. 263–270.

[31] ———, *Problems in additive number theory, II: Linear forms and complementing sets of integers*, J. Théor. Nombres Bordeaux, to appear.

[32] ———, *Supersequences, rearrangements of sequences, and the spectrum of bases in additive number theory*, arXiv preprint 0806.0984, 2008.

[33] M. B. Nathanson, K. O'Bryant, B. Orosz, I. Ruzsa, and M. Silva, *Binary linear forms over finite sets of integers*, Acta Arith. **129** (2007), 341–361.

[34] D. J. Newman, *Tesselation of integers*, J. Number Theory **9** (1977), no. 1, 107–111.

[35] I. Z. Ruzsa, *Appendix in R. Tijdeman, "Periodicity and almost-periodicity"*, 2006.

[36] R. Tijdeman, *Periodicity and Almost-Periodicity*, More Sets, Graphs and Numbers, Bolyai Soc. Math. Stud., vol. 15, Springer, Berlin, 2006, pp. 381–405.

Melvin B. Nathanson
Department of Mathematics
Lehman College (CUNY)
Bronx, NY 10468
USA
and
CUNY Graduate Center
New York, NY 10016
USA
e-mail:`melvyn.nathanson@lehman.cuny.edu`

Chapter 7

Incidences and the spectra of graphs[1]

Jozsef Solymosi

In this contribution we give incidence bounds for arrangements of curves in $\mathbb{F}_q^2$. As an application, we prove a new result that, if $(x, f(x))$ is a Sidon set, then either $A + A$ or $f(A) + f(A)$ should be large. The main goal of the paper is to illustrate the use of graph spectral techniques in additive combinatorics. This is an extended version of the talks I gave in the Additive Combinatorics DocCourse held at the CRM in Barcelona and at the conference "Fete of Combinatorics" held in Keszthely.

7.1 Introduction

The main goal of this contribution is to illustrate the use of graph spectral techniques in additive combinatorics. The problem of finding nontrivial incidence bounds on lines and curves in $\mathbb{F}_q^2$ is closely related to sum-product estimates. In the first section we will prove Garaev's sum-product bound [14] using combinatorial arguments. Such techniques were used in similar context by Vu [27] and by Vinh [26]. Vu gave incidence bounds on polynomial curves and Vinh reproved Garaev's result, an improvement on the Bourgain–Katz–Tao incidence bound for large (larger than q) sets of points and lines in $\mathbb{F}_q^2$.

In Section 7.3 we sketch a spectral proof for Roth's theorem, that every dense set of integers contains three-term arithmetic progressions. There are several ex-

[1]The research was conducted while the author was a member of the Institute for Advanced Study. Funding provided by The Charles Simonyi Endowment. The research was supported by NSERC and OTKA grants and by Sloan Research Fellowship.

amples where one can choose between the Fourier method or a proof based on eigenvalues. A classical example is a discrepancy theorem for arithmetic progressions by Roth [23], who used the Fourier transform. Later, Lovász and Sós proved the theorem using eigenvalues (see [3] or [8] on p. 20).

In the last sections we present new results. We partially answer a question of Bourgain [7], giving incidence bounds similar to Garaev's, but for a more general family of curves. It is a finite field extension of a theorem of Elekes, Nathanson and Ruzsa. Applying Elekes' incidence method [11], Elekes, Nathanson and Ruzsa proved in [13] the following. Let $f\colon \mathbb{R} \to \mathbb{R}$ be a convex function. Then for any finite set $A \subset \mathbb{R}$,

$$\max\{|A+A|, |f(A)+f(A)|\} \geq c|A|^{5/4}. \tag{7.1.1}$$

In the inequality $A+A$ denotes the set of pairwise sums, $A+A = \{a+b : a, b \in A\}$ and $f(A) = \{f(a) : a \in A\}$. We do not have the notion of a convex function in $\mathbb{F}_q$, so we will use a weaker condition on f to get results in $\mathbb{F}_q$ similar to (7.1.1).

7.2 The sum-product problem

An old conjecture of Erdős and Szemerédi states that if A is a finite set of integers, then the sumset or the productset should be large. The sumset of A was defined earlier and the productset is defined in a similar way,

$$A \cdot A = \{ab | a,\, b \in A\}.$$

Erdős and Szemerédi conjectured that the sumset or the productset is almost quadratic in the size of A, i.e.,

$$\max(|A+A|, |A \cdot A|) \geq c|A|^{2-\delta}$$

for any positive δ.

Bourgain, Katz and Tao [6] proved a nontrivial, $|A|^{1+\varepsilon}$, lower bound for the finite field case. Let $A \subset \mathbb{F}_p$ and $p^{\alpha} \leq |A| \leq p^{1-\alpha}$. Then there is an $\varepsilon > 0$ depending on α only, such that

$$\max(|A+A|, |A \cdot A|) \geq c|A|^{1+\varepsilon}.$$

It is important that p is prime, otherwise one could select A being a subring in which case both the product set and the sumset are small, equal to $|A|$. For the case, $\mathbb{F}_q$, where q is a power of an odd prime, the best known bound is due to Garaev [14]. It follows from a construction of Ruzsa, that his bound is asymptotically the best possible in the range $|A| \geq q^{2/3}$. Garaev's proof uses bounds on exponential sums. We are going to derive similar sum-product estimates using spectral bounds for graphs.

Sum-product bounds have important applications, not only to number theory, but to computer science, Ramsey theory and cryptography.

7.2.1 The sum-product graph

The vertex set of the sum-product graph G_{SP} is the Cartesian product of the multiplicative subgroup and the field, $V(G_{SP}) = \mathbb{F}_q^* \times \mathbb{F}_q$ (as before, q is a power of an odd prime). Two vertices, $u = (a, b)$ and $v = (c, d) \in V(G_{SP})$, are connected by and edge, $(u, v) \in E(G_{SP})$, if and only if $ac = b + d$. This multigraph (there are a few loops) has a very special structure which makes it easy to compute the second largest eigenvalue of the graph. The set of eigenvalues are given by the eigenvalues of the adjacency matrix of the graph. The matrix is symmetric, so all $q(q-1)$ eigenvalues are real, we can order them, $\mu_0 \geq \mu_1 \geq \cdots \geq \mu_{q^2-q-1}$. The second largest eigenvalue, λ, is defined as $\lambda = \max(\mu_1, |\mu_{q^2-q-1}|)$. Using λ, one can write isoperimetric inequalities on the graph. In order to do so, we give a bound on λ. First, observe that for any two vertices, $u = (a, b)$ and $v = (c, d) \in V(G_{SP})$, if $a \neq c$ and $b \neq d$, then the vertices have exactly one common neighbor, $N(u, v) = (x, y) \in V(G_{SP})$.

The unique solution of the system

$$\left.\begin{aligned} ax &= b + y \\ cx &= d + y \end{aligned}\right\} \quad \text{is given by} \quad \begin{aligned} x &= (b-d)(a-c)^{-1} \\ 2y &= x(a+c) - b - d. \end{aligned} \tag{7.2.1}$$

If $a = c$ or $b = d$, then the vertices, u, v, have no common neighbors. Let M denote the adjacency matrix of G_{SP}, that is $a_{ij} = 1$ if $(v_i, v_j) \in E(G_{SP})$, and $a_{ij} = 0$ otherwise. M is a symmetric matrix, moreover

$$M^2 = J + (q-2)I - E,$$

where J is the all-one matrix, I is the identity matrix and E is the "error matrix", the adjacency matrix of the graph, G_E, where for any two vertices, $v_i = (a, b)$ and $v_j = (c, d) \in V(G_{SP})$, $(v_i, v_j) \in E(G_E)$ if and only if $a = c$ or $b = d$. As G_{SP} is a $(q-1)$-regular graph, $q-1$ is an eigenvalue of M with the all-one eigenvector, $\overrightarrow{1}$. The matrix M is symmetric, so that eigenvectors of other eigenvalues are orthogonal to $\overrightarrow{1}$. It is a corollary of the Spectral Theorem that there is an orthonormal basis, V, consisting of eigenvectors of M. Let θ denote the second largest eigenvalue of M. The graph, G_{SP}, is connected so the eigenvalue $q-1$ has multiplicity one, and the graph is not bipartite, so for any other eigenvalue, θ, $|\theta| < q-1$. A corresponding eigenvector is denoted by $\overrightarrow{v_\theta}$. Let us multiply both sides of the matrix equation above by $\overrightarrow{v_\theta}$. The "trick" is that $J\overrightarrow{v_\theta} = 0$, as the eigenvectors are orthogonal to the all-one vector, so we get

$$(\theta^2 - q + 2)\overrightarrow{v_\theta} = E\overrightarrow{v_\theta}.$$

Note that E has the same set of eigenvectors as M has. G_E is a $2q-3$-regular graph, so any eigenvalue of E is at most $2q-3$ in absolute value:

$$\theta^2 - q + 2 \leq 2q,$$

$$|\theta| < \sqrt{3q}.$$

For illustration we show the matrices M, M^2 and E for the case when $q = 5$:

$$M = \begin{bmatrix}
0 & 1 & 0 & 0 & 0 & 0 & 0 & 1 & 0 & 0 & 0 & 0 & 0 & 1 & 0 & 0 & 0 & 0 & 0 & 1 \\
1 & 0 & 0 & 0 & 0 & 0 & 1 & 0 & 0 & 0 & 0 & 0 & 1 & 0 & 0 & 0 & 0 & 0 & 1 & 0 \\
0 & 0 & 0 & 0 & 1 & 1 & 0 & 0 & 0 & 0 & 0 & 1 & 0 & 0 & 0 & 0 & 0 & 1 & 0 & 0 \\
0 & 0 & 0 & 1 & 0 & 0 & 0 & 0 & 0 & 1 & 1 & 0 & 0 & 0 & 0 & 0 & 1 & 0 & 0 & 0 \\
0 & 0 & 1 & 0 & 0 & 0 & 0 & 0 & 1 & 0 & 0 & 0 & 0 & 0 & 1 & 1 & 0 & 0 & 0 & 0 \\
0 & 0 & 1 & 0 & 0 & 0 & 0 & 0 & 0 & 1 & 0 & 1 & 0 & 0 & 0 & 0 & 0 & 0 & 1 & 0 \\
0 & 1 & 0 & 0 & 0 & 0 & 0 & 0 & 1 & 0 & 1 & 0 & 0 & 0 & 0 & 0 & 0 & 1 & 0 & 0 \\
1 & 0 & 0 & 0 & 0 & 0 & 0 & 1 & 0 & 0 & 0 & 0 & 0 & 0 & 1 & 0 & 1 & 0 & 0 & 0 \\
0 & 0 & 0 & 0 & 1 & 0 & 1 & 0 & 0 & 0 & 0 & 0 & 0 & 1 & 0 & 1 & 0 & 0 & 0 & 0 \\
0 & 0 & 0 & 1 & 0 & 1 & 0 & 0 & 0 & 0 & 0 & 0 & 1 & 0 & 0 & 0 & 0 & 0 & 0 & 1 \\
0 & 0 & 0 & 1 & 0 & 0 & 1 & 0 & 0 & 0 & 0 & 0 & 0 & 0 & 1 & 0 & 0 & 1 & 0 & 0 \\
0 & 0 & 1 & 0 & 0 & 1 & 0 & 0 & 0 & 0 & 0 & 0 & 0 & 1 & 0 & 0 & 1 & 0 & 0 & 0 \\
0 & 1 & 0 & 0 & 0 & 0 & 0 & 0 & 0 & 1 & 0 & 0 & 1 & 0 & 0 & 1 & 0 & 0 & 0 & 0 \\
1 & 0 & 0 & 0 & 0 & 0 & 0 & 0 & 1 & 0 & 0 & 1 & 0 & 0 & 0 & 0 & 0 & 0 & 0 & 1 \\
0 & 0 & 0 & 0 & 1 & 0 & 0 & 1 & 0 & 0 & 1 & 0 & 0 & 0 & 0 & 0 & 0 & 0 & 1 & 0 \\
0 & 0 & 0 & 0 & 1 & 0 & 0 & 0 & 1 & 0 & 0 & 0 & 1 & 0 & 0 & 0 & 1 & 0 & 0 & 0 \\
0 & 0 & 0 & 1 & 0 & 0 & 0 & 1 & 0 & 0 & 0 & 1 & 0 & 0 & 0 & 1 & 0 & 0 & 0 & 0 \\
0 & 0 & 1 & 0 & 0 & 0 & 1 & 0 & 0 & 0 & 1 & 0 & 0 & 0 & 0 & 0 & 0 & 0 & 0 & 1 \\
0 & 1 & 0 & 0 & 0 & 1 & 0 & 0 & 0 & 0 & 0 & 0 & 0 & 0 & 1 & 0 & 0 & 0 & 1 & 0 \\
1 & 0 & 0 & 0 & 0 & 0 & 0 & 0 & 0 & 1 & 0 & 0 & 0 & 1 & 0 & 0 & 0 & 1 & 0 & 0
\end{bmatrix}$$

$$M^2 = \begin{bmatrix}
4 & 0 & 0 & 0 & 0 & 0 & 1 & 1 & 1 & 1 & 0 & 1 & 1 & 1 & 1 & 0 & 1 & 1 & 1 & 1 \\
0 & 4 & 0 & 0 & 0 & 1 & 0 & 1 & 1 & 1 & 1 & 0 & 1 & 1 & 1 & 1 & 0 & 1 & 1 & 1 \\
0 & 0 & 4 & 0 & 0 & 1 & 1 & 0 & 1 & 1 & 1 & 1 & 0 & 1 & 1 & 1 & 1 & 0 & 1 & 1 \\
0 & 0 & 0 & 4 & 0 & 1 & 1 & 1 & 0 & 1 & 1 & 1 & 1 & 0 & 1 & 1 & 1 & 1 & 0 & 1 \\
0 & 0 & 0 & 0 & 4 & 1 & 1 & 1 & 1 & 0 & 1 & 1 & 1 & 1 & 0 & 1 & 1 & 1 & 1 & 0 \\
0 & 1 & 1 & 1 & 1 & 4 & 0 & 0 & 0 & 0 & 0 & 1 & 1 & 1 & 1 & 0 & 1 & 1 & 1 & 1 \\
1 & 0 & 1 & 1 & 1 & 0 & 4 & 0 & 0 & 0 & 1 & 0 & 1 & 1 & 1 & 1 & 0 & 1 & 1 & 1 \\
1 & 1 & 0 & 1 & 1 & 0 & 0 & 4 & 0 & 0 & 1 & 1 & 0 & 1 & 1 & 1 & 1 & 0 & 1 & 1 \\
1 & 1 & 1 & 0 & 1 & 0 & 0 & 0 & 4 & 0 & 1 & 1 & 1 & 0 & 1 & 1 & 1 & 1 & 0 & 1 \\
1 & 1 & 1 & 1 & 0 & 0 & 0 & 0 & 0 & 4 & 1 & 1 & 1 & 1 & 0 & 1 & 1 & 1 & 1 & 0 \\
0 & 1 & 1 & 1 & 1 & 0 & 1 & 1 & 1 & 1 & 4 & 0 & 0 & 0 & 0 & 0 & 1 & 1 & 1 & 1 \\
1 & 0 & 1 & 1 & 1 & 1 & 0 & 1 & 1 & 1 & 0 & 4 & 0 & 0 & 0 & 1 & 0 & 1 & 1 & 1 \\
1 & 1 & 0 & 1 & 1 & 1 & 1 & 0 & 1 & 1 & 0 & 0 & 4 & 0 & 0 & 1 & 1 & 0 & 1 & 1 \\
1 & 1 & 1 & 0 & 1 & 1 & 1 & 1 & 0 & 1 & 0 & 0 & 0 & 4 & 0 & 1 & 1 & 1 & 0 & 1 \\
1 & 1 & 1 & 1 & 0 & 1 & 1 & 1 & 1 & 0 & 0 & 0 & 0 & 0 & 4 & 1 & 1 & 1 & 1 & 0 \\
0 & 1 & 1 & 1 & 1 & 0 & 1 & 1 & 1 & 1 & 0 & 1 & 1 & 1 & 1 & 4 & 0 & 0 & 0 & 0 \\
1 & 0 & 1 & 1 & 1 & 1 & 0 & 1 & 1 & 1 & 1 & 0 & 1 & 1 & 1 & 0 & 4 & 0 & 0 & 0 \\
1 & 1 & 0 & 1 & 1 & 1 & 1 & 0 & 1 & 1 & 1 & 1 & 0 & 1 & 1 & 0 & 0 & 4 & 0 & 0 \\
1 & 1 & 1 & 0 & 1 & 1 & 1 & 1 & 0 & 1 & 1 & 1 & 1 & 0 & 1 & 0 & 0 & 0 & 4 & 0 \\
1 & 1 & 1 & 1 & 0 & 1 & 1 & 1 & 1 & 0 & 1 & 1 & 1 & 1 & 0 & 0 & 0 & 0 & 0 & 4
\end{bmatrix}$$

$$E = \begin{bmatrix}
0&1&1&1&1&1&0&0&0&0&1&0&0&0&0&1&0&0&0&0\\
1&0&1&1&1&0&1&0&0&0&0&1&0&0&0&0&1&0&0&0\\
1&1&0&1&1&0&0&1&0&0&0&0&1&0&0&0&0&1&0&0\\
1&1&1&0&1&0&0&0&1&0&0&0&0&1&0&0&0&0&1&0\\
1&1&1&1&0&0&0&0&0&1&0&0&0&0&1&0&0&0&0&1\\
1&0&0&0&0&0&1&1&1&1&1&0&0&0&0&1&0&0&0&0\\
0&1&0&0&0&1&0&1&1&1&0&1&0&0&0&0&1&0&0&0\\
0&0&1&0&0&1&1&0&1&1&0&0&1&0&0&0&0&1&0&0\\
0&0&0&1&0&1&1&1&0&1&0&0&0&1&0&0&0&0&1&0\\
0&0&0&0&1&1&1&1&1&0&0&0&0&0&1&0&0&0&0&1\\
1&0&0&0&0&1&0&0&0&0&0&1&1&1&1&1&0&0&0&0\\
0&1&0&0&0&0&1&0&0&0&1&0&1&1&1&0&1&0&0&0\\
0&0&1&0&0&0&0&1&0&0&1&1&0&1&1&0&0&1&0&0\\
0&0&0&1&0&0&0&0&1&0&1&1&1&0&1&0&0&0&1&0\\
0&0&0&0&1&0&0&0&0&1&1&1&1&1&0&0&0&0&0&1\\
1&0&0&0&0&1&0&0&0&0&1&0&0&0&0&0&1&1&1&1\\
0&1&0&0&0&0&1&0&0&0&0&1&0&0&0&1&0&1&1&1\\
0&0&1&0&0&0&0&1&0&0&0&0&1&0&0&1&1&0&1&1\\
0&0&0&1&0&0&0&0&1&0&0&0&0&1&0&1&1&1&0&1\\
0&0&0&0&1&0&0&0&0&1&0&0&0&0&1&1&1&1&1&0
\end{bmatrix}$$

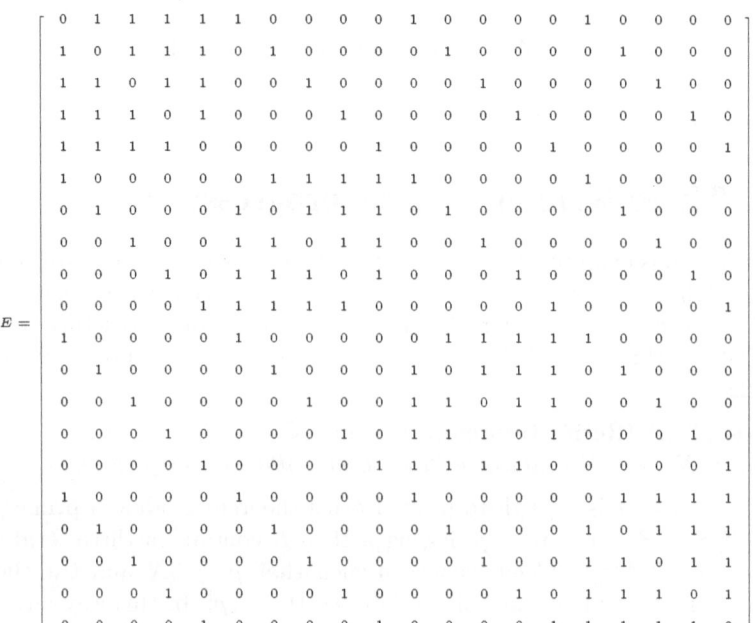

7.2.2 The spectral bound

The small value of the second largest eigenvalue shows us that G_{SP} is a quasi-random graph and we can bound the number of edges between large vertex sets efficiently. We are going to use the following Cheeger-type discrepancy bound. For any two sets of vertices, $S, T \subset V(G_{SP})$,

$$\left| e(S,T) - \frac{|S||T|}{q} \right| \leq \lambda\sqrt{|S||T|}, \tag{7.2.2}$$

where $e(S,T)$ is the number of edges between S and T. (See, e.g., [10] or [2].) Inequality (7.2.2) and the bound on λ imply that

$$e(S,T) \leq \frac{|S||T|}{q} + \sqrt{3q|S||T|}. \tag{7.2.3}$$

From (7.2.3) we can deduct Garaev's sum-product bound [14]. We can suppose that $0 \notin A$, without loss of generality. Set $S = (AA) \times (-A)$ and $T = (A^{-1}) \times (A+A)$. There is an edge between any two vertices $(ab, -c) \in S$ and $(b^{-1}, a+c) \in T$, therefore the number of edges between S and T is at least $|A|^3$. On the other hand,

$$|A|^3 \leq e(S,T) \leq \frac{|S||T|}{q} + \sqrt{3q|S||T|} = \frac{|AA||A+A||A|^2}{q} + \sqrt{3q|AA||A+A||A|^2}.$$

After rearranging the inequality we get the desired sum-product bound:

$$|A+A||AA| \gg \min\left\{q|A|, \frac{|A|^4}{q}\right\}.$$

In particular, if $|A| \approx q^{2/3}$, then $\max\{|AA|, |A+A|\} \gg |A|^{5/4}$.

7.3 Three-term arithmetic progressions

In the previous example it was enough to show that the second largest eigenvalue is small. There are cases where we cannot guarantee that the second eigenvalue is small; however when it is large then we might find some structure in the graph. To illustrate this, we will sketch one of the several possible proofs of Roth's theorem [22].

Theorem 7.3.1 (Roth's theorem). *For any $N \geq 3$, if $S \subset [1,\ldots,N]$ and $|S| \gg N/\log\log N$, then S contains a three-term arithmetic progression.*

Note that it is enough to prove Roth's theorem modulo a prime p. For any $p \geq 3$ if $S \subset \mathbb{F}_p$ and $|S| \gg p/\log\log p$, then S contains a three-term arithmetic progression (3-AP). Indeed, choose p such that $p \geq 3N$ and translate S such that it is in the middle third of the interval $[1,\ldots,p]$. In this way any arithmetic progression modulo q is also a "regular" arithmetic progression.

7.3.1 The 3-AP graph

To prove the "mod p" variant, we define a graph, G_{3AP}, on $2p-1$ vertices. We label the vertices by $v_0, v_1, \ldots, v_{p-1}$, and $v_{-1}, v_{-2}, \ldots, v_{-p+1}$. A way to think of the vertices if they were the $(2p-1)$th roots of unity, assigning v_j to $\exp(\frac{2\pi\imath}{2p-1}j)$. The neighbors of v_0 are defined by the set S in the following way: v_i is connected to v_0 by an edge if and only if $|i| \in S$. (Suppose that $0 \notin S$.) Extend the graph by adding the edges necessary that the mapping, $i \mapsto i+1 \pmod{2p-1}$, is an automorphism of G_{3AP}. Using the roots of unity notation, it means that multiplying the vertices by $\exp(\frac{2\pi\imath}{2p-1}j)$ is an automorphism of the graph for any integer j. (It is the Cayley graph of $\mathbb{Z}/(2p-1)\mathbb{Z}$ with respect to S.)

For graphs with a "nice" automorphism group, finding the eigenvectors and eigenvalues is not a hard task. (See Exercise 8 in [18], Chapter 16 in [4], or [19] for a more detailed description.) In our case it is easy to check that for this circulant graph, $2p-1$ linearly independent eigenvectors are given by the vectors

$$\left[\exp\left(\frac{2\pi\imath k}{2p-1}\right), \exp\left(\frac{4\pi\imath k}{2p-1}\right), \exp\left(\frac{6\pi\imath k}{2p-1}\right), \ldots, \exp\left(\frac{2(2p-1)\pi\imath k}{2p-1}\right)\right]^T,$$

where $0 \leq k \leq 2p-2$. Then the eigenvalues of G_{3AP} are given by the sums

$$\theta_k = \sum_{s\in S} \exp\left(\frac{2\pi\imath sk}{n}\right) + \sum_{s\in S} \exp\left(\frac{-2\pi\imath sk}{n}\right).$$

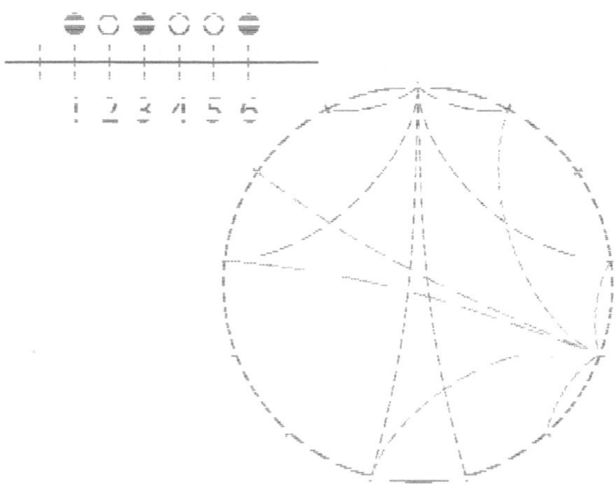

Figure 7.1: A partial drawing of G_{3-AP} for the set shown.

There are two possibilities. Either the second largest eigenvalue is large or all eigenvalues but the largest, $\mu_0 = 2|S|$, are small. In the former case, most of the summands have large positive real part. It implies that there is a long arithmetic progression having a very large intersection with S. We will not explore this case here, instead we show that if all eigenvalues are small, then there is a three-term arithmetic progression in S. The interested reader will find the details for the density increment case in Roth's original paper [22], or in one of the many books discussing Roth's theorem, like [15], [25], or [16]. Our moderate plan here is to show that if $|S|^2/(2p-1) > \lambda$, then S contains a 3-AP.

We can find a relation between the assumption that S has no three-term arithmetic progressions (it is *3-AP-free*) and the structure of the graph G_{3-AP}. We show that if S is 3-AP-free, then there are large vertex sets spanning less than expected edges. For every edge we can define its halving point. Consider the edges as arcs between points on the unit circle. The points are the vertices, represented by the roots of unity and the edges are the shorter circular arcs. The halving point of the edge is the geometric halving point of the circular arc. The number of possible halving points is $4p-2$. The number of edges is $|S|(2p-1)$, so there is a point which is the halving point of at least $\lceil |S|/2 \rceil$ edges. Note that if we had two edges sharing the same halving point, such that there is another edge between the two-two endvertices separated by the halving point, that would imply that there is a 3-AP in S. (See Figure 7.2.)

If S is 3-AP-free, then between the two $\lceil |S|/2 \rceil$-size sets of end vertices, A and B, there are exactly $\lceil |S|/2 \rceil$ edges. Inequality (7.2.2) implies that

$$\left| e(A,B) - \frac{2|S| \lceil |S|/2 \rceil^2}{2p-1} \right| \leq \lambda l \lceil |S|/2 \rceil ,$$

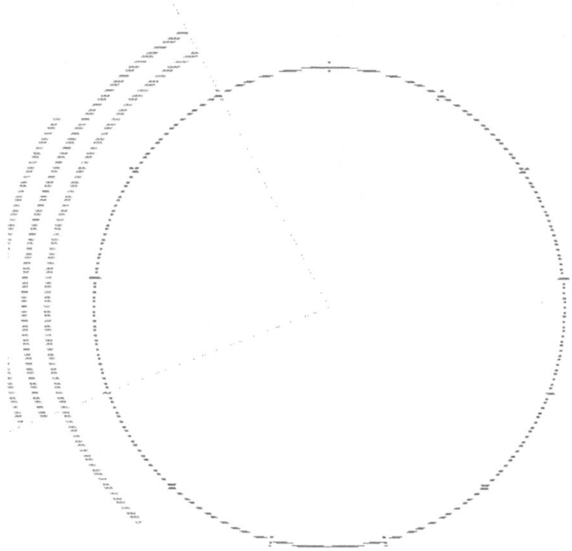

Figure 7.2: If there are two edges sharing the same halving point, h, then the endpoints of the edges can be written as $h+d_1, h-d_1$ and $h+d_2, h-d_2$. If $h+d_2$ and $h-d_1$ are connected by an edge, it means that d_1+d_2 is in S with $2d_1$ and $2d_2$, forming a 3-AP.

from where we get that

$$\frac{|S|^2}{2p-1} \leq \lambda,$$

as we wanted to show.

7.4 Sidon functions

In this section we extend a result of Elekes, Nathanson and Ruzsa [13] to the finite field case.

Theorem 7.4.1 (Elekes, Nathanson and Ruzsa). *Let $f\colon \mathbb{R} \to \mathbb{R}$ be a convex function. Then for any finite set $A \subset \mathbb{R}$,*

$$\max\{|A+A|, |f(A)+f(A)|\} \geq c|A|^{5/4}.$$

7.4.1 Sidon functions

We need a notation which substitutes convexity in finite fields. The graph of a convex function is a Sidon set in $\mathbb{R}^2$, this is the property we are going to use for finite fields. A set $H \subset \mathbb{F}_q \times \mathbb{F}_q$ is a *Sidon set* if for any $h_i, h_j, h_k, h_l \in H$, the equation

$$h_i - h_j \equiv h_k - h_l \pmod{q}$$

implies $i = k$ and $j = l$. A function $f\colon S \to \mathbb{F}_q$, for some $S \subset \mathbb{F}_q$, is said to be a *Sidon function* if its graph $H = \{(x, f(x)) : x \in S\}$ is a Sidon set. Note that the graph of any convex function in $\mathbb{R}^2$ forms a Sidon set.

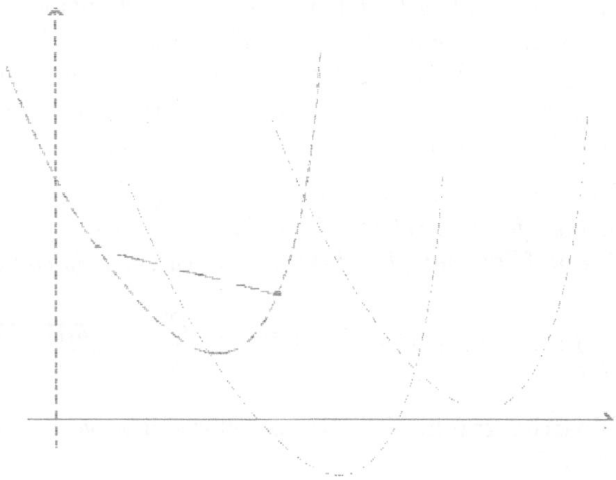

Figure 7.3: The graph of any convex function in $\mathbb{R}^2$ forms a Sidon set.

Theorem 7.4.2. *For any integer, k, and for any $S \subset \mathbb{F}_q, |S| \geq q - k$, if $f\colon S \to \mathbb{F}_q$ is a Sidon function, then for any set $A \subset S$, and sets $B, C \subset \mathbb{F}_q$,*

$$|A + B||f(A) + C| \geq \min\left\{\frac{q|A|}{2}, \frac{|A|^2|B||C|}{8(k+1)q}\right\}.$$

Using the right substitution for C and D, Theorem 7.4.2 gives the following corollaries.

Corollary 7.4.3. *For any integer, k, there is a constant, $c = c(k)$, such that for any $S \subset \mathbb{F}_q, |S| \geq q - k$, if $f\colon S \to \mathbb{F}_q$ is a Sidon function, then for any set $A \subset S$,*

$$|A + A||f(A) + f(A)| \geq c \min\left\{q|A|, \frac{|A|^4}{q}\right\}.$$

It is remarkable that the inequality above matches the Elekes–Nathanson–Ruzsa bound for sets A such that $|A| \approx q^{2/3}$. It has a single-term variant, which we state in a separate statement.

Corollary 7.4.4. *For any integer, k, there is a constant, $c = c(k)$, such that for any $S \subset \mathbb{F}_q, |S| \geq q-k$, if $f\colon S \to \mathbb{F}_q$ is a Sidon function, then for any set $A \subset S$,*

$$|A + f(A)| \geq c \min\left\{\sqrt{q|A|}, \frac{|A|^2}{\sqrt{q}}\right\}.$$

7.4.2 A bipartite incidence graph

The proof of Theorem 7.4.2 is based on the following incidence bound. Let f be a function, $S \to \mathbb{F}_q$, for some $S \subset \mathbb{F}_q$. The *graph of f* is the set of points $\{(x, f(x)) \in \mathbb{F}_q \times \mathbb{F}_q : x \in S\}$. A translate of f by a vector $u = (u', u'') \in \mathbb{F}_q \times \mathbb{F}_q$ is the set $T_u(f) = \{(x + u', f(x) + u'') : x \in S\}$. The translate of the *mirror graph of f* is defined as $T_u(f)^\tau = \{(u' - x, u'' - f(x)) : x \in S\}$.

Lemma 7.4.5. *For any integer, k, and for any $S \subset \mathbb{F}_q, |S| = q - k$ if $f\colon S \to \mathbb{F}_q$ is a Sidon function, then for any set $P \subset \mathbb{F}_q \times \mathbb{F}_q$, the number of incidences between P and s translates of f, the set $\{T_{u_i}(f)\}_{i=1}^s, u_i = (u_i', u_i'')$, is bounded as follows:*

$$\sum_{i=1}^{s} |\{x \in S : (x + u_i', f(x) + u_i'') \in P\}| \leq \frac{|P|s}{q} + \sqrt{2(k+1)q|P|s}.$$

Proof. Define a bipartite graph, $G(A, B)$, as follows. The vertex set of G consists of two copies of $\mathbb{F}_q \times \mathbb{F}_q$.

The edges of $G(A, B)$ are given by the graph of f. Two vertices, $u = (u', u'') \in A = \mathbb{F}_q \times \mathbb{F}_q$ and $v = (v', v'') \in B = \mathbb{F}_q \times \mathbb{F}_q$, are connected by an edge in G if

$$f(v' - u') = v'' - u''. \tag{7.4.1}$$

The neighborhood of a vertex $u \in A$ is given by $N(u) = T_u(f) \subset B$, and neighborhood of a vertex $u \in B$ is described by $N(v) = T_v(f)^\tau \subset A$. The graph, $G(A, B)$, is a $(q-k)$-regular bipartite graph. The spectra of $G(A, B)$ is symmetric. For this graph the second largest eigenvalue is defined as $\lambda = \mu_1$. As the graph is $(q-k)$-regular, the largest and the smallest eigenvalues are $q - k$ and $k - q$. Similarly as we did in the sum-product example, we can bound λ by examining the $q^2 \times q^2$ adjacency matrix of $G(A, B)$, denoted by M. The function f is a Sidon function, therefore the neighborhoods of two vertices in A or in B intersect in at most one vertex. A translate, $T_u(f)$, covers $\binom{q-k}{2}$ vertex pairs. All translates (the neighborhoods of vertices) cover $2\binom{q-k}{2}q^2$ pairs out of the $2\binom{q^2}{2}$ vertex pairs in A and in B. Let us define an error graph, H, which has two components, one in A and one in B, and two vertices, u and v are connected by an edge if and only if there is no vertex connected to both in $G(A, B)$. The error graph, H, has $2(\binom{q^2}{2} - q^2\binom{q-k}{2})$ edges and it is regular of degree $q^2 - 1 - (q-k)(q-k-1)$. Its adjacency matrix is denoted by E,

$$M^2 = \begin{bmatrix} J & 0 \\ 0 & J \end{bmatrix} + (q - k - 1)I - E.$$

As in the first example, we can multiply the equation by an eigenvector of M, which belongs to the second largest eigenvalue,

$$E\overrightarrow{v_\lambda} = (q-k-1-\lambda^2)\overrightarrow{v_\lambda}.$$

We know that H is $(2kq+q-k^2-k-1)$-regular, therefore any eigenvalue of E is less or equal to $2kq+q-k^2-k-1$,

$$|q-k-1-\lambda^2| \le 2kq+q-k^2-k-1,$$

so

$$\lambda < \sqrt{2(k+1)q}.$$

7.4.3 The spectral bound

For bipartite graphs, like $G(A,B)$, inequality (7.2.2) is slightly different. If $G(A,B)$ is an r-regular bipartite graph on n vertices, then for any subsets $A' \subset A$ and $B' \subset B$,

$$\left| e(A',B') - \frac{2r|A'||B'|}{n} \right| \le \lambda\sqrt{|A'||B'|}.$$

Now we are ready to complete the proof of Lemma 7.4.5 to state a bound on incidences between a set of points, P, and s translates, $\{T_{u_i}(f)\}_{i=1}^{s}, u_i = (u_i', u_i'')$. An edge in the graph $G(A,B)$ represents an incidence between a point and a translate,

$$\sum_{i=1}^{s} |\{x \in S : (x+u_i', f(x)+u_i'') \in P\}| \le \frac{|P|s}{q} + \sqrt{2(k+1)q|P|s}. \qquad \square$$

Proof of Theorem 7.4.2. Let us consider the Cartesian product $(A+B) \times (f(A)+C)$. It has $|B||C|$ translates of the smaller product $A \times f(A)$, which contains an $|A|$-element subset of the graph of f. The $|B||C|$ translates determine $|A||B||C|$ incidences in $(A+B)\times(f(A)+C)$. Now we apply Lemma 7.4.5 with substitutions $s = |B||C|$, $|P| = |A+B||f(A)+C|$, and with $|A||B||C|$ incidences. $\square$

Note that Theorem 7.4.2 generalizes Garaev's point-line incidence bound, since the mapping $(x,y) \mapsto (x, y+x^2)$ maps any line, $ax+by=c, a \ne 0$, to a translate of the parabola, $y = x^2$, which is a Sidon function.

For any set $P \subset \mathbb{F}_q \times \mathbb{F}_q$, the number of incidences between P and s lines is bounded by

$$O\left(\frac{|P|s}{q} + \sqrt{q|P|s}\right). \qquad (7.4.2)$$

7.5 Incidence bounds on pseudolines

Incidence bounds in geometries have various applications. The celebrated theorem of Szemerédi and Trotter [24] gives sharp incidence bound for the number of point-line incidences in the Euclidean plane. The Szemerédi–Trotter theorem was extended to pseudolines. For the details about variants of the planar Szemerédi–Trotter theorem we refer the reader to [21].

Our goal here is finding nontrivial incidence bounds for pseudolines in $\mathbb{F}_q^2$. First we give a definition of pseudolines which form a partial geometry in $\mathbb{F}_q^2$. The incidence graph will be a strongly regular graph, therefore we can use standard spectral bounds to estimate incidences.

7.5.1 The incidence bound

The following is a standard (however not the only) definition of pseudolines in the Euclidean plane, see, e.g., [1].

A collection $\mathcal{L}$ of x-monotone unbounded Jordan curves in the plane is called a family of pseudolines if every pair of curves intersects in at most one point.

To find a proper definition of pseudolines in finite fields is not so straightforward. We are going to use one possible definition which has interesting applications. Instead of x-monotone unbounded Jordan curves we consider "lines", $l_i = \{(x, f(x)) : x \in \mathbb{F}_q\}$, where $f\colon \mathbb{F}_q \to \mathbb{F}_q$.

Definition 7.5.1. A collection $\mathcal{L}$ of subsets of $\mathbb{F}_q^2$, $\mathcal{L} = \{l_1, l_2, \ldots, l_k\}$ is called a family of pseudolines if the following conditions hold.

a, For every $a \in \mathbb{F}_q$, any set, l_i, has exactly one point with x-coordinate a.

b, Every pair of sets, l_i and l_j, intersects in at most one point.

c, If $l_i \in \mathcal{L}$, then its y-translates are also in the arrangement, $l_i + (0, a) \in \mathcal{L}$ for any $a \in \mathbb{F}_q$.

The last condition implies that the size of a family of pseudolines is divisible by q.

Theorem 7.5.2. *Let a family of pseudolines, $\mathcal{L}$, and a family of points, P, in $\mathbb{F}_q^2$ be given. Suppose that $|\mathcal{L}| = kq$, and $|P| = n$. Then the number of incidences between m pseudolines and n points is bounded by*

$$I(m, n) \leq n\sqrt{km/q} + \sqrt{qnm}.$$

The incidence bound for pseudolines implies a new bound on point line incidences. It is better than inequality (7.4.2) in line arrangements with a few distinct slopes only.

Corollary 7.5.3. *Let a family of pseudolines, $\mathcal{L}$, and a family of points, P, in $\mathbb{F}_q^2$ be given. Suppose that the lines have no more than k different slopes. Then the number of incidences between the s lines and points of P is bounded by*

$$I(s, |P|) \leq |P|\sqrt{ks/q} + \sqrt{q|P|s}.$$

The bound is better than inequality (7.4.2) if $k < s/q$. To see how Theorem 7.5.2 implies Corollary 7.5.3, observe that the set of all lines with slopes from a given set, forms a family of pseudolines.

The incidence bound in Theorem 7.5.2 is a corollary of the following statement which is proved in the next subsection.

Theorem 7.5.4. *Given a family of pseudolines, $\mathcal{L}$, and two sets of points, P_1 and P_2 in $\mathbb{F}_q^2$. Suppose that $|\mathcal{L}| = kq$, $|P_1| = n_1$ and $|P_2| = n_2$. Then the number of collinear pairs in $P_1 \times P_2$ is bounded as*

$$|\{(p_i, p_j) : p_i \in P_1, p_j \in P_2, \exists \ell \in \mathcal{L} : p_i, p_j \in \ell\}| \leq \frac{kn_1n_2}{q} + q\sqrt{n_1n_2}.$$

Proof of Theorem 7.5.2. Suppose that the m pseudodolines are incident to $t_1, t_2, \ldots, t_m$ points in P. Then, the number of copseudolinear pairs in P is at least $\sum_{i=1}^m \binom{t_i}{2}$. On the other hand, $I(m,n) = \sum_{i=1}^m t_i$, so the number of copseudolinear pairs is at least $m\binom{I(m,n)/m}{2} \sim I(m,n)^2/m$. Using the inequality from Theorem 7.5.4 we have

$$\frac{I(m,n)^2}{m} \leq \frac{kn^2}{q} + qn,$$

concluding the proof of Theorem 7.5.2. □

7.5.2 Strongly regular graphs

In [5] Bose introduced the notation of partial geometries. A set of points and lines is a finite partial geometry if there are integers such that

i. For each two different points p and q, there is at most one line incident with both of them.

ii. Each line is incident with $r + 1$ points.

iii. Each point is incident with $s + 1$ lines.

iv. If a point p and a line L are not incident, then there are exactly t points on L collinear to p.

Lemma 7.5.5. *Any family of pseudolines is a partial geometry.*

The easy proof is left to the reader.

Proof of Theorem 7.5.4. The incidence graph, $G(\mathcal{L})$, of a family of pseudolines is defined as follows. G has q^2 vertices, the elements of $\mathbb{F}_q^2$. Two vertices v and u are connected if and only if the points are collinear, i.e. there is a line, $l \in \mathcal{L}$, such that $v, u \in l$. As we observed earlier, the number of lines is divisible by q. There is an integer, k, $1 \leq k \leq q$, such that $|\mathcal{L}| = kq$.

$G(\mathcal{L})$ is a strongly regular graph, where each vertex has degree $k(q-1)$. Two collinear (adjacent) vertices have $q - 2 + (k-1)(k-2) = q + k^2 - 3k$ common neighbors and non-adjacent vertices have $k^2 - k$ common neighbors. The adjacency matrix of the graph is denoted by A,

$$A^2 = (q + k^2 - 3k)A + (k^2 - k)(J - A - I) + k(q-1)I.$$

The usual trick – multiplying both sides by an eigenvector – helps us to find the eigenvalues. The adjacency matrix of this graph has only three distinct eigenvalues. The largest is $k(q-1)$ and the other two are $q - k$ and $-k$. (For more details about such graphs we refer the reader to [17].) In our applications $q \gg k$, so the second largest eigenvalue is $q - k$. From this, Theorem 7.5.4 follows immediately by applying inequality (7.2.2). □

Bibliography

[1] P.K. Agarwal and M. Sharir, *Pseudoline Arrangements: Duality, Algorithms, and Applications*, manuscript, 2001.

[2] N. Alon and J. Spencer, *The Probabilistic Method*, 3rd. ed., Wiley, 2008.

[3] J. Beck and V.T. Sós, *Discrepancy theory*, Handbook of Combinatorics, Chapter 26, Graham, Grötschel, and Lovász, eds., North–Holland, Amsterdam, 1995, pp. 1405–1446.

[4] N. Biggs, *Algebraic Graph Theory*, 2nd ed., Cambridge University Press, 1993.

[5] R.C. Bose, *Strongly regular graphs, partial geometries and partially balanced designs*, Pacific J. Math. **13** (1963), no. 2, 389–419.

[6] J. Bourgain, N. Katz, and T. Tao, *A sumproduct estimate in finite fields and applications*, Geom. Funct. Anal. **14** (2004), 27--57.

[7] J. Bourgain, personal communication, October 10, 2007.

[8] B. Chazelle, *The Discrepancy Method: Randomness and Complexity*, Cambridge University Press, 2000.

[9] J. Cheeger, *A lower bound for the smallest eigenvalue of the Laplacian*, Problems in Analysis, Papers dedicated to Salomon Bochner, Princeton University Press, Princeton, 1969, 195–199.

[10] F.R.K. Chung, *Spectral Graph Theory*, Amer. Math. Soc., Providence, RI, 1997.

[11] Gy. Elekes, *On the Number of Sums and Products*, Acta Arith. **81** (1997), no. 4, 365–367.

[12] P. Erdős and E. Szemerédi, *On sums and products of integers*, Studies in Pure Mathematics, Birkhäuser, Basel, 1983, pp. 213–218.

[13] Gy. Elekes, M. B. Nathanson, and I. Z. Ruzsa, *Convexity and Sumsets*, J. Number Theory **83** (1999), 194–201.

[14] M.Z. Garaev, *An Explicit Sum-Product Estimate in* $\mathbb{F}_q$, Int. Math. Res. Not. **2007** (2007), 1–11.

[15] R. Graham, B. Rothschild, and J. Spencer, *Ramsey Theory*, Wiley-Intersc. Ser. in Discrete Math. Optim., Wiley, New York, 1990.

[16] A. Granville, M.B. Nathanson, and J. Solymosi, *Additive Combinatorics*, CRM Proceedings & Lecture Notes, vol. 43, Amer. Math. Soc., 2007.

[17] C. Godsil and G. Royle, *Algebraic Graph Theory*, Grad. Texts in Math., vol. 207, Springer, New York, 2001.

[18] L. Lovász, *Combinatorial Problems and Exercises*, AMS Chelsea Publishing, 1979.

[19] ______, *Spectra of graphs with transitive groups*, Period. Math. Hungar. **6** (1975), no. 2, 191–195.

[20] L. Lovász, J. Spencer, and K. Vesztergombi, *Discrepancy of set systems and matrices*, European J. Combin. **7** (1986), 151–160.

[21] J. Pach and P. Agarwal, *Combinatorial Geometry*, Wiley, New York, 1995.

[22] K.F. Roth, *On certain sets of integers, I,* J. London Math. Soc. **28** (1953), 104–109.

[23] ______, *Remark concerning integer sequences*, Acta Arith. **9** (1964), 257–260.

[24] E Szemerédi and W.T Trotter, *Extremal problems in discrete geometry*, Combinatorica **3** (1983), 381–392.

[25] T. Tao and V. Vu, *Additive Combinatorics*, Cambridge Stud. Adv. Math., vol. 105, Cambridge University Press, 2006.

[26] L.A. Vinh, *Szemeredi-Trotter type theorem and sum-product estimate in finite fields*, preprint. arXiv:0711.4427v1.

[27] V.H. Vu, *Sum-product estimates via directed expanders*, Math. Res. Lett. **15** (2008), no. 2, 375--388.

József Solymosi
Department of Mathematics
University of British Columbia
1984 Mathematics Road
Vancouver, BC, V6T 1Z2
Canada
e-mail: `solymosi@@math.ubc.ca`

Chapter 8

Multi-dimensional inverse additive problems

Yonutz V. Stanchescu

We will examine the extremal value of the cardinality of $K \pm K$, for finite sets of lattice points. We will also formulate and discuss various results about the *exact structure* of multi-dimensional sets K, assuming that *doubling constant* $\sigma = \frac{|K+K|}{|K|}$ is very small.

8.1 Direct and inverse problems of additive and combinatorial number theory

Additive number theory is the study of sums of sets and we can distinguish two main lines of research. In a *direct problem* of additive number theory we start with a particular known set A and attempt to determine the structure and properties of the h-folds sumset hA. These are the classical direct problems in additive number theory: Waring's problem, Goldbach conjecture,... (see [27]). As a counterbalance to this direct approach, an *inverse problem* in additive number theory is a problem in which we study properties of a set A, if some characteristic of the h-fold sumset hA is given (see [28]). Sumsets can be defined in any Abelian group G, for example in $\mathbb{Z}$ the group of integers, or in the group $\mathbb{Z}/m\mathbb{Z}$ of congruence classes modulo m, or in the group $\mathbb{Z}^n$ of integer lattice points or in $\mathbb{R}^d$ the d-dimensional Euclidean space.

Let us present a unifying "algorithm" proposed by Freiman [16] for solving inverse additive problems.

- Step 1. Consider some (usually numerical) characteristic of the set under study.

- Step 2. Find an extremal value of this characteristic within the framework of the problem that we are studying.
- Step 3. Study the structure of the set when its characteristic is equal to its extremal value.
- Step 4. Study the structure of the set when its characteristic is near to its extremal value.
- Step 5. ....Continue, taking larger and larger neighborhoods for the characteristic.

In this note we will choose as characteristic the *cardinality of the sumset*

$$2K = K + K = \{x + y : x \in K, y \in K\},$$

or equivalently the "measure of doubling" $\sigma = \frac{|K+K|}{|K|}$. We will examine in detail the *exact structure* of a finite set $K \subseteq G$ in the case of a torsion free Abelian group $G = \mathbb{Z}^n$ or $G = \mathbb{R}^d$, *assuming that the doubling constant is small.* (On the other hand, if σ is an *arbitrary* doubling constant, then Freiman's fundamental result asserts that such a set is a large subset of a multi-dimensional arithmetic progression; see Freiman [14], [15], Bilu [2], Ruzsa [32], Nathanson [28] or Tao and Vu [47].)

8.2 The simplest inverse problem for sums of sets in several dimensions

It is well known that Kneser's theorem [25] for cyclic groups of prime order implies Cauchy–Davenport inequality [5], [7]: $|A + B| \geq \min\{p, |A| + |B| - 1\}$. For the "opposite situation" of a torsion-free Abelian group G, for example $\mathbb{Z}^n$ or $\mathbb{R}^d$, Kneser's result gives the estimate

$$|A + B| \geq |A| + |B| - 1,$$

valid for every two finite sets $A, B \subseteq G$. This is the best possible, since equality is attained when A and B are arithmetic progressions with the same difference.

Freiman showed in [14, p. 24] that it is possible to obtain a much better estimate.

Theorem 8.2.1. *For every finite set $\mathcal{A} \subseteq \mathbb{R}^m$ of affine dimension $\dim \mathcal{A} = d$, one has*

$$|\mathcal{A} + \mathcal{A}| \geq (d + 1)|\mathcal{A}| - \frac{1}{2}d(d + 1). \tag{8.2.1}$$

This is apparently the first result connecting geometry and additive properties of finite sets. The lower bound (8.2.1) cannot be improved (see assertion (i) of Theorem 8.2.2) and thus, Step 2 is solved.

Let us investigate now Step 3: what is the *exact structure* of multi-dimensional sets having the *smallest cardinality* of the sumset? For $d = 1$, inequality (8.2.1) gives $|2\mathcal{A}| \geq 2|\mathcal{A}| - 1$ and one has equality only for an arithmetic progression. We succeeded to extend this result for every dimension $d \geq 2$. More precisely, we obtained in [40] a complete description of the structure of multi-dimensional sets having the smallest cardinality of the sumset.

Theorem 8.2.2. *Let* $\mathcal{A} \subseteq \mathbb{R}^m$ *be a finite set such that* $\dim \mathcal{A} \geq d$ *and*

$$|\mathcal{A} + \mathcal{A}| = (d+1)|\mathcal{A}| - \frac{1}{2}d(d+1). \tag{8.2.2}$$

Then, $\mathcal{A}$ *is a d-dimensional set and the following cases are possible:*

(i) $\mathcal{A}$ *consists of d parallel arithmetic progressions with the same common difference.*

(ii) $\mathcal{A} = \{v_0, v_1, \ldots, v_d\} \cup \{2v_1, v_1 + v_2, 2v_2\}$, *where* v_i *are the vertices of a d-dimensional simplex.*

This result is an analogue of the well-known Vosper's theorem [50], $\mathbb{Z}/p\mathbb{Z}$ being here replaced by the d-dimensional space $\mathbb{R}^d$; it represents the complete solution of Step 3 for the inverse additive problem in several dimensions. Step 4 is described in Section 5 below. Remark that besides "*the standard structure*" of d parallel arithmetic progressions with the same common difference, we may encounter "*the non-standard structure*", when $|\mathcal{A}| = d + 4$.

A natural related problem is to find lower bound for the cardinality of the sum

$$\mathcal{A} + \mathcal{B} = \{a + b : a \in \mathcal{A}, b \in \mathcal{B}\},$$

where $\mathcal{A}$ and $\mathcal{B}$ are finite subsets of $\mathbb{Z}^d$. Ruzsa [33] proved the following result.

Theorem 8.2.3. *If* $\mathcal{A}$ *and* $\mathcal{B}$ *are finite sets in* $\mathbb{Z}^d$ *such that* $|\mathcal{A}| \geq |\mathcal{B}|$ *and the affine dimension* $\dim(\mathcal{A} + \mathcal{B}) = d$, *then*

$$|\mathcal{A} + \mathcal{B}| \geq |\mathcal{A}| + d|\mathcal{B}| - \frac{1}{2}d(d+1). \tag{8.2.3}$$

This generalizes Freiman's inequality (8.2.1), and assertion (i) of Theorem 8.2.2 shows that there is no improvement of (8.2.3) linear in $|\mathcal{A}|$. Nevertheless, under the additional assumption $\dim \mathcal{B} = d$, Gardner and Gronchi obtained in [20] a stronger estimate.

Theorem 8.2.4. *If* $\mathcal{A}$ *and* $\mathcal{B}$ *are finite sets in* $\mathbb{Z}^d$ *such that* $|\mathcal{A}| \geq |\mathcal{B}|$ *and the affine dimension* $\dim(\mathcal{B}) = d$, *then*

$$|\mathcal{A} + \mathcal{B}| \geq |\mathcal{A}| + (d-1)|\mathcal{B}| + (|\mathcal{A}| - d)^{\frac{d-1}{d}}(|\mathcal{B}| - d)^{\frac{1}{d}} - \frac{d(d-1)}{2}, \tag{8.2.4}$$

$$|\mathcal{A} + \mathcal{B}|^{\frac{1}{d}} \geq |\mathcal{A}|^{\frac{1}{d}} + (d!)^{-\frac{1}{d}}(|\mathcal{B}| - d)^{\frac{1}{d}}. \tag{8.2.5}$$

The proof uses the process of the so-called *projection* or *compression* of subsets in $\mathbb{Z}^d$, an idea introduced by Freiman in [14, p. 27]. Moreover, Theorem 8.2.4 – a close discrete analog of the classical Brunn–Minkowski inequality – translates into results for the lattice point enumerator of two convex lattice polytopes (see [20, Corollary 7.1]).

Another interesting development is the Green–Tao inequality (see (8.2.6) below). Note that the lower bound (8.2.1) assumes that the set $\mathcal{A}$ contains some non-degenerate d-dimensional simplex $S = \{e_0, e_0 + e_1, \dots, e_0 + e_d\}$. It turns out that if we replace the *simplex* S with a *parallelopiped* P, then the set $\mathcal{A}$ will have a doubling constant exponential in d. More precisely, we have the following theorem.

Theorem 8.2.5. *If $\mathcal{A} \subseteq \mathbb{R}^m$ is a finite set which contains a non-degenerate d-dimensional parallelepiped $P = e_0 + \{0,1\} \cdot e_1 + \cdots + \{0,1\} \cdot e_d$, then*

$$|\mathcal{A} + \mathcal{A}| \geq 2^{d/2} |\mathcal{A}|. \tag{8.2.6}$$

At the heart of the proof (see [22, Proposition 4.1]) are again a discrete Brun–Minkowski inequality, the machinery of *projection* of sets of lattice points and Plünecke inequality for commutative graphs.

8.3 Small doubling property on the plane

Let us describe some results concerning the structure of finite *planar sets* $\mathcal{K} \subseteq \mathbb{Z}^2$ with small sumset. First we note that Freiman's $3k - 4$ theorem (see [11] or [14, p. 11]) and inequality (8.2.1) easily imply that a planar set of lattice points with

$$|\mathcal{K} + \mathcal{K}| \leq 3|\mathcal{K}| - 4$$

lies on a straight line and is contained in an arithmetic progression of no more than $v = |\mathcal{K} + \mathcal{K}| - |\mathcal{K}| + 1$ terms. Moreover, for doubling $|\mathcal{K} + \mathcal{K}| = 3|\mathcal{K}| - 3$, Theorem 8.2.2 immediately yields the structure of such planar sets: $\mathcal{K}$ consists of two parallel arithmetic progressions with the same common difference or $\mathcal{K}$ is a two-dimensional simplex plus three midpoints of its three edges. Thus, Steps 2 and 3 of Freiman's algorithm are completely solved.

Therefore, a natural problem is to concentrate on the study of Step 4; we ask for the structure of a finite *planar set* of lattice points with small doubling $|\mathcal{K}+\mathcal{K}|$. As one can expect, this question is easier to answer when the cardinality $|\mathcal{K}+\mathcal{K}|$ is close to its minimal possible value $3|\mathcal{K}| - 3$, and becomes much more complicated if we choose bigger values for $|\mathcal{K} + \mathcal{K}|$. Freiman asked in 1973 (see [14] and [17, Section 7]) the following problem:

Problem F. *Find the exact structure of planar sets of lattice points under the doubling hypothesis $|\mathcal{K} + \mathcal{K}| < (4 - \frac{2}{s+1})|\mathcal{K}| - (2s + 1)$.*

Though, the $(2^n - \epsilon)$ theorem ([14, p. 57]), gives a first indication on the structure of $\mathcal{K}$, still this is not so precise as Freiman's result for $s = 2$.

Theorem 8.3.1. *Let $\mathcal{K} \subseteq \mathbb{Z}^2$ be a finite set of dimension* $\dim \mathcal{K} = 2$.

(i) *If $|\mathcal{K}+\mathcal{K}| < \frac{10}{3}|\mathcal{K}| - 5$ and $|\mathcal{K}| \geq 11$, then $\mathcal{K}$ lies on two parallel lines.*

(ii) *If $|\mathcal{K}+\mathcal{K}| < 4|\mathcal{K}| - 6$ and $\mathcal{K}$ lies on two parallel lines, then $\mathcal{K}$ is included in two parallel arithmetic progressions with the same common difference having together no more than $v = |2\mathcal{K}| - 2k + 3$ terms.*

This means that the total number of holes satisfies $h \leq |2\mathcal{K}| - (3k-3)$. For a proof, see [14, p. 28] and [38, p. 135].

In [38] we succeeded to solve the problem for every $s \geq 1$. We obtained the following theorem, which incorporates Theorem 8.3.1(i) as a particular case.

Theorem 8.3.2. *Let $\mathcal{K}$ be a finite set of $\mathbb{Z}^2$ and let $s \geq 1$ be a natural number. If $|\mathcal{K}| \geq k_0(s)$ and*

$$|\mathcal{K}+\mathcal{K}| < \left(4 - \frac{2}{s+1}\right)|\mathcal{K}| - (2s+1), \tag{8.3.1}$$

then there exist s parallel lines which cover the set $\mathcal{K}$.

This is a best possible result, because it cannot be improved by increasing the upper bound for $|\mathcal{K}+\mathcal{K}|$, or by reducing the number of lines that cover $\mathcal{K}$. The theorem is effective and we provide an explicit value for the constant $k_0(s) = O(s^3)$. We also devised a new method of proof of Freiman's $(2^n - \epsilon)$ theorem, for planar sets, i.e., the case of dimension $n = 2$.

Theorem 8.3.3. *Let $s \geq 1$ be an integer and define $\delta = \delta(s) = (8s(s+1))^{-1}$. If $\mathcal{K} \subseteq \mathbb{Z}^2$ is a finite set of lattice points with $|\mathcal{K}+\mathcal{K}| < \left(4 - \frac{2}{s+1}\right)|\mathcal{K}| - (2s+1)$, then there exists a line ℓ in $\mathbb{R}^2$ such that $|\ell \cap \mathcal{K}| \geq \delta|\mathcal{K}|$.*

The estimate $\delta = (8s(s+1))^{-1}$ of Theorem 8.3.3 improves the previous known values of $\delta(s)$, obtained in the general case by Freiman [14], Nathanson [28] and Bilu [2].

Recently, Grynkiewicz and Serra obtained in [23] an exact value for the constant $k_0(s) = 2s^2 + s + 1$. They also succeeded to extend the result for sums of different sets $\mathcal{A} + \mathcal{B}$.

Theorem 8.3.4. *Let $\mathcal{A}, \mathcal{B} \subseteq \mathbb{R}^2$ be finite subsets and let $s \geq 1$ be a natural number.*

(i) *If $\big||\mathcal{A}| - |\mathcal{B}|\big| \leq s+1, |\mathcal{A}| + |\mathcal{B}| \geq 4s^2 + 2s + 1$ and*

$$|\mathcal{A}+\mathcal{B}| < \left(2 - \frac{1}{s+1}\right)(|\mathcal{A}| + |\mathcal{B}|) - (2s+1),$$

then there exist $2s$ (not necessarily distinct) parallel lines which cover the sets $\mathcal{A}$ and $\mathcal{B}$.

(ii) *If $|\mathcal{A}| > |\mathcal{B}| + s, |\mathcal{B}| \geq 2s^2 + \frac{s}{2}$ and*

$$|\mathcal{A}+\mathcal{B}| < |\mathcal{A}| + \left(3 - \frac{2}{s+1}\right)|\mathcal{B}| - (s+1),$$

then there exist $2s$ *parallel lines which cover the sets* $\mathcal{A}$ *and* $\mathcal{B}$.

The next natural question is to consider the doubling coefficient $\sigma = 3.5$; this means that instead of the conditions $|2\mathcal{K}| < 3|\mathcal{K}| - 3$ (i.e., $s = 1$) and $|2\mathcal{K}| < \frac{10}{3}|\mathcal{K}| - 5$ (i.e., $s = 2$), we study now a finite set $\mathcal{K}$ of integer points on a plane, with the following small doubling property $|2\mathcal{K}| < 3.5|\mathcal{K}| - 7$, i.e., the case $s = 3$ of inequality (8.3.1).

The *complete solution* of Step 4 for planar sets under such small doubling hypothesis was obtained in [41]. Take a lattice $\mathcal{L}$ generated by $\mathcal{K}$. We obtain an exact best possible estimate for the number of points of $\mathcal{L}$ that lie in the convex hull of $\mathcal{K}$.

Theorem 8.3.5. *Let* $\mathcal{K}$ *be a finite set of* $\mathbb{Z}^2$ *such that* $|\mathcal{K} + \mathcal{K}| < 3.5|\mathcal{K}| - 7$.

(i) *If* $|\mathcal{K}| \geq 1344$, *then the set* $\mathcal{K}$ *lies on no more than three parallel lines.*

(ii) *If* $\mathcal{K}$ *is not contained in any two parallel lines, then the convex hull of* $\mathcal{K}$ *is included in three compatible arithmetic progressions having together no more than* v *terms, where*

$$v = |\mathcal{K}| + \frac{3}{4}\Big(|\mathcal{K} + \mathcal{K}| - \frac{10}{3}|\mathcal{K}| + 5\Big) = \frac{3}{4}\Big(|\mathcal{K} + \mathcal{K}| - 2|\mathcal{K}| + 5\Big). \tag{8.3.2}$$

The paper [44] is devoted to the generalization of Theorem 8.3.5 to the case $s \geq 4$. We shall consider now a finite set $\mathcal{K}$ of lattice points on a plane having the *small doubling property*

$$|2\mathcal{K}| < (4 - \frac{2}{s+1})|\mathcal{K}| - (2s + 1).$$

We wish to obtain a reasonable estimate for the number of lattice points of a "minimal" parallelogram that covers the set $\mathcal{K}$; more precisely, if $\mathcal{L}$ is a lattice generated by $\mathcal{K}$, we are interested in precise upper bounds for the number of points of $\mathcal{L}$ that lie in the convex hull of $\mathcal{K}$. Our main result asserts that $\mathcal{K}$ is located inside a parallelogram that lies on a few lines which are well filled.

Theorem 8.3.6. *Let* $s \geq 19$ *be an integer and let* $\mathcal{K}$ *be a finite subset of* $\mathbb{Z}^2$ *that lies on exactly* s *parallel lines. If*

$$|2\mathcal{K}| < (4 - \frac{2}{s+1})|\mathcal{K}| - (2s + 1),$$

then there is a lattice $\mathcal{L} \subseteq \mathbb{Z}^2$ *and a parallelogram* $\mathcal{P}$ *such that*

$$|\mathcal{P} \cap \mathcal{L}| \leq 24(|\mathcal{K} + \mathcal{K}| - 2|\mathcal{K}| + 1) \tag{8.3.3}$$

and $\mathcal{K} \subseteq (\mathcal{P} \cap \mathcal{L}) + v$, *for some* $v \in \mathbb{Z}^2$.

This gives an accurate description for the structure of planar sets having *doubling coefficient* $\sigma = \frac{|2K|}{|K|} < 4$: the set $\mathcal{K}$ can be covered by s parallel lines (the best possible result). Moreover, a suitable affine isomorphism maps $\mathcal{K}$ into a set of lattice points that lies inside a parallelogram of bounded area. The reader will notice that Chang's quantitative estimate $c(\sigma) \leq \exp(C\sigma^2 log^3\sigma)$ is extremely weak (see [6]) when compared to the new upper bound (8.3.3) implied by Theorem 8.3.6. This estimate is nearly sharp and it is far better than the bounds arising from the known proofs of Freiman's theorem.

We found that a similar inequality can be formulated for planar sets that lie on $s \geq 3$ parallel lines and have small sumset

$$|2\mathcal{K}| < (5 - \frac{2}{s-1})|\mathcal{K}| - (2s+1) + \frac{2}{s-1}.$$

We have examples showing that this inequality cannot be further relaxed. Moreover, we believe that for a best possible result, the constant factor 24 of Theorem 8.3.6 should be replaced by $\frac{1}{2}(1 + \frac{1}{s-1})$, if instead of a covering parallelogram $\mathcal{P}$ we consider the convex hull of $\mathcal{K}$. We suggest the following conjecture.

Conjecture. Let $\mathcal{K}$ be a finite subset of $\mathbb{Z}^2$ that lies on exactly $s \geq 2$ parallel lines. If

$$|\mathcal{K} + \mathcal{K}| < \max\left\{4|\mathcal{K}| - 6, (5 - \frac{2}{s-1})(|\mathcal{K}| - 1) - 2s + 4\right\},$$

then the convex hull of $\mathcal{K}$ can be covered by $2s - 2$ compatible arithmetic progressions having together no more than $v = \frac{s}{2(s-1)}\left(|\mathcal{K} + \mathcal{K}| - 2|\mathcal{K}| + 2s - 1\right)$ terms.

So far, this estimate has been proved only for $s = 2$ (Theorem 8.3.1) and $s = 3$ (Theorems 8.3.2 and 8.3.5).

Various applications of such results have surfaced: Theorems 8.3.1 and 8.3.2 are used in [4] in order to obtain a structure result for sets of integers with small upper density; connections to finite Beatty sequences have been found in [29].

8.4 Planar sets with no three collinear points on a line

In this section $r_k(n)$ denote the maximal number of *integers* that can be selected from the interval $[1, n]$ without including an arithmetic progression of k-terms, $k \geq 3$.

Let $\mathcal{A} \subseteq \mathbb{Z}^2$ be a finite set, not containing any three collinear points. Freiman asked in 1973 for a lower bound for $|\mathcal{A} + \mathcal{A}|$ (see [14, p. 27]). As a first step in the investigation of this problem, we showed in [43] that if $\mathcal{A}$ is a finite set of lattice points not containing any three-term arithmetic progression, then $\frac{|\mathcal{A} \pm \mathcal{A}|}{|\mathcal{A}|}$ is unbounded, as $\lim |\mathcal{A}| = \infty$. More precisely, we have the following theorem.

Theorem 8.4.1. *Let $\mathcal{A} \subseteq \mathbb{Z}^2$ be a finite set of n lattice points. If $\mathcal{A}$ does not contain any three collinear points, then there is a positive absolute constant $\delta > 0$ such* *that*

$$|\mathcal{A} \pm \mathcal{A}| \gg n(\log n)^{\delta}. \tag{8.4.1}$$

The constant δ can be easily computed: for instance, any positive δ smaller than 0.125 will do. This lower bound provides an answer to Freiman's question.

The proof shows that there is an intimate connection between two seemingly unrelated problems: planar sets with no three points on a line and non-averaging sets of integers (see also [30]).

Definition. *A finite set of integers $\mathcal{B} \subseteq \mathbb{Z}$ is called a non-averaging set of order t, if for every $1 \leq m, n \leq t$, the equation*

$$mX_1 + nX_2 = (m+n)X_3$$

has no nontrivial solutions with $X_i \in \mathcal{B}$; we say that (X_1, X_2, X_3) is a nontrivial solution, if X_1, X_2, X_3 are distinct elements of $\mathcal{B}$.

Let $s_t(n)$ be the maximal cardinality of a *non-averaging set of order* t included in the interval $[1, n]$. It is clear that a non-averaging set of order $t = 1$ is simply an integer set containing no arithmetic progressions and thus $s_t(n) \leq s_1(n) = r_3(n)$. Recently Bourgain obtained the estimate $r_3(n) \ll \frac{n}{(\log n)^{\frac{1}{2}}}(\log\log n)^{\frac{1}{2}}$, and therefore

$$s_t(n) \ll \frac{n}{(\log n)^{\frac{1}{2}}}(\log\log n)^{\frac{1}{2}}.$$

Remark. We obtained a *more exact* inequality, valid for sets $\mathcal{A} \subseteq \mathbb{Z}^2$ containing no $k-$terms arithmetic progressions: for every integer $t \geq 1$ we have

$$|\mathcal{A} \pm \mathcal{A}| \geq \frac{1}{2}|\mathcal{A}|\left(\frac{n}{s_t(n)}\right)^{\frac{1}{4t}}. \tag{8.4.2}$$

We formulate the following problem.

Problem S. *Suppose that $t \geq 1$ is a fixed, positive, but rather large integer. Is it true that $s_t(n) \ll \frac{n}{(\log n)^{4t}}$, or at least $s_t(n) \ll \frac{n}{(\log n)^{c}}$, for a positive absolute constant $c \geq \frac{1}{2}$?*

Note that Freiman's question asks for a nontrivial lower estimate of $|\mathcal{A} + \mathcal{A}|$ for a set $\mathcal{A} \subseteq \mathbb{Z}^2$ containing no three collinear points and in Problem S we want to estimate the density of a sequence of natural numbers $\mathcal{B}$, assuming that t *linear equations* do not hold for $\mathcal{B}$. Inequality (8.4.2) shows that any upper bound for $s_t(n)$, *better* than the trivial one $r_3(n)$, will lead to a corresponding sharpening of (8.4.1).

As regards lower bounds, we obtained in [43] the following

Theorem 8.4.2. (i) *For every $t \geq 1$, there is a positive constant c_t such that for every n one has $s_t(n) \geq n\exp(-c_t\sqrt{\log n})$.*

(ii) *There is no* $\epsilon_0 > 0$ *such that the inequality* $|\mathcal{A}+\mathcal{A}| \gg |\mathcal{A}|^{1+\epsilon_0}$ *holds for every finite set* $\mathcal{A} \subseteq \mathbb{Z}^2$ *containing no three collinear points.*

The proof uses Freiman's fundamental concept of isomorphism [14], Behrend's method [1] and a result of Ruzsa [31] about sets of integers containing no nontrivial three-term arithmetic progressions.

A recent improvement of the lower bound (8.4.1) was obtained by Sanders [35]:

$$|\mathcal{A}+\mathcal{A}| \gg_\epsilon |\mathcal{A}|(\log |\mathcal{A}|)^{\frac{1}{3}-\epsilon}.$$

8.5 Exact structure results for multi-dimensional inverse additive problems

A natural question (see [16, p. 16] and [17, p. 249]) is to generalize Theorem 8.3.1 to the multi-dimensional case $d = \dim(\mathcal{K}) \geq 3$. Assume that the doubling coefficient of the sumset $2\mathcal{K}$ is not much exceeding the minimal one, i.e.,

$$d+1 \leq \sigma = \frac{|2\mathcal{K}|}{|\mathcal{K}|} < \rho_d.$$

What can be said about the *exact structure* of $\mathcal{K}$? The expected result is: If $\rho_d = d+1+\frac{1}{3}$, then the set K is contained in d "short" arithmetical progressions.

In [45] we solved the above question for the first open case $d = 3$.

Theorem 8.5.1. *Let* $\mathcal{K}$ *be a finite subset of* $\mathbb{Z}^3$ *of affine dimension* $\dim \mathcal{K} = 3$.

(i) *If* $|\mathcal{K}+\mathcal{K}| < \frac{13}{3}|K| - \frac{25}{3}$ *and* $|\mathcal{K}| > 12^3$, *then* $\mathcal{K}$ *lies on three parallel lines.*

(ii) *If* $\mathcal{K}$ *lies on three parallel lines and* $|\mathcal{K}+\mathcal{K}| < 5|\mathcal{K}| - 10$, *then* $\mathcal{K}$ *is contained in three arithmetic progressions with the same common difference, having together no more than* $v = |\mathcal{K}+\mathcal{K}| - 3|\mathcal{K}| + 6$ *terms.*

In [46] we completely describe the structure of $\mathcal{K}$ for doubling coefficient $\sigma < d + \frac{4}{3}$.

Theorem 8.5.2. *Let* $\mathcal{K}$ *be a finite subset of* $\mathbb{Z}^d$ *of affine dimension* $\dim \mathcal{K} = d \geq 2$. *If* $k = |\mathcal{K}| > 3 \cdot 4^d$ *and*

$$|\mathcal{K}+\mathcal{K}| < (d+\frac{4}{3})|\mathcal{K}| - c_d,$$

where $c_d = \frac{1}{6}(3d^2+5d+8)$, *then* $\mathcal{K}$ *lies on* d *parallel lines.*

Moreover, under the additional assumption that $\mathcal{K}$ lies on d parallel lines, we give a sharp estimate for the number of points of the convex hull of $\mathcal{K}$.

Theorem 8.5.3. *Let $\mathcal{K}$ be a d-dimensional finite subset of $\mathbb{Z}^d$ that lies on d parallel lines. If*

$$|\mathcal{K}+\mathcal{K}| < (d+2)|\mathcal{K}| - \frac{1}{2}(d+1)(d+2),$$

then the convex hull of $\mathcal{K}$ is contained in d parallel arithmetic progressions with the same common difference, having together no more than $v = |\mathcal{K}+\mathcal{K}| - d|\mathcal{K}| + \frac{1}{2}d(d+1)$ terms.

These results are best possible and cannot be sharpened by reducing the quantity v or by increasing the upper bounds for $|\mathcal{K}+\mathcal{K}|$.

We found that a similar inequality can be formulated for d-dimensional sets that have a small doubling coefficient $\sigma = d+2-\frac{2}{s-d+3}$ (where $s \geq d$ is a positive integer). In this case we prove that $\mathcal{K}$ lies on no more than s parallel lines.

These results can be used to make Freiman's main theorem more precise. In a joint work with Freiman [19] we studied the *exact structure* of d-dimensional sets satisfying the small doubling property

$$|2K| < (d+2-\epsilon)|K|.$$

8.6 Difference sets

We will present now some results on difference sets in a d-dimensional Euclidean space. The need for lower estimates for $|\mathcal{A}-\mathcal{A}|$ in terms of $|\mathcal{A}|$ has been raised by Uhrin in [48], where the trivial inequality $|\mathcal{A}-\mathcal{A}| \geq 2|\mathcal{A}|-1$ is used to prove theorems sharpening the classical theorem of Minkowski–Blichfeldt in geometry of numbers. It can be stated that the sharper estimation for $|\mathcal{A}-\mathcal{A}|$ we have, the sharper results in geometry of numbers can be proved (see also [49]). Let $\mathcal{A} \subseteq \mathbb{R}^d$ be a finite set and (as Step 1 of Freiman's algorithm requires) we choose as numerical characteristic the cardinality of the difference set $\mathcal{A}-\mathcal{A}$.

A first result was obtained in 1989, when Freiman, Heppes and Uhrin proved in [18] an inequality analogous to (8.2.1):

$$|\mathcal{A}-\mathcal{A}| \geq (d+1)|\mathcal{A}| - \frac{1}{2}d(d+1). \tag{8.6.1}$$

This immediately yields that if $d=1$ and $\mathcal{A} \subseteq \mathbb{R}$, then $|\mathcal{A}-\mathcal{A}| \geq 2|\mathcal{A}|-1$ and if $d=2$ and $\mathcal{A} \subseteq \mathbb{R}^2$, then $|\mathcal{A}-\mathcal{A}| \geq 3|\mathcal{A}|-3$. These two inequalities cannot be strengthened. However, the lower bound (8.6.1) is not exact for $d=3$. Freiman, Heppes and Uhrin in [18] and Ruzsa in [33] conjectured that the "correct" lower bound for $\dim \mathcal{A} = 3$ is

$$|\mathcal{A}-\mathcal{A}| \geq 4.5|\mathcal{A}| - 9. \tag{8.6.2}$$

In [39] (see also [36]) we completely solved the above conjecture and showed that (8.6.2) is a best possible lower bound for $|\mathcal{A}-\mathcal{A}|$.

Theorem 8.6.1. *Let $\mathcal{A}$ be a finite set of $\mathbb{R}^3$ and let $\{e_1, e_2, e_3\}$ be the standard basis of $\mathbb{R}^3$.*

(i) *If the affine dimension* $\dim \mathcal{A} = 3$, *then* $|\mathcal{A} - \mathcal{A}| \geq 4.5|\mathcal{A}| - 9$.

(ii) *Equality is attained if and only if $\mathcal{A}$ is a union of four parallel arithmetic progressions:* $\mathcal{A} = \{0, e_1, e_2, e_1 + e_2\} + \{0, e_3, 2e_3, \ldots, ke_3\}, \quad k \geq 1$.

Theorem 8.6.1 solves Steps 2 and 3 of Freiman's algorithm: it gives the structure of three-dimensional sets having the smallest cardinality of the difference set; for two-dimensional sets the situation is similar.

Theorem 8.6.2. *Let $\mathcal{D}$ be a finite set in $\mathbb{R}^2$ of affine dimension* $\dim \mathcal{D} = 2$. *Then* $|\mathcal{D} - \mathcal{D}| = 3|\mathcal{D}| - 3$ *if and only if $\mathcal{D}$ consists of two parallel arithmetic progressions with the same number of elements and the same common difference.*

Let us give now a short description of the multi-dimensional case $d \geq 4$. Let $s(d)$ be the maximal positive number for which the inequality

$$|\mathcal{A} - \mathcal{A}| \geq s(d)|\mathcal{A}| - t(d)$$

holds for every finite set $\mathcal{A}$ of affine dimension $\dim \mathcal{A} = d$. What can one say about $s(d)$? The exact value of $s(d)$ is known only for $d = 1, 2$ and 3. Ruzsa conjectured in [33] that $s(d) = 2d - 2 + \frac{2}{d}$, for every $d \geq 4$. The aim of the paper [42] is to prove the following upper bound for $s(d)$.

Theorem 8.6.3. *For every integer $d \geq 2$ one has* $s(d) \leq 2d - 2 + \frac{1}{d-1}$.

This readily disproves Ruzsa's conjecture. Moreover, in view of inequality (8.6.2) and Theorem 8.6.3, it seems that the equality $s(d) = 2d - 2 + \frac{1}{d-1}$ is true for every $d \geq 2$. Thus, we suggest the following conjecture.

Conjecture. For every finite set $\mathcal{A}$ of affine dimension $\dim \mathcal{A} = d \geq 2$, one has

$$|\mathcal{A} - \mathcal{A}| \geq (2d - 2 + \frac{1}{d-1})|\mathcal{A}| - (2d^2 - 4d + 3).$$

Of course, in view of Theorem 8.6.3, if the above inequality is true, then it is best possible.

8.7 Finite Abelian groups

Similar questions can be asked for any group G. A short and incomplete list of results for $G = \mathbf{F}_p, (\mathbf{F}_2)^d, \mathbb{Z}/n\mathbb{Z}$ will show that additive questions in finite abelian groups are generally more difficult than analogous problems in $\mathbb{Z}$.

(i) Consider for the beginning *sums of congruence classes modulo a prime p*. Take two finite sets A and B in $\mathbf{F}_p$ and choose as characteristic the *cardinality of the sum* $A + B = \{a + b : a \in A, \, b \in B\}$. Then the solution of Step 2 is Cauchy–Davenport theorem $|A + B| \geq \min\{p, |A| + |B| - 1\}$. The answer to Step 3 is given by Vosper's theorem [50], which classify those pairs A, B of sets of residues for which equality holds in Cauchy–Davenport inequality.

The next natural question is to consider Step 4 and to analyze the case when the cardinality of the sum is not much exceeding its extremal value. Freiman [14, p. 46] generalized Vosper's theorem for sumsets of the form $A+A$ in $\mathbf{F}_p$, by describing the structure of A in the case $|2A| < c|A| - 3$, with $c < 2.4$; either $|A|$ is large or the set A is located in a short arithmetic progression. This has been recently extended to any c by Green and Ruzsa [21], using the rectification principle of Freiman and Bilu–Lev–Ruzsa [3].

(ii) For sumsets in *vector spaces over finite fields*, Eliahou and Kervaire proved in [10] that

$$|A+B| \geq \min\left\{p^t\left(\lceil\frac{|A|}{p^t}\rceil + \lceil\frac{|B|}{p^t}\rceil - 1\right) : \ t = 0, 1, \ldots, d\ \right\},$$

for every two sets A and B included in $(\mathbf{F}_p)^d$. Step 2 is solved. Deshouillers–Hennecart–Plagne gave in [9] an answer to Steps 3 and 4 by obtaining a structure theorem under the assumption $|A+A| = c|A|$, with $1 \leq c < 4$. In this instance the set A is contained in a coset $a+H$ of order at most $\frac{|A|}{u(c)}$ where $u(c) > 0$ is an explicit function depending only on c. Recently Step 5 was solved by Ruzsa and Green [34], not only for $G = \mathbf{F}_p^d$, but also for commutative torsion groups: if A is a subset of a commutative group G of exponent r and if $|A+A| < k|A|$, then A is contained in a coset of a subspace of size no more than $k^2 r^{2k^2-2}$.

(iii) When G is an *arbitrary Abelian group*, Kneser [25] gave a deep generalization of Cauchy–Davenport's theorem: Let A and B be two finite subsets of an Abelian group G. One has $|A+B| \geq |A| + |B| - |H|$, where H is the stabilizer of $A+B$. Important results concerning the equality case in Kneser's theorem are due to Kemperman [24] and Lev [26]. In a step beyond Kneser's theorem, Deshouillers and Freiman [8] proved a structural result for the cyclic group $G = \mathbb{Z}/n\mathbb{Z}$ assuming that $|A+A| < 2.04|A|$ and $|A|$ sufficiently small.

Bibliography

[1] F.A. Behrend, *On sets of integers which contain no three terms in arithmetic progression*, Proc. Nat. Acad. Sci. USA **32** (1946), 331–332.

[2] Y.F. Bilu, *Structure of sets with small sumset*, Astérisque **258** (1999), 77–108.

[3] Y.F. Bilu, V.F. Lev and I.Z. Ruzsa, *Rectification principles in additive number theory. Dedicated to the memory of Paul Erdös*, Discrete Comput. Geom. **19** (1998), no. 3 (Special Issue), 343–353.

[4] G. Bordes, *Sum-sets of small upper density*, Acta Arith. **119** (2005), no. 2, 187–200.

[5] A. Cauchy, *Recherches sur les nombres*, Jour. École polytechn. **9** (1813), 99–116.

[6] M.C. Chang, *A polynomial bound in Freiman's theorem*, Duke Math. J. **113** (2002), no. 3, 399–419.

[7] H. Davenport, *On the addition of residue classes*, Jour. London Math. Soc. **10** (1935), 30–32 and **22** (1947), 100–101.

[8] J.-M. Deshouillers and G.A. Freiman, *A step beyond Kneser's theorem for abelian finite groups*, Proc. London Math. Soc. **86** (2003), no. 1, 1–28.

[9] J.-M. Deshouillers, F. Hennecart and A. Plagne, *On small sumsets in* $(\mathbb{Z}/2\mathbb{Z})^n$, Combinatorica **24** (2004), no. 1, 53–68.

[10] S. Eliahou and M. Kervaire, *Sumsets in vector spaces over finite fields*, J. Number Theory , **71**, (1998), no. 1, 12–39.

[11] G.A. Freiman, *On the addition of finite sets I*, Izv. Vyssh. Zaved. Math. **13** (1959), no. 6, 202–213.

[12] ———, *Inverse problems of additive number theory. On the addition of sets of residues with respect to a prime modulus*, Soviet Math. Dokl. **2** (1961), 1520–1522.

[13] ———, *Inverse problems of additive number theory VI, On addition of finite sets III*, Izvest. Vuz. Mathem., **3** (1962), no. 28, 151–157.

[14] ______, *Foundations of a Structural Theory of Set Addition*, Transl. Math. Monogr., vol. 37, Amer. Math. Soc., Providence, RI, 1973.

[15] ______, *What is the structure of K if $K + K$ is small?*, Lecture Notes in Math., vol. 1240, Springer, New York, 1987, pp. 109–134.

[16] ______, *Structure theory of set addition, Astérisque* **258** (1999), 1–33.

[17] ______, *Structure theory of set addition II. Results and problems*, Paul Erdös and his mathematics, I Budapest, 1999, Bolyai Soc. Math. Stud., vol. 11, János Bolyai Math. Soc., Budapest, 2002, pp. 243–260.

[18] G.A. Freiman, A. Heppes and B. Uhrin, *A lower estimation for the cardinality of finite difference sets in $\mathbb{R}^n$*, Coll. Math. Soc. J. Bolyai, Budapest **51** (1989), 125–139.

[19] G.A. Freiman and Y.V. Stanchescu, preprint.

[20] R.J. Gardner and P. Gronchi, *A Brunn-Minkowski inequality for the integer lattice*, Trans. Amer. Math. Soc. **353** (2001), no. 10, 3995–4024.

[21] B. Green and I.Z. Ruzsa, *Sets with small sumset and rectification*, Bull. London Math. Soc. **38** (2006), no. 1, 43–52.

[22] B. Green and T. Tao, *Compressions, convex geometry and the Freiman–Bilu theorem*, Q. J. Math. **57** (2006), no. 4, 495–504.

[23] D.J. Grynkiewicz and O. Serra, *Properties of two dimensional sets with small sumset*, arXiv:0710.3127 v1.

[24] J.H.B. Kemperman, *On small sumsets in an abelian group*, Acta Math. **103** (1960), 63–88.

[25] M. Kneser, *Abschätzungen der asymptotischen Dichte von Summenmengen*, Math. Z. **58** (1953), 459–484.

[26] V.F. Lev, *On small sumsets in abelian groups. Structure theory of set addition*, Astérisque **258** (1999), xv, 317–321.

[27] M.B. Nathanson, *Additive Number Theory: The Classical Bases*, Grad. Texts in Math., vol. 164, Springer, New York, 1996.

[28] ______, *Additive Number Theory: Inverse Problems and the Geometry of the Sumsets*, Grad. Texts in Math., vol. 165, Springer, New York, 1996.

[29] J. Pitman, *Sumsets of finite Beatty sequences. In honor of Aviezri Fraenkel on the occasion of his 70th birthday.* Electron. J. Combin. **8** (2001), no. 2, Research Paper 15, 23 pp.

[30] K.F. Roth, *On certain sets of integers*, J. London Math. Soc. **28** (1953), 104–109.

[31] I.Z. Ruzsa, *Arithmetical progressions and the number of sums*, Period. Math. Hungar. **25** (1992), no. 1, 105–111.

[32] ______, *Generalized arithmetical progressions and sumsets*, Acta Math. Hung. **65** (1994), no. 4, 379–388.

[33] ______, *Sums of sets in several dimensions*, *Combinatorica* **14** (1994), no. 4, 485–490.

[34] I.Z. Ruzsa and B. Green, *Freiman's theorem in an arbitrary abelian group*, J. London Math. Soc. (2) **75** (2007), no. 1, 163–175.

[35] T. Sanders, *Three term arithmetic progressions in sumsets*, arXiv:0611304 v1, to appear in Proc. Edinb. Math. Soc.

[36] Y.V. Stanchescu, *On finite difference sets in* $\mathbb{R}^3$, Proceedings of the Conference on the Structure Theory of Set Addition, CIRM, Marseille, 1993, pp. 149–162.

[37] ______, *On addition of two distinct sets of integers*, Acta Arith. **75** (1996), no. 2, 191–194.

[38] ______, *On the structure of sets with small doubling property on the plane (I)*, Acta Arith. **83** (1998), no. 2, 127–141.

[39] ______, *On finite difference sets*, Acta Math. Hungar. **79** (1998), no. 1–2, 123–138.

[40] ______, *On the simplest inverse problem for sums of sets in several dimensions*, Combinatorica **18** (1998), no. 1, 139–149.

[41] ______, *On the structure of sets of lattice points in the plane with a small doubling property*, *Astérisque* **258** (1999), 217–240.

[42] ______, *An upper bound for d-dimensional difference sets*, Combinatorica **21** (2001), no. 4, 591–595.

[43] ______, *Planar sets containing no three collinear points on a line and non-averaging sets of integers*, Discrete Mathematics **256** (2002), no. 1-2, 387–395.

[44] ______, *On the structure of sets with small doubling property on the plane (II)*, Integers **8**, (2008), no. 2, A10, 1–20.

[45] ______, *Three dimensional sets with small sumset*, Combinatorica **28** (2008), no. 3, 343–355.

[46] ______, *The structure of d-dimensional sets with small sumset*, submitted.

[47] T. Tao and H.V. Vu, *Additive Combinatorics*, Cambridge Stud. Adv. Math., vol. 105, Cambridge University Press, 2006.

[48] B. Uhrin, *Some estimations useful in geometry of numbers*, Period. Math. Hungar. **11** (1980), 95–103.

[49] ______, *On a generalization of Minkowski convex body theorem*, J. Number Theory, **13** (1981), 192–209.

[50] A.G. Vosper, *The critical pairs of subsets of a group of prime order*, J. London Math. Soc. **31** (1956), 200–205; Addendum, 280–286.

Yonutz V. Stanchescu
Department of Mathematics
Afeka Academic College
218 Bney Efraim
Tel Aviv 69107
Israel
e-mail: yonis@afeka.ac.il

Advanced Courses in Mathematics CRM Barcelona

Edited by
Manuel Castellet

BIRKHÄUSER

Since 1995 the Centre de Recerca Matemàtica (CRM) in Barcelona has conducted a number of annual Summer Schools at the post-doctoral or advanced graduate level. Sponsored mainly by the European Community, these Advanced Courses have usually been held at the CRM in Bellaterra.
The books in this series consist essentially of the expanded and embellished material presented by the authors in their lectures.

Argyros, S. / Todorcevic, S.
Ramsey Methods in Analysis (2005)
ISBN 978-3-7643-7264-4

Audin, M. / Cannas da Silva, A. / Lerman, E.
Symplectic Geometry of Integrable Hamiltonian Systems (2003)
ISBN 978-3-7643-2167-3

Bertoluzza, S. / Falletta, S. / Russo, G. / Shu, C.-W.
Numerical Solutions of Partial Differential Equations (2008)
ISBN 978-3-7643-8939-0

Brady, N. / Riley, T. / Short, H.
The Geometry of the Word Problem for Finitely Generated Groups (2006)
ISBN 978-3-7643-7949-0

The origins of the word problem are in group theory, decidability and complexity, but, through the vision of Gromov and the language of filling functions, the topic now impacts the world of large-scale geometry, including topics such as soap films, isoperimetry, coarse invariants and curvature.
The first part introduces van Kampen diagrams in Cayley graphs of finitely generated, infinite groups; it discusses the van Kampen lemma, the isoperimetric functions or Dehn functions, the theory of small cancellation groups and an introduction to hyperbolic groups. The second part is dedicated to Dehn functions, negatively curved groups, in particular, CAT(0) groups, cubings and cubical complexes. In the last part, filling functions are presented from geometric, algebraic and algorithmic points of view. Many examples and open problems are included.

Brown, K.A. / Goodearl, K.R.
Lectures on Algebraic Quantum Groups (2002)
ISBN 978-3-7643-6714-5

Catalano, D. / Cramer, R. / Damgård, I. / Di Creszenso, G. / Pointcheval, D. / Takagi, T.
Contemporary Cryptology (2005)
ISBN 978-3-7643-7294-1

Christopher, C. / Li, C.
Limit Cycles of Differential Equations (2007)
ISBN 978-3-7643-8409-8

Cohen, R.L. / Hess, K. / Voronov, A.A.
String Topology and Cyclic Homology (2006)
ISBN 978-3-7643-2182-6

Da Prato, G.
Kolmogorov Equations for Stochastic PDEs (2004)
ISBN 978-3-7643-7216-3

Drensky, V. / Formanek, E.
Polynomial Identity Rings (2004)
ISBN 978-3-7643-7126-5

Dwyer, W.G. / Henn, H.-W.
Homotopy Theoretic Methods in Group Cohomology (2001)
ISBN 978-3-7643-6605-6

Markvorsen, S. / Min-Oo, M.
Global Riemannian Geometry: Curvature and Topology (2003)
ISBN 978-3-7643-2170-3

Mislin, G. / Valette, A.
Proper Group Actions and the Baum-Connes Conjecture (2003)
ISBN 978-3-7643-0408-9

Myasnikov, A. / Shpilrain, V. / Ushakov, A.
Group-based Cryptography (2008)
ISBN 978-3-7643-8826-3

This book is about relations between three different areas of mathematics and theoretical computer science: combinatorial group theory, cryptography, and complexity theory. It is explored how non-commutative (infinite) groups, which are typically studied in combinatorial group theory, can be used in public key cryptography. It is also shown that there is a remarkable feedback from cryptography to combinatorial group theory because some of the problems motivated by cryptography appear to be new to group theory, and they open many interesting research avenues within group theory.
Then, complexity theory, notably generic-case complexity of algorithms, is employed for cryptanalysis of various cryptographic protocols based on infinite groups, and the ideas and machinery from the theory of generic-case complexity are used to study asymptotically dominant properties of some infinite groups that have been applied in public key cryptography so far.
Its elementary exposition makes the book accessible to graduate as well as undergraduate students in mathematics or computer science.